BOTANICAL RESEARCH AND PRACTICES

FUSARIUM

EPIDEMIOLOGY, ENVIRONMENTAL SOURCES AND PREVENTION

BOTANICAL RESEARCH AND PRACTICES

Additional books in this series can be found on Nova's website under the Series tab.

Additional e-books in this series can be found on Nova's website under the e-book tab.

BOTANICAL RESEARCH AND PRACTICES

FUSARIUM

EPIDEMIOLOGY, ENVIRONMENTAL SOURCES AND PREVENTION

TONY FERARRI RIOS
AND
EMILIA REYES ORTEGA
EDITORS

Nova Science Publishers, Inc.
New York

LIBRARY OF CONGRESS CATALOGING-IN-PUBLICATION DATA

Fusarium : epidemiology, environmental sources and prevention / editors: Tony Ferarri Rios and Emilia Reyes Ortega.
p. cm.
Includes index.
ISBN 978-1-61942-539-2 (hardcover)
1. Fusarium diseases of plants--Epidemiology. 2. Fusarium diseases of plants--Prevention. I. Ferarri Rios, Tony. II. Reyes Ortega, Emilia.
SB741.F9F875 2012
632'.4677--dc23
2011050656

Published by Nova Science Publishers, Inc. † New York

Contents

PREFACE

Fusarium is commonly associated with higher plants and are among the most ubiquitous fungi in terrestrial ecosystems. Many fusarium species are serious plant pathogens, causing symptoms such as necrotic lesions, rot, and wilt. In this book, the authors present current research in the study of the epidemiology, environmental sources and prevention of fusarium. Topics include the development of natural products from fusarium including chemotherapeutic agents for human diseases and agricultural applications; the molecular basis of fusarium wilt in tomatoes and disease management; lectin preparations for the biocontrol of fusarium species and fusarium head blight and DON contamination management in soft and durum wheat cultivation.

Chapter 1 - *Fusarium* spp. are commonly associated with higher plants and are among the most ubiquitous fungi in terrestrial ecosystems. Many *Fusarium* species are serious plant pathogens, causing symptoms such as necrotic lesions, rot, and wilt. Although *Fusarium* species are being regarded more important in plant pathology, several species also produce a diverse range of toxic metabolites which may contaminate foods and animal feeds of plant origin but only a small number are considered to be important in context of human health. Several *Fusarium* species isolated have been reported to produce natural products. Both pharmaceutical and agricultural industries are interesting for new products, from natural source to be developed into new industrial products. The natural products are adapted to a specific function in nature; the search for novel secondary metabolites should concentrate on organisms that inhabit different ecosystems. In this context, *Fusarium* species are a promising group of organisms capable to produce a wide variety of metabolite with interesting biological activities such as, antibacterial, antifungal, antiviral, immunosuppressant, antitumour and antimalarial. Many works describe yet the biological activities from crude extracts from *Fusarium* against pathogens of the humans. *Fusarium* produce a wide variety of chemical compounds from different classes, such as, terpenoids, alkaloids, and others. This review attempts to summarize the current status of the importance of natural products from *Fusarium* for drug discovery, lead candidates for the development of new chemotherapeutic agents that can be used for treatment of human diseases and agricultural applications.

Chapter 2 - *Fusarium* is one of the most ubiquitous, abundant, and important genera of soil microfungi, and about 1000 species have been described. Identification of *Fusarium* species is difficult due to great variation in morphological and nonmorphological characteristics, including virulence. The genus contains many species of environmental, agricultural and human health importance. Fusaria have long been recognized as pathogens of many plant species. Wheat and other small grain cereals may be attacked by a wide range of *Fusarium* spp. and on different plant organs. However, infestation of the ears appears to be the most critical, leading to Fusarium head blight (FHB), also known as scab. *F. graminearum* and *F. culmorum* are the predominant *Fusarium* species infecting wheat. Predominance of a specific species in a region is mostly influenced by climatic conditions especially temperature and moisture requirements. Plant diseases caused by *Fusarium* spp. include vascular wilts, seedling damping-off, and rots of stems, crowns, and roots. Some species of *Fusarium* also produce mycotoxins resulting in food contamination. In general, biotic and abiotic stresses (heat, water and insect damage) cause plant stress and predispose plants in the field to mycotoxin contamination. The main mycotoxin classes of concern produced by *Fusarium* species include the trichothecenes, moniliformin, fumonisins and zearalenone. Traditionally crop rotations have been proven to be an effective strategy to control many soil-borne diseases, but because most of these pathogens can survive for a long period of time, the effectiveness of crop rotation is limited once a disease outbreak occurs. Biological control and natural products (like essential oil) were an alternative strategy to control *Fusarium* disease.

Chapter 3 - Fusarium wilt is a widespread and serious chickpea disease caused by the soil-borne fungus *Fusarium oxysporum* f.sp. *ciceris*. Eight physiological races of the pathogen (0, 1A, 1B/C, 2, 3, 4, 5 and 6) have been reported. Persistence of the pathogen in soil and its capacity to survive there for years even in the absence of host renders its control difficult. The use of resistant cultivars is the most effective, economical and environmentally friendly method of managing the disease. Therefore breeding for resistance to Fusarium wilt has been an important objective of many chickpea research programmes. Systematic screening of germplasm collections have identified useful resistance sources that have been used successfully to breed resistant cultivars. Besides, genetic studies have revealed that resistance genes to Fusarium wilt is governed by one two or three genes depending on the race. Genetic linkage analyses and quantitative trait loci (QTL) mapping have identified molecular markers that could be useful for marker-assisted selection (MAS) and gene pyramiding. On the other hand, comparative genomics and synteny studies with model species and closely related legumes promises to further advance the knowledge of the chickpea genome and provide breeders with additional genes and selectable markers for use in MAS. In this chapter the authors will review the current state of molecular breeding for Fusarium wilt in chickpea. Special emphasis is given on the application of marker technology, and QTL analysis for MAS in this species.

Chapter 4 - The commercial tomato, *Lycopersicon esculentum* is one of the most popular vegetables in the world. It is cultivated for its culinary, industrial and ornamental values. It has also served as a model plant for various scientific studies. Tomato is affected by several economically important plant diseases, notably vascular wilt caused by *Fusarium oxysporum* f. sp. *lycopersici* (Fol). *Fusarium* is a soil borne fungus which infects the plant root and colonises the xylem vessels, resulting in wilting, stunting and death. *Fusarium* wilt of tomato continues to be a major cause of yield losses in endemic regions. As a result of its growth within the confinement of the vascular tissues of the host plant, chemical and cultural attempts at protecting tomato against *Fusarium* wilt have been rendered ineffective or impractical. The most effective method of control is breeding for resistant cultivars. Resistance is mediated by the I, I-2 and I-3 resistance genes in tomato against Fol. There is remarkably little knowledge available about the molecular mechanism or/and pathogenicity genes required by Fol to cause disease and how hosts combat or tolerate the pathogen. Understanding the process by which Fol infects, colonises and interacts with tomato host plant could lead to more efficient ways of controlling it. More so, the interaction between Fol and tomato, which has become a model system for the study of the molecular mechanism of disease susceptibility and resistance, could give a better insight into the control of wilt in many agriculturally important crops caused by other species of *Fusarium*. This chapter highlights the nature and molecular basis of *Fusarium* wilt in tomato, as well as disease management.

Chapter 5 - Crown and root rot is the most important disease of asparagus worldwide. Symptoms include rots and vascular discoloration of roots, rhizomes and stems, necrotic lesions of roots and stems, yellowing and senescence of stems and spears and dwarfing and death of plants. This brings about reductions in plant vigour and individual plants yield related to decline of plants and replant problem. The complex etiology of Fusariumcrown and root rot of asparagus (FCRR), with several species of *Fusarium* reported as causal agents, depend on soil and climate conditions whereby prevalence of the species involved differs among crop areas. *Fusarium oxysporum* f. sp. *asparagi*, *F. proliferatum* and *F. verticillioides* (syn. *F. moniliforme*) are the pathogens most commonly associated to FCRR. *F. solani*, less widespread in asparagus fields, include some isolates highly pathogenic to asparagus, at least in Spain. Other *Fusarium* species are often recovered although with much lower frequencies than *F. oxysporum* f. sp. *asparagi* or *F. proliferatum*. Among those, *F. subglutinans*, *F. redolens* f. sp. *asparagi* and *F. culmorum* seem to play a pathogenic role in the disease complex. Asparagus plants stressed either by biotic or abiotic factors, such as drought or old crops, are more prone to infections by different *Fusarium* spp. in the complex. With regard to optimal temperature for fungal growth and disease development, roughly ranging 20-30ºC, *F. solani* is favoured by higher temperature than *F. oxysporum* f. sp. *asparagi* and *F. proliferatum*. Within *Liliaceae*, *F. oxysporum* f. sp. *asparagi*and *F. solani* show a certain degree of pathogenic specialisation which contrasts with the lack of specificity of the other species above mentioned. The genetic intra-specific diversity, mainly in *F. oxysporum* f. sp. *asparagi*, is related to the host range, as studies of populations have commonly shown these to be composed of many different vegetative compatibility groups (VCG) which frequently include a reduced or single number of isolates; these also differ among studies conducted in different countries. Disease management can not be achieved by resistant cultivars because of the very narrow genetic diversity of *Asparagus officinalis*. The old practice of applying rock salt to asparagus soils proved effective in reducing populations of *Fusarium* spp. and infections, and increasing plant vigour, possibly because of a shift of microbial populations towards a disease suppressiveness composition. Except for the use of methyl isothiocianate as a soil fumigant and thiabendazol as a fungicide, which costs seem unaffordable to growers, no chemical control measures are available. However, use of organic soil amendments followed by polyethylene tarping of soil seems to be effective to control FCRR under some conditions. Incorporation of fresh residues of broccoli or grass into infested soils reduces Foa densities due to anaerobiosis and the accumulation of toxic compounds. The latter could also be the main mechanism by which fresh poultry manure application achieves reduction of Foa propagules in soil, disease levels, and increases fresh weight of plants and yields. In addition, promising biological control agents such as some *Trichoderma* isolates or non-pathogenic isolates of *F. oxysporum* that proliferate in the presence of toxic asparagus residues encourage long-term experiments aimed at implementing such control practice, which is compatible with Ecological Agriculture requirements.

Chapter 6 - Lectins are proteins from non-immune origin, containing two or more binding sites to mono or oligosaccharides. Plant lectins, found in different plant tissues, show a marked heterogeneity with respect to their molecular structure and sugar-binding specificity. Fungi are frequently pathogens of cultivated plant species representing one of the main causes of crop losses. Growth of fungi has been considered a major factor in biodeterioration of materials such as wood, stored grains, paintings, sculptures, leather, and oil. *Fusarium* is a filamentous fungi genus widely distributed in soil and in association with plants. *Fusarium* is a pathogen of bananas, corn, rice, wheat, tomatoes, peanuts, among other crops. Some *Fusarium* species produce mycotoxins that have been found infecting breakfast and infant cereals. *Fusarium* species can affect human health causing cutaneous infections and keratitis, for example. The intensive use of synthetic fungicides for plant disease control is cause of public concern because of the presence of residues in water and crops for animal and human consumption. Biodegradable fungicides derived from plants have been searched aiming to minimize environmental damage. Antifungal effects on *Fusarium* species have been reported for lectins isolated from different plant tissues. The role of lectins in fungi plant defense mechanisms may have evolved from its ability to recognize and agglutinate cells or glycoconjugates in sites for potential invasion by infectious agents, with ability to inhibit fungal growth and germination. Antifungal activity has been reported mainly for chitin-binding lectins which are able to bind chitin present in fungal cell wall and interfere in spore germination. In summary, this chapter discusses the optimistic perspectives for lectin preparations that can be applied in agriculture as cheap and biodegradable antifungal agents preventing pollution of the environment without damages to the biodiversity.

Chapter 7 - Some *Fusarium* species are human pathogens while other *Fusarium* species are phytopathogens.A diversity of antifungal proteins displays inhibitory activity toward *Fusarium* species. They include chitin binding proteins, chitinases, cyclophilin-like proteins, defensins, embryo-abundant protein-like proteins, glucanases, lectins, lipid transfer proteins, osmotins, peroxidases, proteases, ribonucleases, storage 2S albumins, thaumatin-like proteins,and other antifungal proteins. The antifugal mechanisms of chitinases, defensins, glucanases, and thaumatin-like proteins are well known but those of the other aforementioned proteins await elucidation.

Chapter 8 - The annotation of complete genome sequences of phytopathogenic fungi is an ongoing initiative for plant pathologists and different *Fusarium* species, such as *F. solani, F. verticillioides,* and *F. oxysporum* have been the focus of a number of these studies. *F. graminearum* was one of the initial filamentous fungi to have its whole genome sequenced and the resulting data have since served as an invaluable tool to study fungal development, gene expression and function, and population genetics. Techniques involving expressed sequence tags, DNA microarrays, site directed mutagenesis, gene replacement, and large scale DNA sequencing, among others, have enabled a more comprehensive characterization of genes intimately involved in pathogenesis, including virulence factors, such as mycotoxin-encoding genes. Genomics has facilitated the study of relatively large gene sets such as polyketide synthases (PKS) among *Fusarium* species and the determination of the functions of a number of PKS genes in *F. graminearum* was made possible using gene replacement techniques. Additionally, DNA microarrays have been used to identify differentially expressed genes in *F. graminearum* and *F. verticillioides* under diverse conditions while site directed mutagenesis has aided in the characterization of fungicide resistant *F. graminearum*. These as well as other examples of how genomics has contributed to deciphering the pathogenicity of *Fusarium* species are examined. This chapter provides an overview of the techniques commonly utilized in genomics-based studies as well as a review of published research highlighting key genomic resources relevant to effective management of diseases caused by different *Fusarium* species. Potential future applications of the burgeoning field of genomics to *Fusarium* research as well as possible consequences of these studies are also discussed.

Chapter 9 - Fusarium Head Blight (FHB), also known as scab, is a devastating disease of enormous economic importance throughout the world that attacks all classes of barley and wheat. Every year, all of the most important cereal producers in the world are affected by this disease. *Fusarium graminearum* and *Fusarium culmorum* are considered the most pathogenic and widespread agents of FHB in wheat. Both *F. graminearum* and *F. culmorum* produce deoxynivalenol (DON, also known as vomitoxin), a mycotoxin of the trichotecenes group, one of the most widespread mycotoxins in cereals, which can be synthesised in the field in small grain cereals and in maize. This mycotoxin has a great impact on the health of animals and humans, due to their cytotoxic activity and immunosuppressive effects. As a consequence, FHB and DON contamination are responsible for serious direct and indirect economic losses. Economic losses can be attributed to yeld loss due to fungal disease, reduced technological and nutritional quality of grain, reduced crop value due to mycotoxin contamination, reduced animal productivity, human health costs. Prevention, monitoring, sampling, chemical analysis, litigation, mitigation, and research costs also need to be taken into account. Moreover, FHB and DON limits, which have been established in many parts of the world for animal feeding and human consumption, have made the worldwide cereal market much more selective. *F. graminearum* and F. culmorum inoculum is usually abundant and its production, dispersal and deposition are influenced by weather conditions. Warm, humid weather, frequent rainfall and heavy dew favour spore germination. Temperature and water activity of the host are the fundamental factors that affect fungal growth inside grain. DON synthesis is mainly influenced by temperature and water ativity and its production seems to play a decisive role on the aggressiveness of *F. graminearum* and *F. culmorum*: it has been suggested that in the presence of DON, plant defence mechanisms are not triggered fast enough, thereby leading to an increased aggressiveness of the pathogen. The accumulation of DON in the grain is intimately related to the development of the disease in the field. Prevention strategies through pre-harvest agronomic management can achieve the quality and safety standars required by the market and international regulations. Among the agricultural practices, FHB development and DON contamination in wheat grains are mainly affected by tillage, crop rotation, varieties and fungicide application. None of these strategies on their own are able to significantly reduce the impact of this disease. Nevertheless, careful planning of all the different management decisions canl lead to the designing of crop management systems able to reduce FHB and mycotoxin incidence and severity. In other words, it is necessary for producers to make use of the available wheat varieties that are known to be more resistant and to apply all of the good agricultural practices tuned to a determined cultivation area. Useful tools for prevention are represented by FHB and DON predictive models: the use of a model to predict the outcome of a disease is desirable to enhance and trigger management opportunities with the aim of reaching high technological, nutritional and productivity quality and safety of the production.

Chapter 10 - A microcosm studies was carried out to determine the effect of *Aspergillus niger*-treated mixture of two agro-industrial waste (AIW) materials, dry olive wastes and sugar beet press mud, on tomato (*Lycopersicon esculentum*) plants grown in a soil inoculated with *Fusarium oxysporum* f. sp. *lycopersici* (*Fol*). These waste materials were selected as they constitute a major environmental problem especially for Mediterranean countries. Agro-wastes were treated in conditions of solid-state fermentation in the presence of Morocco apatite (RP) and further applied at a rate of 50 g/kg soil. Soil-plant systems were additionally inoculated or not with the arbuscular mycorrhizal (AM) fungus *Glomus intraradices*. Plant growth and nutrition, symbiotic developments and soil enzymatic activities were stimulated in *Fol*-free soil supplemented with treated agro-wastes and significantly greater in treatments where AM fungus was introduced compared with the non-amended control. The introduction of *Fol* into the soil-plant system adversely influenced all studied parameters. AM fungus alone reduced the effect of the plant pathogen on tomato plants and significantly decreased (by 2.3×10^3) the number of *Fol*-CFU compared with the non-mycorrhizal control grown in non-amended soil. Higher levels of pathogen control were achieved in *A. niger*/AIW/RP-amended mycorrhized plant-soil system. In separate, *in vitro* studies, *A. niger* demonstrated a strong suppressive effect on *Fol* with a 5-fold reduction of the pathogen colony diameter compared to the control. Production of siderophore-like metabolites was also detected in chrome azurol S plate assays with *A. niger*.

In: Fusarium
Editors: Tony Ferarri Rios and Emilia Reyes Ortega
ISBN: 978-1-61942-539-2

Chapter 1

NATURAL PRODUCTS FROM *FUSARIUM*

***Fernanda F. Campos**[*,1], **Ezequias P. Siqueira**[2], and **Betania B. Cota**[2]*

[1]Departamento de Microbiologia, Instituto de Ciências Biológicas, Universidade Federal de Minas Gerais, Belo Horizonte, MG, Brazil

[2]Laboratório de Química de Produtos Naturais, Centro de Pesquisas René Rachou, Fiocruz, Belo Horizonte, MG, Brazil

ABSTRACT

Fusarium spp. are commonly associated with higher plants and are among the most ubiquitous fungi in terrestrial ecosystems. Many *Fusarium* species are serious plant pathogens, causing symptoms such as necrotic lesions, rot, and wilt. Although *Fusarium* species are being regarded more important in plant pathology, several species also produce a diverse range of toxic metabolites which may contaminate foods and animal feeds of plant origin but only a small number are considered to be important in context of human health. Several *Fusarium* species isolated have been reported to produce natural products. Both pharmaceutical and agricultural industries are interesting for new products, from natural source to be developed into new industrial products. The natural products are adapted to a specific function in nature; the search for novel secondary metabolites should concentrate on organisms that inhabit different ecosystems. In this context, *Fusarium* species are a promising group of organisms capable to produce a wide variety of metabolite with interesting biological activities such as, antibacterial, antifungal, antiviral, immunosuppressant, antitumour and antimalarial. Many works describe yet the biological activities from crude extracts from *Fusarium* against pathogens of the humans. *Fusarium* produce a wide variety of chemical compounds from different classes, such as, terpenoids, alkaloids, and others. This review attempts to summarize the current status of the importance of natural products from *Fusarium* for drug discovery, lead candidates for the development of new chemotherapeutic agents that can be used for treatment of human diseases and agricultural applications.

* Author for correspondence: Departamento de Microbiologia, Instituto de Ciências Biológicas, Universidade Federal de Minas Gerais, Belo Horizonte, MG, P. O. Box 486, CEP 31270-901, Brazil. E-mail: ffcmicro@gmail.com, Phone: +55-31-3409 2741 Fax: +55-31-3409 2730.

1. Introduction

1.1. Fusarium, the Genus, Habitats and Distribution

The generic concept of *Fusarium* was first studied by Link (1263) with the primary character being the presence of the distinctive canoe- or banana-shaped conidia well known to all that work with the genus. Yet Seifert (1935) also notes that the production of the distinctive macroconidia is not necessarily the sole realm of *Fusarium*, as a number of coelomycetes produce spores that could easily be identified as belonging to a *Fusarium* spp., if these spores were the only criterion that was used for identification. Many of the *Fusarium* species currently used in taxonomic systems are poorly defined and in many cases type specimens no longer exist. If the descriptions of many of these species were published today they would be unlikely to meet the current requirements of the International Code of Botanical Nomenclature (Greuter et al., 2000) because the description is vague or poorly defined.

This gender displays a high degree of diversity in relation to the attributes morphological, physiological and ecological and it is not surprising that representatives of this genus occur in most ecological niches in various geographic regions of the world (Burgess et al., 1997). It is estimated that since 1900, nearly a thousand species of Fusarium have been described, most of the examination of their structures (sporodochia) directly in the natural hosts or substrates, which led to many of these species are synonyms, in view of the great variability Fusarium in different environments and substrates (Burgess et al., 1994).

The complex microbial community that inhabits the rhizosphere interact with soil pathogens, and influence, consequently, infection by fungi. In the rhizosphere, in addition to pathogens, there are fungi that have antagonistic activity against pathogens. They can adversely affect population densities and metabolic activities of soil-borne pathogens from the production of primary and secondary metabolites with antimicrobial effect, such as the production of extracellular enzymes, proteins and mycotoxin (Raaijmakers et al., 2009). On the other hand the pathogens of the soil can produce primary and secondary metabolites useful for pharmaceutical, food and cosmetics industry.

Fusarium species are distributed across many geographical regions and substrates, also widely distributed in soils, plants, and air (Booth, 1971; Burgess et al., 1994; Nelson et al., 1994; Summerell et al., 2003). This fungus can survive for long periods in soil by the formation of structures called chlamydospores, which can colonize branches, leaves, flowers, and fruits through their conidia that are spread by air or by water. Some species are cosmopolitan and can be found in most geographical region whereas some species occur predominantly in tropical and subtropical regions, or cool to warm temperate regions (Table 1) (Burgess et al., 1994). Moreover, *Fusarium* species are even found in the enclosed buildings such as offices and hospitals (Reddy et al., 2010).

The types of vegetation are a important factor for the occurrence of *Fusarium* species such as rice, beans, wheat. Temperature in different climatic regions also affects the species distribution and virulence (Sangalang et al., 1995; Saremi et al., 1999). For example, when the temperature is low, the *Fusarium* disease that affecting alfalfa increased (Richard et al., 1982).

Table 1. The occurrence of some Fusarium species in relation to climate adapted from Burgess et al. (1994)

Species which occur in most climatic regions	Species which occur in temperate regions	Species which occur in subtropical and tropical regions
F. chlamydosporum	F. acmuminatum	F. beomiforme
F. equiseti	F. avenaceum	F. compactum
F. proliferatum	F. crookwellense	F. decemcellulare
F. oxysporum	F. culmorum	F. longipes
F. poae	F. graminearum	
F. semitectum	F. sambucinum	
F. solani	F. sporotrichioides	
F. tricinctum	F. subglutinans	

Fusarium species recovered from both natural and agricultural ecosystems have distinct climatic preferences (Backhouse et al., 2001; Saremi et al., 1999). The climate, and even local variations in weather, can limit the range of species observed (de Wolf et al., 2003; Rossi et al., 2002; Vigier et al., 1997). In some cases these differences have been used to establish disease forecasting systems for *Fusarium* diseases. In broad terms there are species that prefer tropical climates, hot arid climates or temperate climates, and a fourth group that has a cosmopolitan range. A concern with many of the species in the fourth group is that these may be sets of sibling species, in other words, species that are morphologically indistinguishable but genetically distinct, that have yet to be adequately resolved. For example, maize stalk rot from plants in relatively warm dry areas, Similar to their counterparts in temperate regions, *Fusarium* species in the tropics are quite diverse in terms of numbers of species, distribution, host range and virulence (Gordon, 1960). However, there are some notable differences between species found in tropical and temperate regions. Plant diseases caused by *Fusarium* spp. in the tropics are becoming more significant with the introduction of intensive, high-yielding production systems and genetically uniform cultivars (Waller and Brayford, 1990). *F. oxysporum* and *F. solani* are widely dispersed saprophytic soil inhabitants throughout the tropics.

Fusarium is known as soil-borne fungi because the genus is commonly found in soils and very widely distributed in soils across geographical region (Burgess et al., 1988; Burgess et al., 1994; Sangalang et al., 1995). Physical and chemical properties of the soil also affect the abundance of *Fusarium* species. For instance, the numbers of *F. solani F. phaseoli* are lower when soil pH decreased and the levels of Ca, Mg, K, and P reduced. Furthermore, the physical and chemical properties of the soil are correlated with decrease of *Fusarium* wilt of banana in Central American banana plantations (Smith and Snyder, 1971). By performing changes in soil, soil pH, and soil water supply, banana wilt caused by *F. oxysporum* f. sp. *cubense* can be suppressed (Peng et al., 1999). Temperature and availability of water also affect the distribution and the population of *Fusarium* species in soils (Sangalang et al., 1995).

Fusarium species can present two modes of nutrition saprotrophs and facultative pathogens with saprotrophic phases. Plant debris in soils plays a very important role as nutrient reservoir for *Fusarium* species to continue living in soils as saprotrophs (Burgess, 1981; Burgess et al., 1988). A fungus needs three attributes to be consistently isolated from soils, the spores must be able to commence activity, the mycelium must be able of make vegetative growth, and the fungus must be able to survive in minimal conditions. There are two phases of existence in the soil for fungi an active growth phase and a survival phase (Sangalang et al., 1995). An active growth phase is when the soil environment and the remained substrates are suitable with enough nutrients. On the other hand, a survival phase is when the soil conditions and environments are harsh with fewer nutrients. In the survival phase, soil fungi such as *F. oxysporum* will form dormant structures which are chlamydospores. Other dormant form is the multicellular resting bodies known as sclerotia. During this dormant stage, *Fusarium* species implies minimal respiration rate and reserve nutrients accumulated in the mycelium those results in maximum longevity of survival (Garrett, 1981). Some *Fusarium* species produces no resting bodies and survives by continuing through slow saprophytic activity within the colonized substrate. In addition, the survival of Fusarium in plant in the soils continues in the residues left after harvest of a diseased crop (Garrett, 1981).

We try to provide here a brief introduction to the genus Fusarium. We know that there are excellent review articles that deal specifically on taxonomy, distribution and habitat of this genus of fungi. Our goal was to introduce the subject showing some amounts of the gender and then highlight the importance of natural products of Fusarium. Below follows an extensive review on the subject.

2. Natural Products from Fungi

Products of natural origins can be called "natural products". Natural products include: an entire organism (for example, a plant, an animal, or a microorganism) that has not been subjected to any kind of processing or treatment other than a simple process of preservation (for example, drying), part of an organism (for example, leaves or flowers of a plant, an isolated animal organ), an extract of an organism or part of an organism, and exudates, and pure compounds (for example, alkaloids, coumarins, flavonoids, glycosides, lignans, steroids, sugars, terpenoids, etc.) isolated from plants, animals, or microorganisms (Samuelsson, 1999). However, in most cases the term natural products refers to secondary metabolites, small molecules (<2000 Da) produced by an organism that are not strictly necessary for the survival of the organism. Concepts of secondary metabolism include products of overflow metabolism as a result of nutrient limitation, shunt metabolism produced during idiophase, defense mechanism regulator molecules, etc (Cannell, 1998). Natural products can be from any terrestrial or marine source: plants (for example, paclitaxel-Taxol from *Taxus brevifolia*), animals (for example, vitamins A and D from cod liver oil), or microorganisms (for example, doxorubicin from *Streptomyces peucetius*).

Humans have always relied on natural products and have continually explored their application to improving various aspects of our lives. Thus, we have sought to use natural products such as spices, flavoring agents, perfumes, cosmetics, and dyes to control and overcome diseases. Some were recognized as poisons and others were used as insecticides and pesticides. This interest in, and fascination with, natural products continues today and, despite the fact that periodic claims pronounce that the era of natural products chemistry is over, this area of research is still attracting considerable interest and offering significant promise (Bode et al., 2005).

In recent years, a significant revival of interest in natural products as a potential source for new medicines has been observed among academia as well as pharmaceutical companies. Several modern drugs (~40% of the modern drugs in use) have been developed from natural products. More precisely, according to Cragg et al. (1997), 39% of the 520 new approved drugs between 1983 and 1994 were natural products or their derivatives, and 60–80% of antibacterial and anticancer drugs were from natural origins. In 2000, approximately 60% of all drugs in clinical trials for the multiplicity of cancers had natural origins. In 2001, eight (simvastatin, pravastatin, amoxycillin, clavulanic acid, azithromycin, ceftriaxone, cyclosporin, and paclitaxel) of the 30 top-selling medicines were natural products or their derivatives, and these eight drugs together totaled US $16 billion in sales. Apart from natural product-derived modern medicine, natural products are also used directly in the ''natural'' pharmaceutical industry, which is growing rapidly in Europe and North America, as well as in traditional medicine programs being incorporated into the primary health care systems of Mexico, the People's Republic of China, Nigeria, and other developing countries. The use of herbal drugs is once again becoming more popular in the form of food supplements, nutraceuticals, and complementary and alternative medicine.

3. Primary Metabolites from Fusarium

Microorganisms are capable of growing on a wide range of substrates and can produce a remarkable spectrum of products. The relatively recent advent of *in vitro* genetic manipulation has extended the range of products that may be produced by microorganisms and has provided new methods for increasing the yields of existing ones. The commercial exploitation of the biochemical diversity of microorganisms has resulted in the development of the fermentation industry and the techniques of genetic manipulation have given this well-established industry the opportunity to develop new processes and to improve existing ones. The term fermentation is derived from the Latin verb *fervere,* to boil, which describes the appearance of the action of yeast on extracts of fruit or malted grain during the production of alcoholic beverages (Raimbault, 1998).

The growth of a microorganism may result in the production of a range of metabolites but to produce a particular metabolite the desired organism must be grown under precise cultural conditions at a particular growth rate. If a microorganism is introduced into a nutrient medium that supports its growth, the inoculated culture will pass through a number of stages and the system is termed batch culture. Initially, growth does not occur and this period is referred to as the lag phase and may be considered a period of adaptation. Following an interval during which the growth rate of the cells gradually increases, the cells grow at a constant, maximum rate and this period is referred to as the log or exponential phase (Gottelieb and Etten, 1964; Calvo et al., 2002).

The kinetic description of batch culture may be rather misleading when considering the product-forming capacity of the culture during the various phases, for, although the metabolism of stationary phase cells is considerably different from that of logarithmic ones, it is by no means stationary. Bu'Lock et al. (1986) proposed a descriptive terminology of the behaviour of microbial cells which considered the type of metabolism rather than the kinetics of growth. The term 'trophophase' was suggested to describe the log or exponential phase of a culture during which the sole products of metabolism are either essential to growth, such as amino acids, nucleotides, proteins, nucleic acids, lipids, carbohydrates, etc., or are the by-products of energy-yielding metabolism such as ethanol, acetone and butanol. The metabolites produced during the trophophase are referred to as primary metabolites (Figure1). Some examples of primary metabolites of produced by *Fusarium* species are described below.

Bu'Lock et al. (1986) suggested the term 'idiophase' to describe the phase of a culture during which products other than primary metabolites are synthesized, products which do not have an obvious role in cell metabolism. The metabolites produced during the idiophase are referred to as the secondary metabolites. The interrelationships between primary and secondary metabolism are illustrated in figure 1, from which it may be seen that secondary metabolites tend to be synthesized from the intermediates and end-products of primary metabolism.

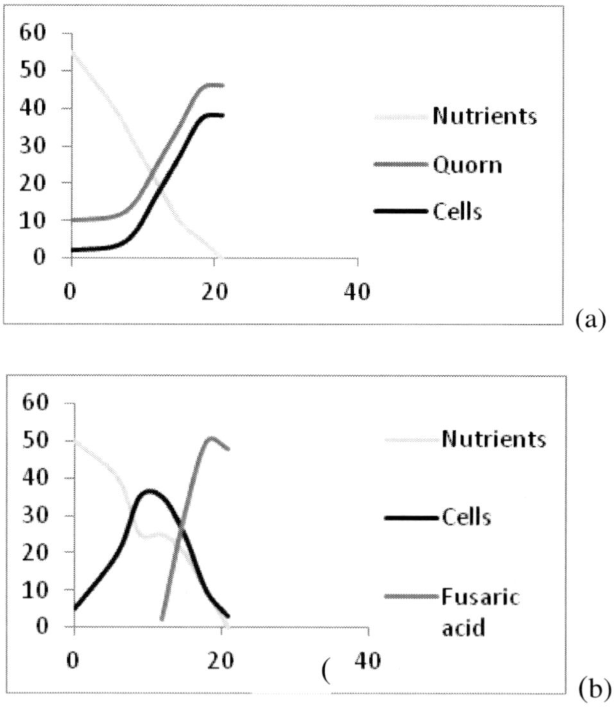

Figure 1. Contrast between production of primary (trophophase) and secondary (idiophase) metabolites. (a) Quorn production by *Fusarium venenatum* – an example of a primary metabolite. (b) Fusaric acid production by *Fusarium* species – an example of a secondary metabolite. Note how Fusaric acid is not made until after mid-log phase.

3.1. Proteins from Fusarium

The production of mycoprotein provides an excellent example of the use of filamentous fungi to provide high-class protein. Mycoprotein is a fibre-rich food produced by the continuous fermentation of a selected strain of *Fusarium venenatum*, first isolated from a field. It was originally deposited as a strain of *Fusarium graminearum*. However, thorough investigations of the taxonomy by O'Donnell et al. (1998) and Yoder and Christianson (1998) have revealed, that the strain belongs to the species *Fusarium venenatum* established by Nirenberg (1995).

Fusarium venenatum A3/5 (ATCC PTA-2684) was first investigated as a potential protein source for human consumption during the late 1960s by the British company Rank Hovis McDougall (RHM). RHM was seeking to develop a microbial protein source which would be cheap and easy to produce from starch- or glucose-based media, but which would also be palatable. They found that filamentous fungi provided a suitably textured product and *F. venenatum* A3/5 was selected as the best fungus for further product development after 3 years of screening approximately 3,000 different fungi (Anderson and Solomons 1984: Wiebe et al., 2002).

In order to bring mycoprotein from *F. venenatum* A3/5 onto the market, it was necessary for RHM to invest 12 years in researching the safety of the organism (as a potential plant pathogen) and of the final product (Edelman et al., 1983; Anderson and Solomons 1984; Solomons 1986), making mycoprotein the most carefully tested food product on the European market. Mycoprotein produced from *F. venenatum* A3/5 was approved by the Ministry of Agriculture, Fisheries and Food (MAFF) for sale in the United Kingdom in 1984, but toxicity testing has continued (Miller and Dwyer 2001), as have investigations into allergic reactions. Some strains of *F. venenatum* (e.g. NRRL 22198) are known to produce type A trichothecenes, including diacetoxyscirpenol (DAS), scirpentriol, 15-acetoxyscirpenol and 4 monoaceytoxyscirpenol (O'Donnell et al., 1998). Isotrichodermin, isotricodermol, sambucinol, apotrichothecen, culmorin, culmorone and enniatin B have also been detected in cultures of *F. venenatum*. However, the strain used for myco-protein production, ATCC PTA-2684 (formerly NRRL 26139 = ATCC 20334), did not produce mycotoxins when grown in conditions which induced mycotoxin production in the strain NRRL 22198 (O'Donnell et al. 1998), and the growth conditions used for production are not suitable for mycotoxin production.

Mycoprotein is produced as a cake with about 30% solids which have textures and flavoured that allow to make a range of food products such as chicken substitute. Mycoprotein is also used as an extender in pies, burgers and cold slicing meats. Mycoprotein is bland and so can easily be flavoured; it also has a texture that provides 'bite'. Its composition is similar to that of lean beef. Mycoprotein has been approved for human use in the UK and applications for certification have been made elsewhere in the world. It is now marketed in the UK under the trade name Quorn (Anderson and Solomons 1984; Wiebe et al., 2002). One of the advantages that mycoprotein has over soya, yeasts and bacterial proteins is its filamentous nature, which gives it a texture and bite similar to that of meat. Although filamentous proteins can be produced by spinning bacterial and plant cells, such a step is unnecessary with mycoprotein which is naturally filamentous. Quorn is produced by growing *F. graminearum* in a 1300-litre continuous culture fermenter. The organism is grown at 30°C at pH 6 on a sterilized medium containing glucose syrup, mineral salts, trace metals, choline and biotin. Oxygen is supplied as compressed air to maintain dissolved oxygen concentrations in the 0.05-0.3 atmosphere range. Ammonia gas is used to control the pH of the medium and also to provide the nitrogen source (Wiebe et al., 2002).

Immediately after fermentation the RNA content of the mycelium is around 10 %, which is too high for animal or human consumption. The RNA content is therefore reduced, using thermal shock and endogenous ribonucleases, to a value less than 2 %. After RNA reduction, the mycelium is continuously harvested on an horizontal filter bed, which forms a filter cake made up of a mat of interwoven fungal hyphae, which can be stored at minus 18 °C and will keep almost indefinitely. Mycoprotein typically contains 44.3 % protein, 18.3 % dietary fibre and only 13 % fat. It also contains an ideal blend of aminoacids and has a low sodium content. It is therefore an ideal food for diet-conscious consumers in the developed world. Quorn is free of mycotoxins, which hitherto limited the use of *Fusarium* species in biotechnology. However, it has yet to be produced on an economic scale, which would require a large fermenter to be in continuous use producing some 20000-30000 tonnes of biomass per year, and involving processing techniques similar to those used in breweries or pharmaceutical industries (Anderson and Solomons 1984; Wiebe et al., 2002).

3.2. Enzymes from Fusarium

Nowadays, enzymes have become an important tool to obtain various industrial products. Enzymes of different origin and activity are utilized in food, chemical, textile and pharmaceutical industries. Various species of filamentous fungi are the main organisms that produce industrial enzymes (Alkorta et al., 1998).

Cellulases are a group of hydrolytic enzymes capable of hydrolysing cellulose to smaller sugar components like glucose units. Cellulolytic enzymes play an important role in nature's biodegradation processes where plant lignocellulosic material are efficiently degraded by cellulolytic fungi and bacteria. In industry, these celluloytic enzymes have found novel applications in the production and processing of chemicals, foods and manufactured goods such as paper, rayon and cellophane. Cellulases, for instance have been extensively utilized for extraction of valuable components from plant cells, improvement of nutritional values of animal feed and the preparation of plant protoplasts in genetic research (Mandels, 1985). Generally, fungi produce three major types of cellulolytic enzyme: endoglucanase, exoglucanase and cellobiohydrolase (Klyosov, 1990). These enzymes are extracellular and inductive in nature (Enari, 1983). The ability to produce cellulase are widespread among fungi and this has become the subject of extensive investigation.

Microbial lipases have already established their vast potential regarding to their usage in different industries. The interest in microbial lipase production has increased in the last decades, because of its large potential in industrial applications as additives for foods (flavor modification), fine chemicals (synthesis of esters), waste water treatment (decomposition and removal of oil substances), cosmetics (removal of lipids), pharmaceuticals (digestion of oil and fats in foods), leather (removal of lipids from animal skins) and medical (blood triglyceride assay) (Pandey et al., 1999; Burkert et al., 2004; Kumar et al., 2005). However, the biggest market of their use is in the detergent formulation (Prazeres et al., 2006). Regarding detergents applications, new challenges for lipases producers can be pointed (Rathi et al., 2001). The high variation in the triglyceride content of fat stains, requiring lipases with low substrate specificity; the relatively harsh washing conditions (pH 10-11 and 30-60ºC), requiring stable enzymes; and the effects of chemical denaturation and/or proteolytic degradation caused by detergents additives such as surfactant linear alkyl benzene sulfonate (LAS) and proteases (Rathi et al., 2001).

Prazeres et al. (2006) evaluated the effect of different commercial detergents and surfactants on enzymatic activity of lipase from *Fusarium oxysporum* through *p*-nitrophenylpalmitate (*p*NPP) assay. The enzyme was compatible with various ionic and non-ionic surfactants as well as commercial detergents. Lipase activity was strongly inhibited by Sodium Dodecyl Sulphate (SDS), but not by Triton X-100 and Triton X-114. The best assay conditions observed for this lipase were pH 8.0 and 50ºC. The enzyme was stable at alkaline pH and remained 93% of residual activity during 1 h incubation at 60ºC. The highest lipase activity was measured with triglycerides of middle and long chain fatty acids (C8-C18). According Prazeres et al. (2006) this enzyme showed a variable specifity/hydrolytic activity towards various fats and oils. All these properties and its resistance towards various surfactants and tolerance to commercial detergents make this lipase a potential additive for detergent formulation.

Kwon et al. (2007) evaluated thirty seven species of *Fusarium* for their ability of producing extracellular enzymes using chromogenic medium containing substrates such as starch, cellobiose, CM-cellulose, xylan, and pectin. Among the tested species *F. mesoamericanum*, *F. graminearum*, *F. asiaticum*, and *F. acuminatum* showed high β-glucosidase acitivity. Xylanase activity was strongly detected in *F. proliferatum* and *F. oxysporum*. Strong pectinase activity was also found in *F. oxysporum* and *F. proliferatum*. Amylase activity was good in *F. oxysporum*. No clear activity in cellulase was found from all the *Fusarium* species tested.

Maia et al. (1999) studied a Brazilian strain of *Fusarium solani* for ability to produce extracellular lipase. The authors evaluated the production in peptone-olive oil medium. The fungus produced 10,500 U.L of lipase after 72 hours of cultivation at 25ºC in shake-flask at 120 rpm in a medium containing 3% (w/v) peptone plus 0.5% (v/v) olive oil. The optimum lipase activity was achieved at pH 8.6 and 30ºC and a good enzyme stability (80% activity retention) was observed at pH ranging from 7.6 to 8.6, and the activity rapidly dropped at temperatures above 50ºC. Lipase activity was stimulated by the addition of n-hexane to the culture medium supernatants, in contrast to incubation with water-soluble solvents.

Nwagu and Okolo (2011) observed the production of α- amylase by the fungi *Fusarium* sp. isolated from the soil. Growth and enzyme production occurred at 30, 45 and 55 ºC. Soybean meal at 1% concentration, supplemented with 0.2% NH4Cl and 2.5% corn starch elicited the highest amylase yield. Optimum pH for the enzyme was pH 6.5 which retained over 60% of its activity after 24 h incubation at the pH range of 4.5-7.0. The enzyme showed high activity from 40-70 ºC with optimal activity at 50 ºC and 78% activity was retained after incubation at 70 ºC for 30min. Catalytic function of the crude amylase was stimulated by Mg (136%), Ca (118%) and Zn (118%) at 2mM concentrations. The enzyme hydrolyzed cassava, potato and yam starches effectively. According to Nwagu and Okolo (2011), large amounts of amylase could be produced by the *Fusarium* sp. on cheap and easily accessible substrates. Further studies and optimization could prove the amylase to be very useful in industrial processes requiring amylase or its hydrolytic products.

Ueda et al. (2007) isolated *Fusarium* sp., which produces a strongly fibrinolytic enzyme, from plant leaf (Hibiscus). Fibrinolytic alkaline protease was purified from a culture filtrate of

Fusarium sp. by precipitation with (NH4)2SO4 and column chromatography with CM-Toyopearl 650M and Superdex 75. Maximum activity of protease was observed at pH 9.5 and 50°C. Purified protease was active between pH 2.5 and 11.5 and was found to be stable up to 50°C. The enzyme derived from *Fusarium* sp. is useful for thrombolytic therapy because this enzyme showed pH resistance. The *Fusarium* protease was able to maintain the activity for long periods of time *in vivo* because of its high pH resistance. The *Fusarium* protease shows a significant potential for food fortification and nutraceutical applications, such that their use could effectively prevent cardiovascular diseases. The detailed protein and gene structure of the *Fusarium* protease are currently under investigation.

Sridevi and Charya, (2011), studied the production of the extracelular enzymes by 30 fungal species. They found *Fusarium* sp., with high xylanase activity, low cellulase production and biomass, showed promising industrial applicable results. These results are in consistence with the results obtained by Gupta et al. (2009), where they showed the xylanase production ability by *Fusarium* species for the first time. They believe that *Fusarium* could be a useful organism for the large-scale production of enzymes for paper and pulp industry (Gupta et al., 2009; Wood and Mc Crae, 1986).

4. Secondary Metabolites from Fusarium

The gibberellins (GA), oxygenated tetracyclic diterpenoids, are known as plant growth regulators with hormonal functions (Lang, 1956). Gibberellins have a carboxylic group at C-7 and nearly all have a $\Delta^{16,17}$ double bond. The structures of the C19 and C20-gibberellins are based on the 20-norgibberellane and on the gibberellane skeletons, respectively (Ávalos et al., 2007).

GAs have been identified as secondary products, mainly in *Fusarium fujikuroi*, the pathogenic fungus of the rice which causes overgrowth symptoms ("bakanae" disease) that synthesizes high levels of gibberellic acid (GA_3) (Troncoso et al., 2010). Others strains (*F. verticillioides* and *F. proliferatum*) were isolated from infected rice seedlings in different geographic regions (Desjardins et al., 1997). Malonek et al. (2005) showed that from 50 *Fusarium* species, most of the sexually fertile species (B, D, E, F, G, and I) and anamorphs within the *G. fujikuroi* species complex, have the entire GA biosynthetic gene cluster or at least some of the genes. However the expression of GA biosynthetic genes and GA production were detected only in *F. fujikuroi* and one isolate of *F. konzum*. They speculate that GAs play an important role in the process of infecting and/or invading rice seedlings.

GAs were identified as diterpenoid compounds by incorporation of ^{14}C-labelled acetate and mevalonate into GA_3 in fungal cultures (Birch et al. 1958, 1959). Fungal gibberellins are synthesized through the mevalonate pathway. Mevalonate, produced from 3-hydroxymethylglutaryl CoA, is the precursor of two C5 compounds, isopentenyl pyrophosphate and dimethylallyl pyrophosphate. Thus, a major difference between the GA pathways in *F. fujikuroi* and plants is the stage at which the hydroxyl groups are introduced, and the different enzymes responsible for the 3-β-hydroxylation (Hedden et al., 2002; Tudzynski, 2005).

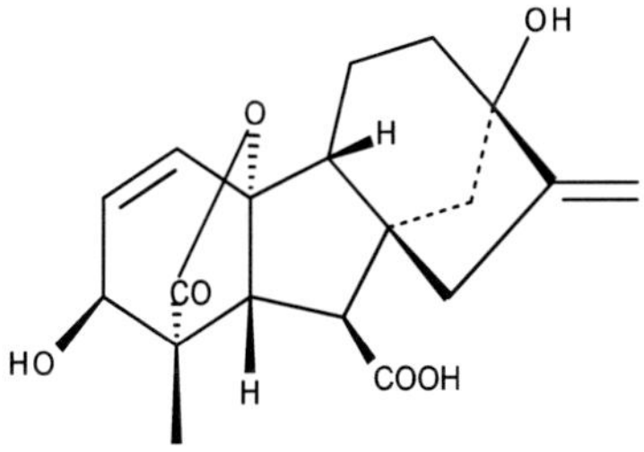

GA3

Ent-kaurenoid derivatives, as kaurenolides (diterpenoid lactones) and fujenoic acids (e.g. seco-ring B kaurenoids) are biosynthetically related to the GAs and produced by *Gibberella fujikuroi* (Cross et al., 1963a,b; Hedden et al., 1973). Kaurenolides and seco-ring B kaurenoids are formed by the P450-1 monooxygenase (GA14 synthase) of *G. fujikuroi* and are, probably, side products from stabilization of radical intermediates involved in GA14 synthesis (Rojas et al., 2004).

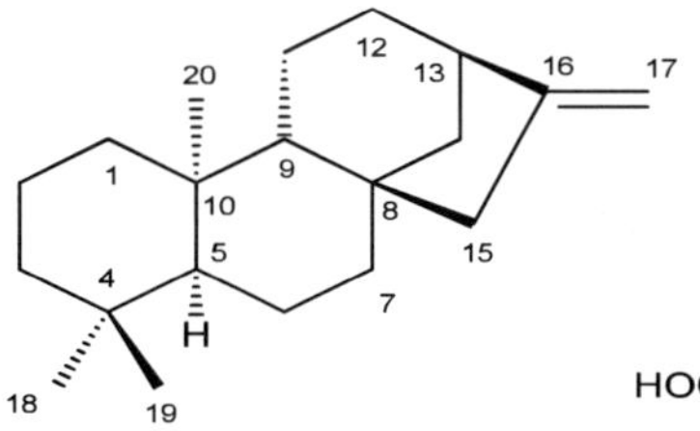

Ent-kaurene

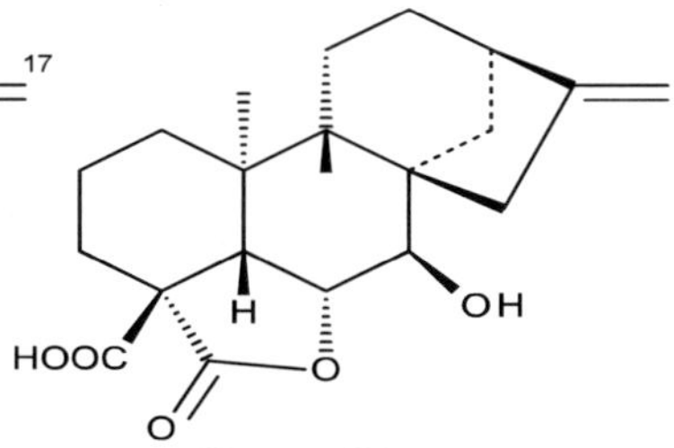

Kaurenolide

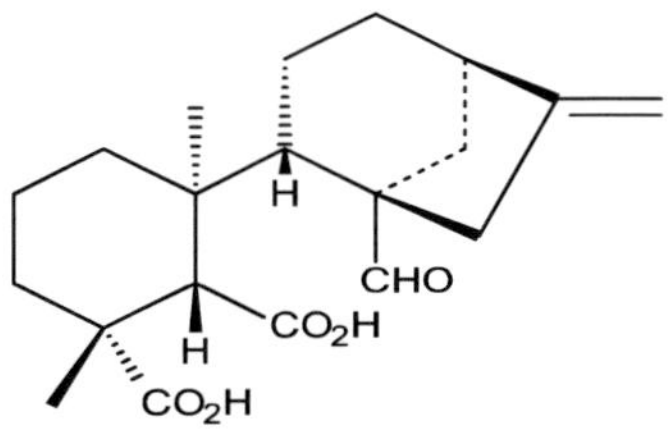

Fujenoid acid

Six gibepyrones A-F, *α*-pyrones, were isolated from *Gibberella fujikuroi* (IMI 58289; *F. fujikuroi*) grown on calcium lactase as the carbon source. Gibepyrones A and B exhibited weak inhibitory activity against *S. aureus* (MIC 200 μg/ml) and *S. cerevisiae* (MIC 100 μg/ml) (Barrero et al., 1993).

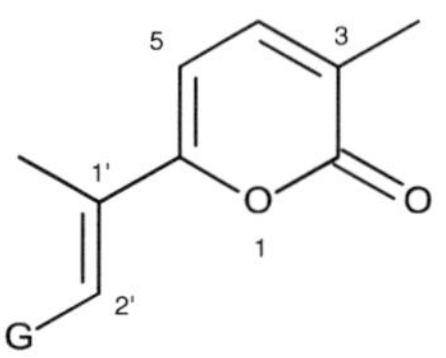

Gibepyrone A = CH_3
Gibepyrone B = CH_2OH
Gibepyrone C = CHO
Gibepyrone D = COOH
Gibepyrone G = $COOCH_3$

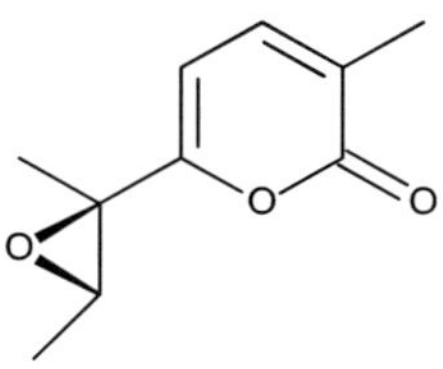

Gibepyrone E

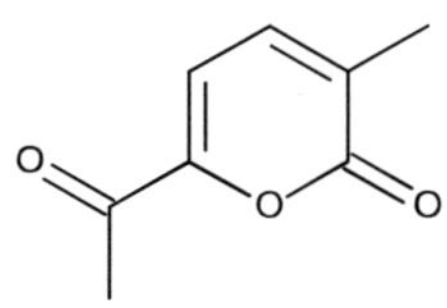

Gibepyrone F

Several species of *Fusarium* cause symptoms of *Fusarium* head blight (FHB) in wheat (*Triticum spp.*), barley (*Hordeum vulgare*) and oat (*Avena sativa*) (Rudd et al., 2001) and deoxynivalenol production has been associated with the virulence of *Fusarium culmorum* (Scherm et al., 2011). Trichothecene mycotoxins can induce toxicoses in animals and humans, to cause deaths and can contribute to the ability of some fusaria to cause plant disease (Ciegler and Bennett, 1980).

Trichothecenes have a tetracyclic sesquiterpenoid 12,13-epoxytrichothec-9-ene ring in common, and the 12,13-epoxy ring is responsible for the toxicological activity. They are derived from farnesyl-pyrophosphate followed by cyclization and oxygenations steps (Kimura et al., 2007). On the basis of the absence or presence of characteristic functional groups, the trichothecenes can be classified into four types (A, B, C, D) (Table 2). Type B trichothecenes typically have a C-7 hydroxyl group only in *Fusarium* genus and the degree of toxicity of trichothecenes does not necessarily correlate with chemical classification (McCormick et al., 2011).

Table 2. Classification of trichothecenes types, functional characteristics and representative compounds from each group

Type	Functional characteristic	Compounds
A	without carbonyl group at the C-8 position	T-2 toxin and HT-2 toxin
B	carbonyl group at the C-8 position	deoxynivalenol and nivalenol
C	epoxy ring between C-7 and C-8 or between C-9 and C-10	crotocin and baccharin
D	macrocyclic ring between C-4 and C-15with two ester linkages	satratoxin and roridin

Li et al., 2011.

Currently approximately 190 members of the trichothecene family have been identified (Li et al., 2011). They are stable under different environmental conditions because when exposed to air, light, or submitted to autoclaving, they are not inactivated. During the 1970s and 1980s, the trichothecene mycotoxins were used by the Soviet Union as warfare against Southeast Asia and Afghanistan in attacks known as "yellow rain" (Haig, 1982; Mirocha, 1982). Wannemacher and Wiener (1997) describe mechanism of action, metabolism and relative acute parenteral toxicity of the most abundant trichothecene mycotoxins. Other recent review covers the types of trichothecenes, their complexity, and proposed biosynthetic pathways (McCormick et al., 2011).

Li et al. (2011) describe on the acute and chronic toxicity, apoptosis studies in the immune system and metabolism of T-2 toxin (T-2). T-2 has the highest toxicity of all the trichothecenes, although it is less frequently detected compared to the other trichothecenes. T-2 as like as moniliformin, have been associated with the Kashin-Beck disease. T-2 inhibits aggrecan synthesis and promotes aggrecanase and pro-inflammatory cytokine production in cultured chondrocytes. Wistar rats fed with a diet containing T-2 toxin (100 ng/kg chow), for six and ten months, showed chondrocyte degeneration/necrosis and loss, chondrocyte clones, and loss of proteoglycan staining of articular cartilage (Wang et al., 2011). Recently, a strain of *Fusarium* sp. (UFMGCB 551), isolated from the plant *Piptadenia adiantoides* J.F. Macbr (Fabaceae), furnished T2-toxin and a mixture of 8-*n*-butyrylneosolaniol and 8-isobutyrylsolaniol as antifungal compounds against strains of *Paracoccidioides brasiliensis* ranging from 75 to 640 nmol (Campos et al., 2011).

Members of the A-type trichothecenes are represented by scirpentriol (3α,4β, 15-trihydroxy-12,13epoxytrichothec-9-ene) and its acetylated and unacetylated derivatives. They occur in cereals and non-grains (soybean, banana and potato). Diacetoxyscirpenol an others trichothecenes were found in dust samples from ventilation systems of office buildings spaces of the Montreal urban area, reportedly affected by the "sick buildings syndrome" (Smoragiewicz et al., 1993). From 79 bulk samples of moldy interior finishes, fifteen samples (19%) contained trichothecenes inclunding diacetoxyscirpenol (Tuomi et al., 2000).

Scirpentriol, 3-, 4- and 15-monoacetoxyscirpenol (MAS), 3,4-, 3,15-, and 4,15-diacetoxyscirpenol (DAS), and 3,4,15-triacetoxyscirpenol were compared for their acute oral lethality in broiler chicks, lethality in brine shrimp, and dermal toxicity in guinea pigs. 4,15-diacetoxyscirpenol was the most toxic in the three assays and about 3 times more toxic in chicks than aflatoxin. In general, a free hydroxy group at the 3-position was a primary determinant of toxicity, but scirpenol family did not follow precisely one pattern of toxicity (Richardson and Hamilton, 1990). LD_{50} values of 4,15-DAS, 15-MAS, T-2 and deoxynivalenol to mice, by intraperitoneal administration, were 15.3, 4.5, 9.1 and 43 mg/kg body weight, respectively (Thompson and Wannemacher, 1986; Schollenberger et al., 2007).

Anguidine (4,15-DAS) is produced by strains of *F. equiseti* and has activity (0.05-1.6 mg/kg/day) *in vivo* against the transplantable mouse tumor P-388 lymphatic leukemia (Claridge et al., 1978). Anguidine is highly cytotoxic (<1 μg/ml for HeLa and L929) in several continuous tumor cell lines and irreversibly blocks protein synthesis by inhibiting protein chain initiation through degradation of polyribosomes. It blocks the G_1, into S and G_2 into mitosis transitions human adenocarcinoma cell line (Dosik et al, 1978). Anguidine reached Phase II trials in patients with colorectal adenocarcinoma (DeSimone et al., 1986), central nervous-system tumors (Goodwin et al., 1983), gastrointestinal malignancies (Bukowski et al., 1982), sarcomas unresponsive to prior chemotherapy (Thigpen et al., 1981) but resulted in unsuccessfully therapies because, in part, of its toxicity (thrombocytopenia, leukopenia, nausea and vomiting, hypotension, fever) and pathophysiologic effects (Bukowski et al., 1982; DeSimone et al., 1986).

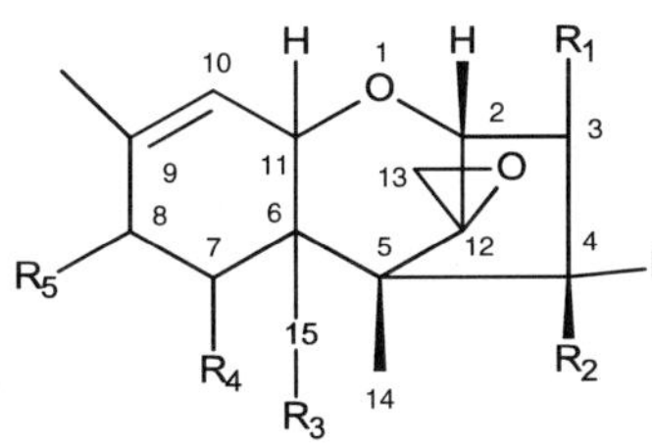

Name	R1	R2	R3	R4	R5
Nivalenol	-OH	-OH	-OH	-OH	=O
Deoxynivalenol	-OH	-H	-OH	-OH	=O
Scirpentriol	-H	-H	-OH	-OH	-OH
15-acetoxyscirpentriol	-H	-H	-OAc	-OH	-OH
8-Isobutyrylneosolaniol	-OH	$-OCOCH_3$	-OAc	-H	$-OCOCHMe_2$
8-*n*-butyrylneosolaniol	-OH	$-OCOCH_3$	-OAc	-H	$-OCOCH_2CH_2Me$
Anguidine	-OH	$-OCOCH_3$	-OAc	-H	-H

Fusaproliferin (FUS, 18-[1-(acetoxymethyl)-2-methylethyl]-10,17-dihydroxy-3,7,11,15-tetramethylbicyclo[13.3.0]octadeca-trans,*trans,trans,cis*-2,6,11,17-tetraen-16-one or proliferin), a a bicyclic sesterterpene, was isolated by Randazzo et al. (1993) from *Fusarium proliferatum*, a fungal pathogen of maize. It derived from the isoprenoid pathway via common terpen intermediates originating from geranylfarnesyl pyrophosphate starting from five isoprene units linked head-to-tail in a linear fashion (Jestoy, 2008; Santini et al., 1996). FUS is produced by several closest *Fusarium* species pathogens of maize and cereal grains worldwide as *F. subglutinans* and *F proliferatum*. This compound can induce teratogenic effects, e.g., cephalic dichotomy, macrocephaly and limb asymmetry in chicken embryos (Ross et al., 1993). Fusaproliferin production by 71 strains, isolated from various geographic areas (Europe, North America, Argentina, and Africa), ranged from 30 to 1,500 μg/g of dried ground substrate (Logrieco et al., 1996). It is toxic to *Artemia salina* (LC_{50} 53.4 μM) and to the lepidopteran cell line SF-9 (IC_{50} 70 μM). Fusaproliferin was also toxic to the human nonneoplastic B-lymphocyte cell line IARC/LCL 171 (IC_{50} 55 μM) (Logrieco et al., 1996) but did not show activity against *S. aureus* and *E. coli* at concentrations ranged from 1 ng/ml to 10 μg/ml by turbidimetry assay (Meca et al., 2009).

Marine-derived *Fusarium* sp., obtained from a driftwood sample in a mangrove habitat located at Bahanas, furnished rearranged tetracarbocyclic sesterpenoids that include the halogen containing neomangicols (Renner et al., 1998) and the related mangicols (Renner et al., 2000). Neomangicols A-C are class of C_{25} rearranged sesterterpenes, which the tetracyclic ring system appears to be produced by atypical terpenoid pathways resulting in irregular "head to tail" isoprene configurations. They are produced by *Fusarium* (strain CNC477), but A and B have halogen substituent at C-23, a chlorine or a bromine atom, respectively. It suggests the presence of halogenating enzymes, a feature in the marine microorganisms, indicative that this strain is adapted to the marine environment. Neomangicols A and B were active against a panel of cell lines. Neomangicol A was the most active against MCF-7 (human breast carcinoma) and CACO-2 (human colon carcinoma) cell lines, displaying a mean of the IC_{50} value of 10 μM and although neomangicol B was less active (mean of the IC_{50} value of 27 μM), it was activity against *Bacilus subtilis* at 50 μg/disc (Renner et al., 1998).

Further, from the same strain, *Fusarium heterosporum* CNC-477, cultured in a seawater-based medium, were obtained spirotricyclic structures named mangicols A-G. They showed GI_{50} values ranging from 18 to 36 μM in the National Cancer Institutes 60 cell line panel (mangicol A, 24.5 μM; B, 20.4 μM; C, 17 μM; D, 24.5 μM; E, 17.8 μM; F, 36.3 μM; G, 25.1 μM).

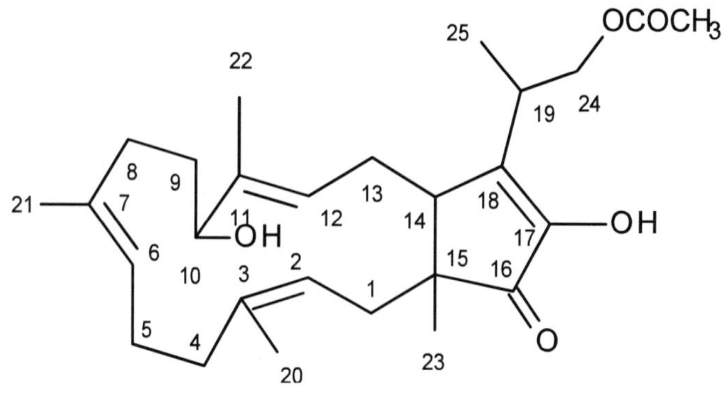

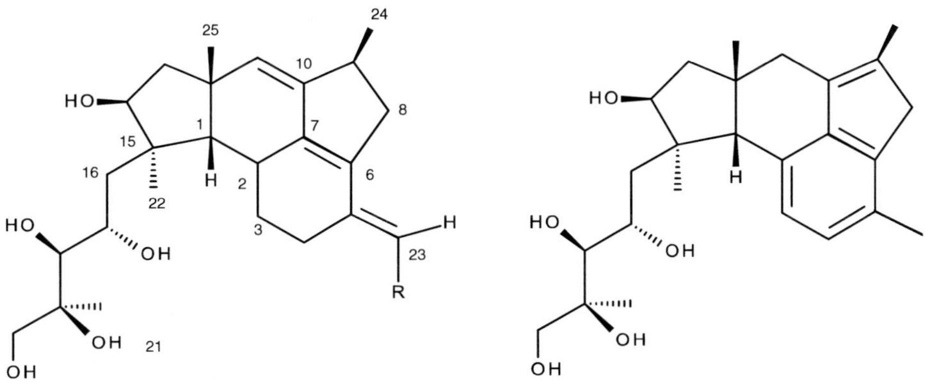

Neomangicol A R = Cl

Neomangicol B R = Br

Neomangicol C

Mangicols A and C showed significant antiinflammatory activity in the phorbol myristate acetate-induced mouse ear edema model (81 and 57% reduction in edema, respectively) at the standard testing dose of 50 μg per ear; values with the potencies of the antiinflammatory agents as indomethacin (71% reduction) (Renner et al., 2000).

Feeding studies (sodium [1,2-^{13}C]acetate and sodium [1-^{13}C]acetate) allowed to propose that three of the five isoprene units of mangicols were incorporated intact, where C-17 through C-21, comprise the tail and the second (C-4 through C-7 plus C-23) and third (C-8 through C-11 plus C-24) intact isoprene units appeared to constitute the other end of the C_{25} unit. The eleven-membered ring is formed, followed by two Meerwein (1,2-alkyl) shifts, simple cation-induced ring closures and 1,2-hydride transfers (Renner et al., 2000).

Culmorins are tricyclic sesquiterpene diols derived from farnesyl pyrophoshpate (Hanson and Nyfeler, 1976) that have been reported from *Fusarium culmorum, F. graminearum* and *Fusarium venenatum* (Langseth et al., 2001). The parent culmorin (2,6,6,9-tetramethyltricyclo[5,4,0^{1,7},0^{2,9}]undecane-8,11-diol) was first isolated in 1937 from *F. culmorum* by Ashley et al. (1937) and its structure was elucidated only in 1967 (Barton and Werstiuk, 1967, 1968).

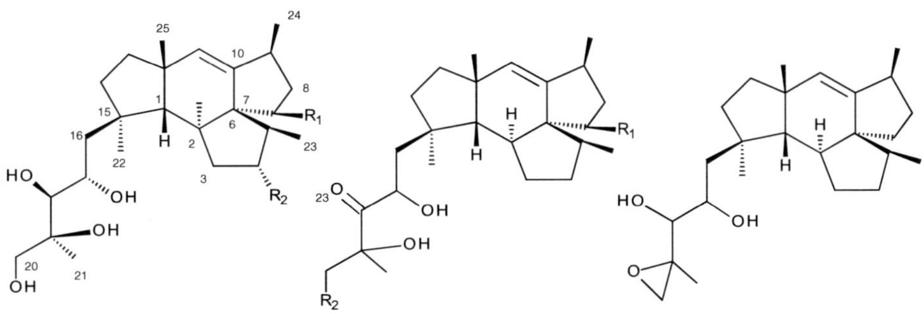

Mangicol A R₁ = OH, R₂ = H

Mangicol B R₁ = H, R₂ = OH

Mangicol C R₁ = H, R₂ = H

Mangicol D R₁ = OH, R₂ = H

Mangicol E R₁ = H, R₂ = OH

Mangicol F R₁ = H, R₂ = H

Mangicol G

Culmorin has antifungal activity (Strongman et al., 1987) but it is not mutagenic in the Ames test (2.5 μmol^{-1} by plate) (Pedersen and Miler, 1999). The first hydroxyculmorins; 5-, 12, and

15-hydroxy-culmorins, were isolated and characterized from *F. culmorum* (Kasitu et al., 1992). From 23 *Fusarium culmorum* and 21 *F. graminearum* isolates, analysed by gas chromatography mass spectrometry (GC-MS), 15-hydroxy-culmorin followed by 5-hydroxy-culmorin were the main other metabolites produced *F. culmorum*, while 5- and 12- hydroxy-culmorin, were the main metabolites of *F. graminearum* (Langseth et al., 2000).

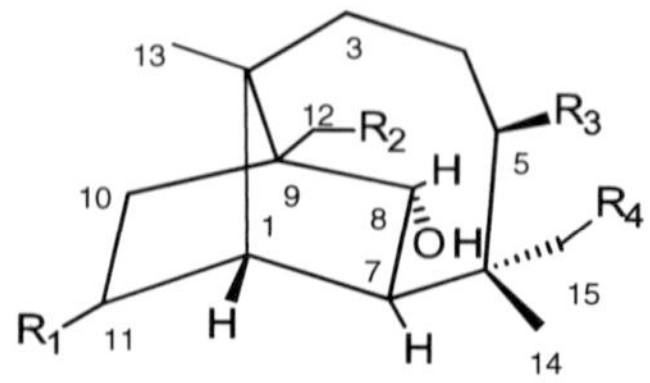

Name	R1	R2	R3	R4
Culmorin	α-OH	–H	–H	–H
5-Hydroxyculmorin	α-OH	–H	-OH	–H
15-Hydroxyculmorin	α-OH	–H	–H	-OH
12-Hydroxyculmorin	α-OH	-OH	–H	–H

Steroid sulfates were recognized as secondary metabolites from *Fusarium* sp. in 1989 and named as integracides, members of the oxygenated tetracyclic triterpenoids of the 4,4-dimethylergostane family. These compounds were isolated from *Fusarium* sp. (L-681,512, ATCC 20858), a strain from soil microorganism, as elastase inhibitors (IC_{50} 1.6-2.5 and μM) to be useful in the treatment of inflammatory and degenerative diseases (Richard and Yujin, 1989; Ashe and Fletcher, 1989).

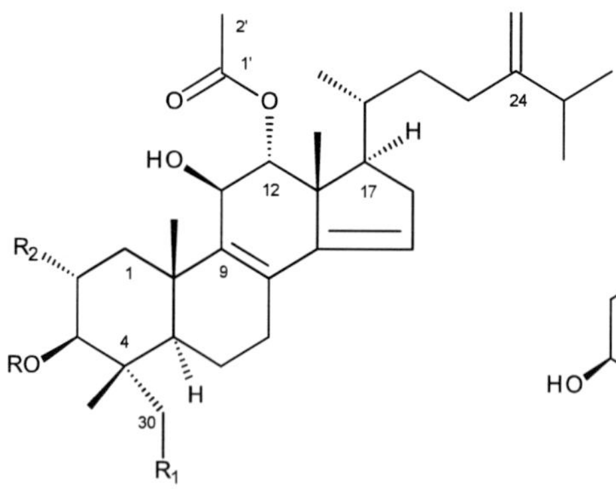

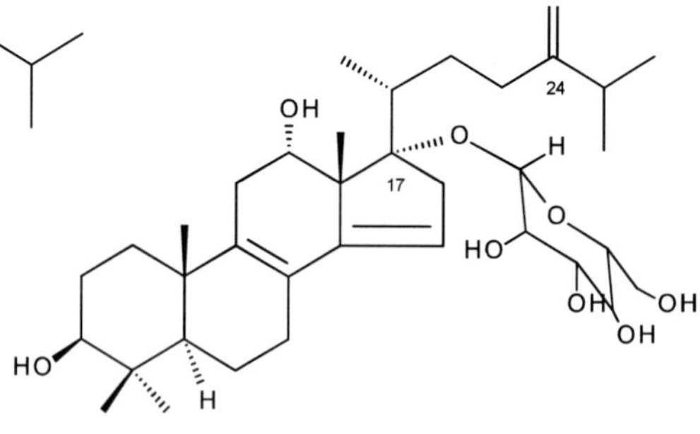

Integracide D

R	R1	R2	Name
$-SO_3H$	-H	-OH	Integracide A
-H	-H	-OH	Integracide B
-H	-OH	-OH	Integracide C

One of these compounds, integracide A [4,4,24-trimethylcholesta-8,14,24(28)-trien-2α,3β,11α,12β-tetrol 12- acetate, 3-sulfate], was found in large amounts as phytotoxic compound from methanol extract of white corn grits contamined with *F. graminearum* NRRL. Burmeistert and Vesonder (1990) confirmed the presence of integracide A in 40 *Fusarium* strains belonging to *F. sporotrichioides, F. chlamydosporum, F. acuminatum, F. sambucinum, F. culmorum* and *F. graminearum* by TLC and UV spectrum analyses. Integracide A demonstrated antiinsectan properties against *Spodoptera frugiperda* (armyworm) and *Carpophilus hemipterus* (beetle) larvae at 2500 and 4000 ppm, respectively (Dowd et al., 1992). It showed phytotoxic activity on germinating of wheat and tomato seeds (25-200 μg/ml) and antimicrobial activity against fungi (*Candida, Aspergillus, Cryptococcus, Mucor, Penicillium and Fusarium*), two Gram-positive bacteria (two *B. subtilis* strains and *S. aureus*) and one alga (*Chlorella pyrenoidosa* NRRL YB-3399) at 160 μg/ml (Burmeister and Vesonder, 1990).

A-108835, diastereomeric compound from integracide A, which differs only by the stereochemistry at the C-11 and C-12 positions, was isolated as major bioactive compound from *Fusarium compactum* (AB 21941-103). The compound had no antifungal and antibacterial activities (Gram-negative) at 100 μg/ml, in spite of it had been presented weak activity against some *S. aureus* and *Streptococcus* strains in the concentrations ranging from 6 to 50 μg/ml. This metabolite also inhibits proteolytic activity of rhinovirus 3C protease on a fluorogenic peptide substrate (IC_{50} 48 μg/ml) (Brill et al., 1996).

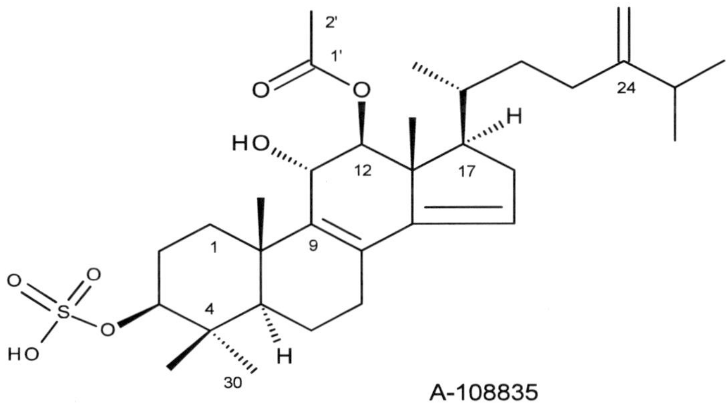

A-108835

Others members of the group, integracides B-D, were isolated together integracide A from *Fusarium* strains (ATCC74469 and FO-6651), obtained from soil collected in Tanzania and Japan, respectively (Singh et al., 2003a, Tabata et al., 1999). The desulfated compounds from integracide A (B, C and D) inhibit HIV-1 integrase enzyme, responsible by integration of viral DNA into host cell DNA, (IC_{50} 82, >100, 50 μM, respectively), but were significantly less active in comparison to integracide A (IC_{50} 4 μM) (Singh et al., 2003b). Integracides A (L681,512-1) and compounds L681,512-2, L681,512-3 and L681,512-4 are inhibitors of cholesteryl ester transfer protein (IC_{50} 8.2-34 μg/ml) (Tabata et al., 1999), of strand transfer reactions of recombinant HIV-1 integrase (IC_{50} 5.6-15 μM), of 3'-end processing activity (IC_{50} 2-10 μM), of preintegration complex (IC_{50} 30-50 μM) and have antiviral activity (CIC_{95} 6-50 μM) in the multiple cycle H9 viral spread assay. Unfortunately, they also showed toxicity to Hella cells at 12-50 μM (Singh et al., 2003a, b).

Carotenoids are terpenoid pigments widely distributed in nature, and provide red, orange, or yellow colours to many organisms. Light promotes orange pigmentation in cultures of *Fusarium fujikuroi* and induces accumulation of neurosporaxanthin, a β-apo-4'-carotenoic acid (Ávalos and Cerda´-Olmedo, 1987). Neurosporaxanthin is produced from phytoene backbone through five desaturations, an end-cyclization by carotene cyclases, an oxidative cleavage reaction and a final oxidation step by different oxidases (Prado-Cabrero et al., 2009).

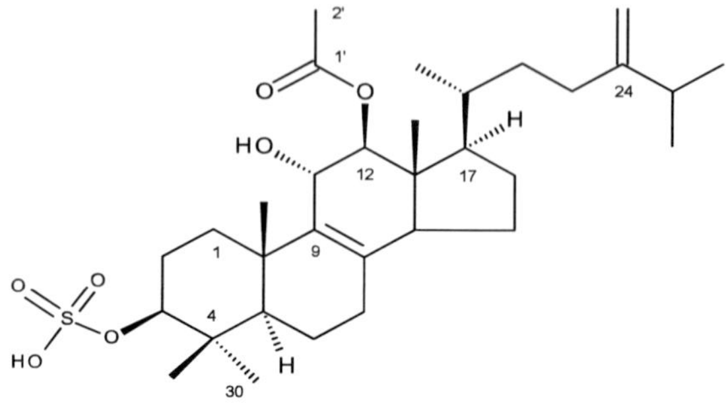

L681,512-2

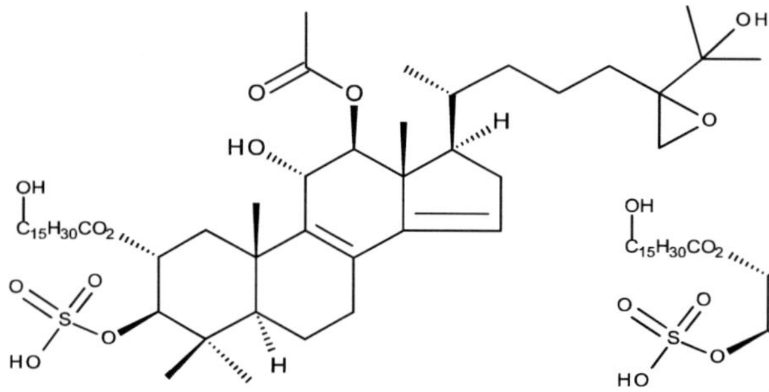

L681,512-3

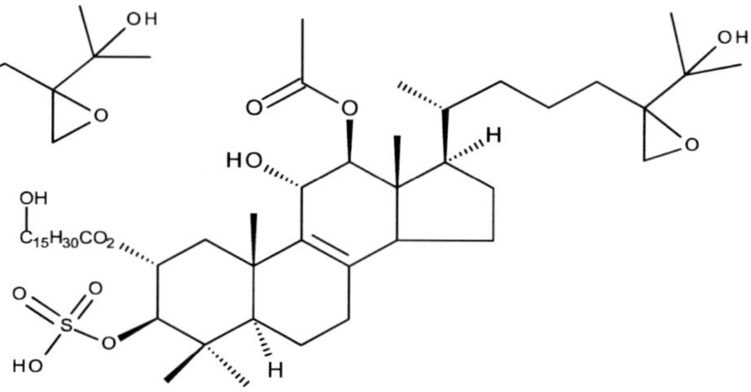

L681,512-4

Cultured cells of a marine *Fusarium* sp. (strain T-1), isolated from the surface of seawater in Japan, yielded a carotenoid glycosyl ester, neurosporaxanthin β-D-glucopyranoside (4'-apo-β-caroten-4'-oic acid β-D-glucopyranoside) together with neurosporaxanthin, α-carotene, β-carotene and torulene (Sakaki et al., 2002). Carotenoids are extremely rigid backbone due to the linear chain of conjugated C=C bonds in *trans*-configuration that confers mechanical properties and retaining their membrane-reinforcing function in some fungi. Moreover, the linear system of conjugated C=C bonds provides high reducing potential, making them potent antioxidants (Vershinin, 1999).

Papulacandins are a class of natural products that has lipids linked to saccharides and cinnamyl ester at the C-6'. Fusacandins A and B, isolated from *Fusarium sambucinum* (strain AB 1900A-1314), are papulacandins that contains two units of galactose and one of glucose, the latter connected as a *C*-glycoside to an aromatic moiety. They have additional galactose unit comparison to papulacandins (Hochlowski et al., 1995; Jackson et al., 1995). Fusacandin A has no significant antibacterial activity (MIC > 100 μg/ml) against Gram-negative and Gram-positive bacteria, but has antifungal activity (MIC 3.12-6.25 μg/ml) against strains of *C. albicans* (Jackson et al., 1995). It inhibits *in vitro* (1,3)-β-glucan enzyme and reduces incorporation of radiolabel in (1,3)-β-glucan and (1,6)-β-glucan in whole cell studies.

PF-1042/BE-29602 and saricandin are antifungal papulacandins produced by *Fusarium* sp. (AB 2202W), obtained from a soil sample collected in Nepal. They inhibit the growth of *C. albicans* (MIC 7.8 and 0.49 μg/ml, respectively) and inhibit glucan synthase (IC_{50} 242 and 6.5 μg/ml, respectively). Broth diluition assay with sorbitol indicated that these agents, like as the fusacandin A, act on the fungal cell wall (Chen et al., 1996; Jackson et al., 1995).

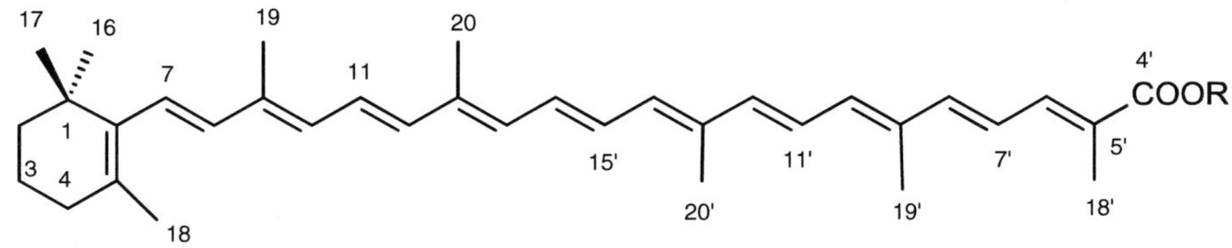

Neurosporaxanthin R = H

Neurosporanxanthin-β-D-glucopyranoside R =

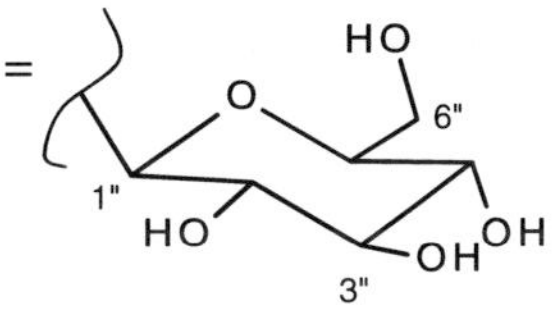

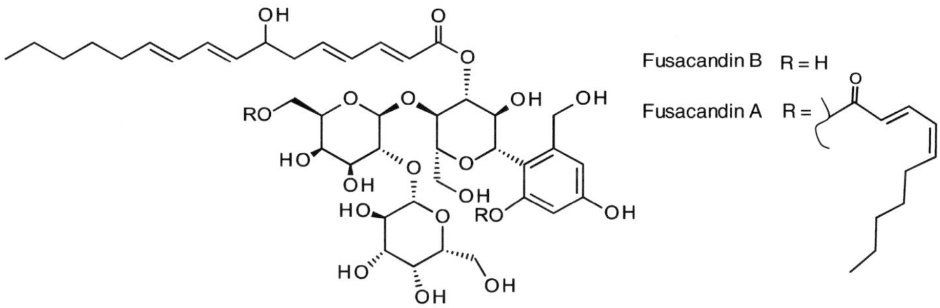

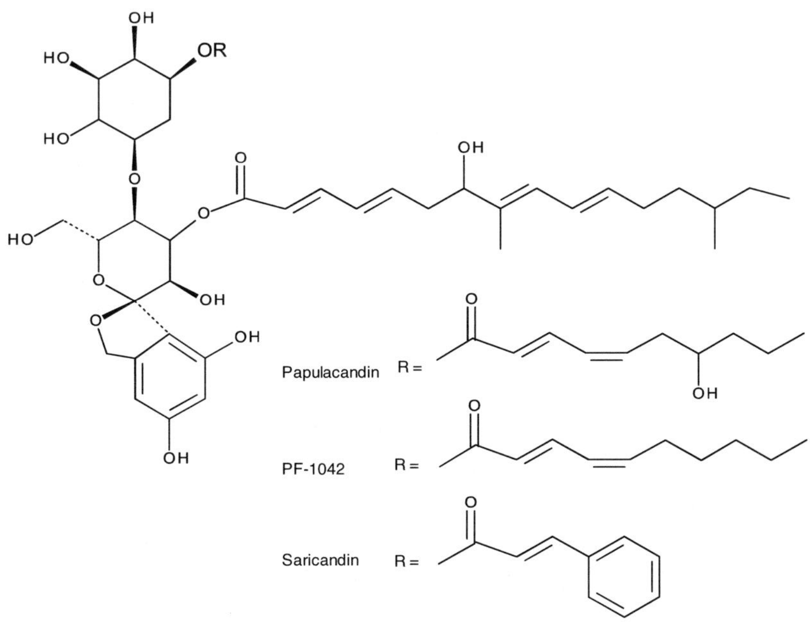

Polyhydroxydecanoate tetramers, halymecins A-C, were identified from *Fusarium* (FK-N30), a strain isolated of a marine alga (*Halymenia dilatata*) collected in Japan. Halymecin A showed antimicroalgal activity against *Skeletonema costatum* (MIC 4 μg/ml) and exhibited MIC values ranging from 68 to 500 μg/ml against others algae (*Oscillaroria amphinia, Brachiomonas submarina, Prorocentrum micans*). It was cytotoxic against HeLa S3 (IC_{50} 6 μg/ml) and BALB3T3 (IC_{50} 7 μg/ml) and active against Gram-positive and negative bacteria (MIC 10 μg/ml) (Chen et al., 1996).

Zearalenone [6-(10-hydroxy-6-oxo-*trans*-1-undecenyl)-*β*-resorcylic acid lactone; ZEA) is a mycotoxin produced by several *Fusarium* spp. (Zinedine et al., 2007). This toxin may occurs post-harvest storage of cereals under poor storage conditions (moisture>22%). It was isolated for the first time in 1962 from the mycelia of the fungus *Giberella zeae* (*Fusarium graminearum*) (Hodge et al., 1966a,b). ZEA is, also known as F-2 toxin, do not degrade at high temperatures (Kuiper-Goodman et al., 1987). In mammals, ZEA is rapidly absorbed after oral administration and can be metabolised in intestinal cells where it is keto group at C-8 is reduced to transformed into *α*- or *β*-isomers (*α*- or *β*-zearalenol-ZOL) which are conjugated with glucuronic acid (Minervini and Dell'Aquila, 2008). ZEA exhibits low acute toxicity (oral LD_{50} values of >2,000–20,000 mg/kg b.w.) after oral administration in mice, rats and guinea pigs (Flannigan, 1991).

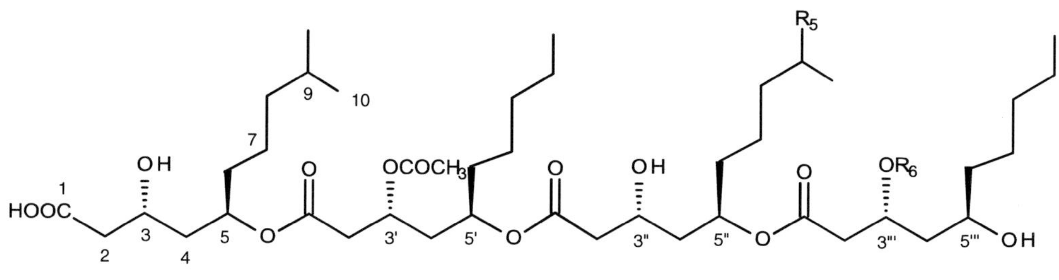

Halymecin A R_6 = H

Halymecin B R_6 = *β*-D-mannosyl

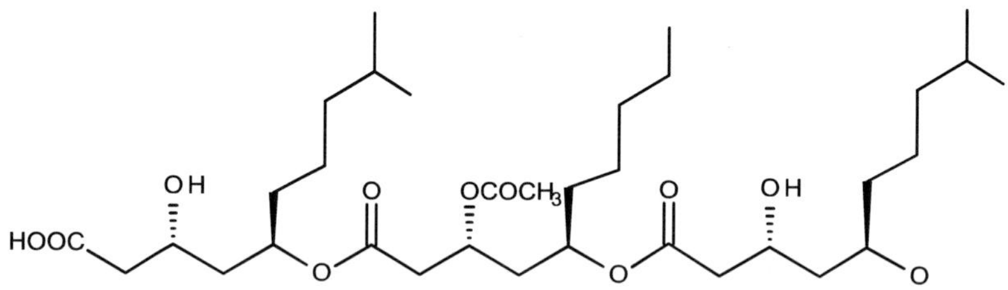

Halymecin C

ZEA was regarded as a causative agent in epidemics of premature thelarche in girls; breast development in girls below the age of eight years without others signs of premature sexual development (Saenz de Rodriguez and Toro-Sola, 1982; Schoental, 1983). ZEA induced dose-dependent premature uterine growth (ED_{50} = 1.3 mg/kg BW), stimulating uterine weight gains in neonatal rats (Sheehan et al., 1984). This mycotoxin binds to human estrogen receptors (ER*α* and ER*β*) and stimulates the transcriptional activity of both ER subtypes at concentrations of nanomolar. ZEA is a full agonist for ERα and a mixed agonist-antagonist for ERβ (Kuiper et al., 1998). According to Takemura et al. (2007) the IC_{50} values of ZEA for binding affinity to human ERα and ERβ proteins were 240.4 and 165.7 nM, respectively. Alcoholic metabolites of ZEA exhibited lower toxicity, but they were estrogenic compounds, being α-ZOL more active (Richardson et al., 1985; Celius et al., 1999) than ethynyl oestradiol (Le Guevel and Pakdel, 2001). This was probably due to the relative binding affinities to estrogen receptors which were in decreasing order: α-ZOL, ZEA and β-ZOL (Miksicek, 1994; Fitzpatrick et al., 1989). ZEA has structure flexible enough to allow to occupy the active site of the human ER in a strikingly similar manner as 17β-estradiol, such that the phenolic rings of ZEA occupy the same receptor region as occupied by the A-ring of 17β-estradiol. But this requires considerable conformational alterations upon binding to the estrogen receptors and confers to ZEA lower affinity than the natural estrogens (Takemura et al., 2007).

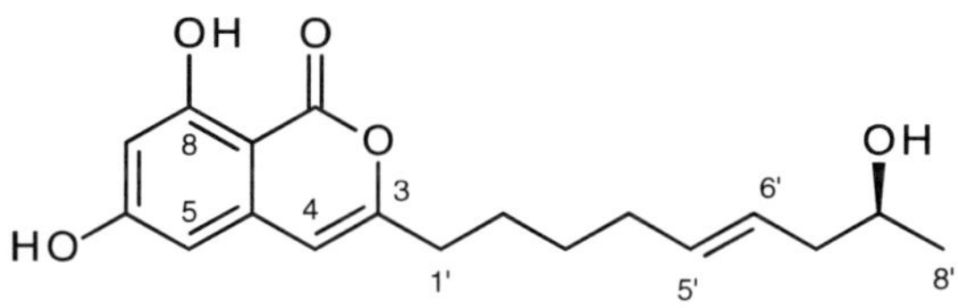

Fusariumin

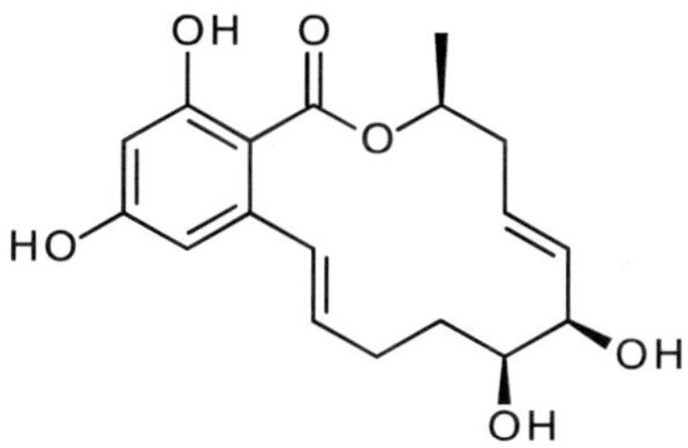

Aigilomycin D

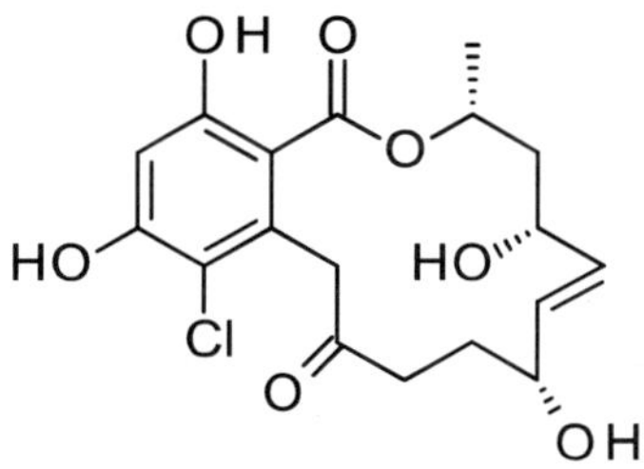

Pochonin N

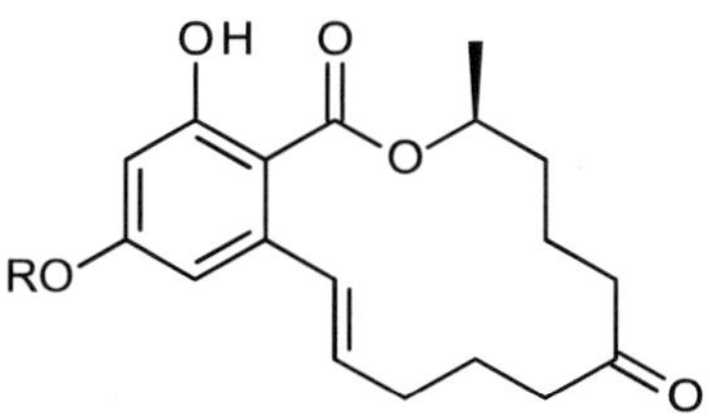

Zearalenone R = H

Zearalenone-4-sulfate R = -SO_3

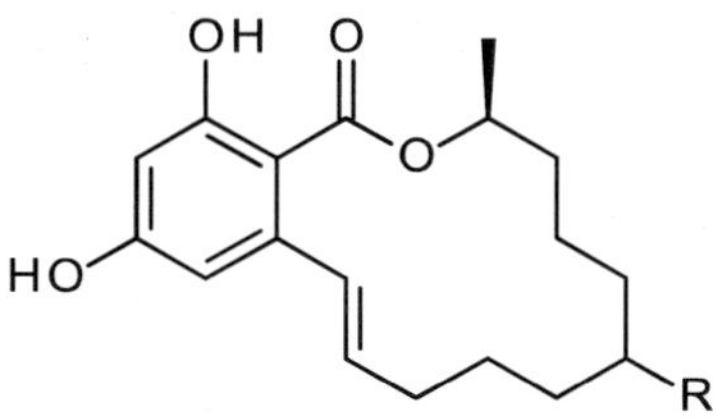

α-Zearalenol R = α-OH

β-Zearalenol R = β-OH

Plasencia and Mirocha (1991) reported the isolation of zearalenone-4-sulfate from cultures of *Fusarium graminearum* grown on a rice substrate and found this metabolite in *Fusarium equiseti, Fusarium sambucinum,* and *Fusarium roseum*. The rat uterus enlargement bioassay, by oral administration, showed that ZEA-4-sulfate has estrogenic properties of the ZEA.

There is an association between endometrial cell proliferation and the presence of ZEA in human endometrium. It was found in 22 adenocarcinoma samples (endometrial adenocarcinomas n = 27, endometrial hyperplasia n = 11) and in 3 endometrial hyperplasia samples from a total of 49 women (Tomaszewski et al, 1998). α- and β-Zearalenol have cytotoxicity activity by inhibiting cell viability (IC_{50} 84 μM and 50 μM), protein (IC_{50} 117 μM and 60 μM) and DNA (110 μM and 90 μM, 9 μM for ZEA) syntheses. They induce oxidative damage (lipid peroxidation) and increase the level of expression of stress proteins (Hsp 70, Hsp 27) in Vero cells (Othmen et al., 2008).

Immunosuppressing effect has been found in human peripheral blood mononuclear cells exposed to ZEA or α-ZOL (Berek et al., 2001). ZEA inhibited totally T and B lymphocyte proliferation from the stimulation with phytohemagglutinin and pokeweed mitogen at 30 μg/ml. The inhibitory effects of ZEA were further related to cell necrosis/apoptosis. Necrosis was observed in both the lymphocyte and monocyte/granulocyte gate by flow cytometry analysis (Vlata et al., 2006). Marin et al. (2011) demonstrated that ZEA and its derivatives altered several parameters of the innate immune response of swine neutrophils or polymorphonuclear cells. They induced at 1 μM an increase of intracellular O_2^-synthesis, decreased IL-8 synthesis at 10 μM, but they did not affect polymorphonuclear cells phagocytosis.

Hydrogenation of ZEA with a Raney nickel catalyst was able to reduce both the 1’,2’-double bond and the 6’-ketone, resulting in zearalanol that has four times the estrogenicity of ZEA and has been marketed as an anabolic agent under the name Ralgrow® to promote growth of sheep and cattle (Hodge et al., 1966).

Recently, Yang et al. (2011) isolated fusariumin, an isocoumarin derivative and two resorcylic acid lactones; aigialomycin D and pochonin N. These metabolites were isolated from *Fusarium* sp. (LN-10), an endophytic fungus from the leaves of *Melia azedarach* and demonstrated to inhibit *A. salina* in 78.2%, 76.7% and 82.8%, respectively, at a concentration of 10 μg/ml.

Antibiotic y (5-hydroxy-9-methoxy-2-methyl-4H,6H-pyrano[3,4-g]-1-benzopyran-4,6-dione-8-carboxilic acid, avenacein Y), a lateropyrone, was isolated from *F. lateritium* This fungus has been isolated from *Tsuga heterophylla* (Raf.) Sarg. trees suffering from dieback, and produces visible quantities of crystals when cultured on malt extract at room temperature. The presence of antibiotic Y can correlate with the formation of macroconidia of atypical shape due to high levels of yellow crystals (Bushell et al., 1984). *Fusarium avenaceum* was identified as causal agent of the wet apple core rot disease in apple orchards such as Gloster, Jona Gold, and Fuji. From 20 naturally infected apples sampled, eleven contained antibiotic Y in concentration levels ranging from 0.08 to 51.3 μg g^{-1} (Golinski et al., 1986; SØrensen et al., 2009). It has inhibitory effect on *S. aureus* growth and the highest activity occurs at pH 6.2 (MIC 0.37 ug/ml). Antibiotic Y was found to be nontoxic to 3T3 mouse cells in concentrations up to 5,000 ng/ml and nontoxic to in the chicken embryo bioassay (Bushell et al., 1984).

Chlamydosporol (*trans*-7,8-di hydro-7- hydroxy-4-m ethoxy- 7,8-dimethyl -2*H*,5*H*-pyrano [4,3-*b*] pyran-2-one) was isolated from three different *Fusarium* species (*F. chlamydosporum, F. acuminatum* and *F. culmorum*). The isomers 6α and 6β of chlamydosporol were obtained from *Fusarium chlamydosporum*, an endophytic fungus isolated from rice (Savard et al., 1990). The mixture was active in *A. salina* and in HeLa cell assays (LC_{50} 400 μg/ml) (Savard et al., 1990). Chlamydosporol frequently occurs in *Fusarium* genus (23 of 40 isolates, 58%) in amounts as high as 0.8% (wt/wt) dried culture medium (Shier and Abbas, 1992). It causes food refusal and weight loss in rats and has cytotoxic effects in cultured mouse (IC_{50} 2.5 μg/ml), and in human fibroblast cells (IC_{50} 7.5 μg/ml) (Abbas et al., 1992). The addition of $CaBr_2$ to the fermentation of a marine-derived, *Fusarium tricinctum*, resulted in production of chlamydosporol (as isomers α and β), fusarielin A and halogenated chlamydosporol analogues: bromomethylchlamydosporols A and B.

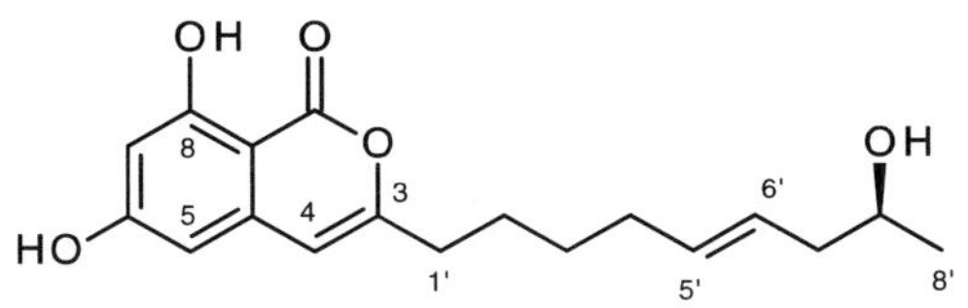

Fusariumin

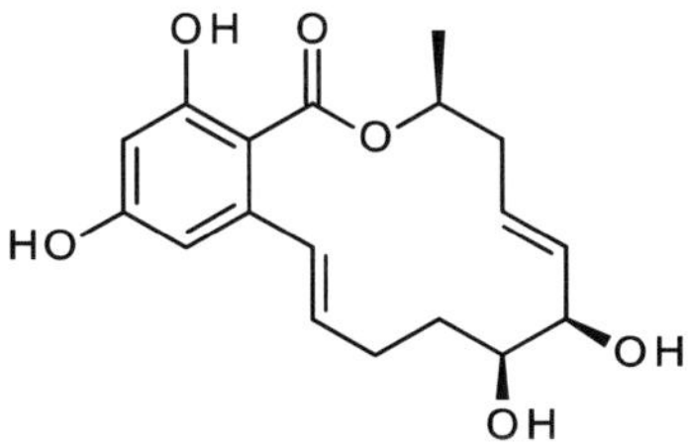

Aigilomycin D

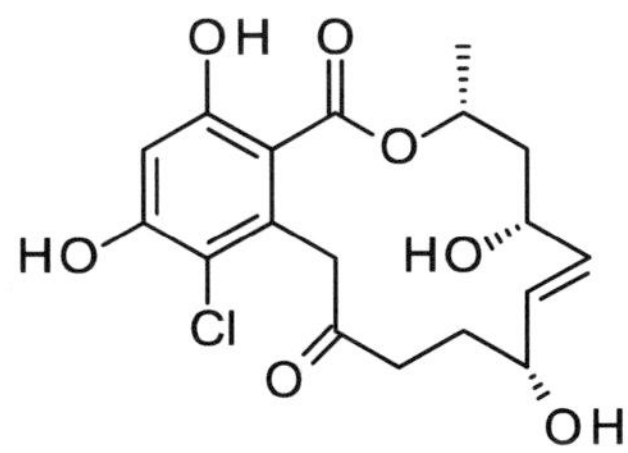

Pochonin N

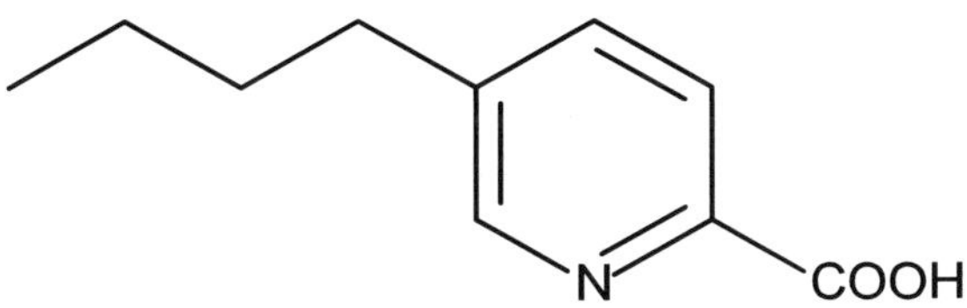

Fusaric acid

Chlamydosporol and fusarielin A exhibited MIC values of 31.5 μg/ml against *S. aureus* and methicillin-resistant *S. aureus* and MIC values of 62.5 μg/ml against multidrug-resistant *S. aureus*, nevertheless, halogenated derivates (MIC 15.6 μg/ml) were more active in both assays (Nenkep et al., 2010).

Naphthoquinone pigments occur in several species of the genus *Fusarium* and some of them are summarized in review of Medentsev and Akimenko (1998). Rubrofusarin, a nafto-γ-pirone and, aurofusarin, a homodimeric naphthoquinone, were first described as *Fusarium culmorum* pigments by Ashley et al. (1937). Javanicin and oxyjavanicin, naphthaquinones compounds, were described as metabolites from *F. javanicum* (Anstein, 1946a, b). Others quinones, as lateritiin I, norjavanicin and avenacein and fructigenin were, further, isolated from *Fusarium* spp. (Cook et al., 1947; Chilton, 1968). Medentsev and Akimenko (1998) suggest that antibiotic and phytotoxic effects of quinones are due to the interaction with the oxidative systems, considering the fact that they accept the reducing equivalents from the redox enzymes and transferred them directly to oxygen.

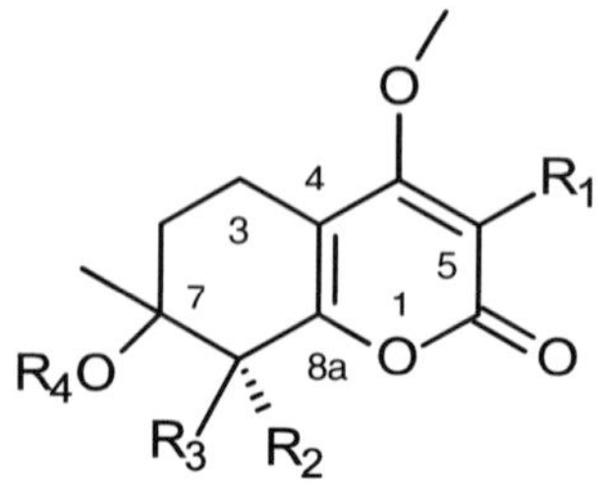

Name	R1	R2	R3	R4
Bromomethylchlamydosporol A	-H	-Br	$-CH_3$	-α-OCH_3
Bromomethylchlamydosporol B	-Br	-Br	$-CH_3$	-α-OCH_3
α-Chlamydosporol	-H	$-CH_3$	-H	-α-OH
β-Chlamydosporol	-H	$-CH_3$	-H	-β-OH

Fusarubin and dihydrofusarubins A are quinones that show cytotoxic activity (IC_{50} 3.1 and 1.7 µg/ml, respectively) against proliferation of mouse leukemia L1210 culture cells and activity against Gram-positive (MIC 50-100 and 12.5-50 µg/ml, respectively) and against *C. albicans* (MIC 25-50 µg/ml) (Kurobane et al., 1986). Branco et al. (2008) demonstrated that rubrofusarin (5,6-dihydroxy-8-methoxy-2-methylbenzo[g]cromen-4-one) shows total inhibition of human DNA topoisomerase II-α at 120 µM, result comparable to etoposide at 100 µM.

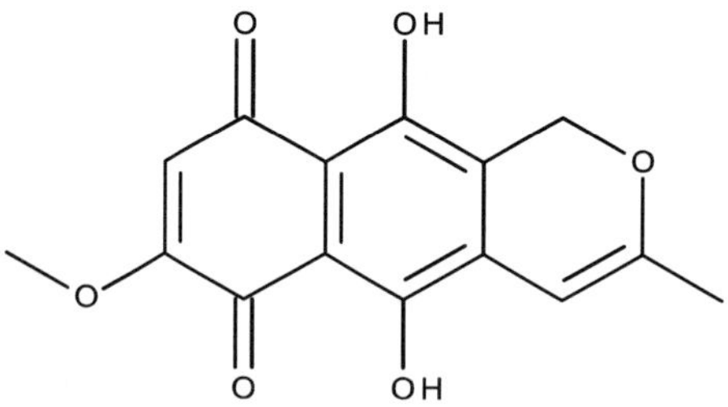

Fusarubin

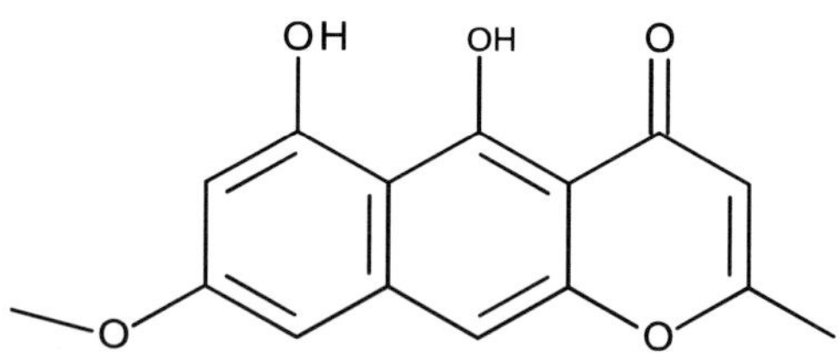

Rubrofusarin

Bikaverin has been encountered in a number of microorganisms including several *Fusarium* species (Chelkowski et al., 1992; Jones et al., 1987; Brewer and Arsenault, 1973; Cornforth et al., 1971; Balan et al., 1970) and 6-deoxybikaverin has previously been obtained only as a product in the synthesis of bikaverin (Kjaer et al., 1983). Bikaverin was previously isolated as mycogonin, the red pigment of *Mycogone jaapii* Lindau, as passiflorin, a pigment from *F. oxysporum f. sp. passiflorae* and as lycopersin, a metabolite from *F. oxysporum* (Brewer and Arsenault, 1973).

Bikaverin, after 48 hours, led to the complete death of the *Leishmania brasiliensis* at concentration of 0.15 μg/ml (Balan et al., 1970). It is cytotoxic to Erlich ascites carcinoma, leukemia L5178, and sarcoma 37 cells probably by inhibiting nucleic acid and protein synthesis (Fusca et al., 1975). From all metabolites isolates by Zhang et al. (2007), only beauvericin (IC_{50} 1.41-2.29 μM) and bikaverin (IC_{50} 0.26 -0.43 μM) were active toward NCI-H460 (human non-small-cell lung cancer), MIA Pa Ca-2 (human pancreatic carcinoma), MCF-7 (human breast cancer) and SF-268 (human CNS CANCER). 6-deoxybikaverin was completely devoid of activity, suggesting the requirement of the C-6 hydroxy group for cytotoxic activity (Zhan et al., 2007). Newly, beauvericin was isolated from *F. oxyporum*, obtained from the bark of *Cinnamomum kanehirae*, an endemic plant of Taiwan and exhibited anti-bacterial activity towards methicillin-resistant *Staphylococcus aureus* (MIC 3.13 μg/ml) and *Bacillus subtilis* (MIC 3.13 μg/ml) (Wang et al., 2011).

The marine-derived fungi *Fusarium* spp. (PSUF14 and PSU-F135) were isolated from a gorgonian sea fan (*Annella* sp.). Fusaranthraquinone (octahydroanthraquinone), fusarnaphthoquinones A-C (hydronaphthoquinones) and, fusarone (cyclopentanone), were isolated as new compounds from both fungi. Others 13 known compounds were also isolated, but none of them was active in antibacterial assays (*S. aureus*, MIC values >350 μM) and in antifungal assays (*C. neoformans* and *M. gypseum*, MIC values >200 μM). 9α-Hydroxyhalorosellinia A, bostrycin, 11-acetoxy bostrycin, nigrosporin B, javanicin and anhydrofusarubin displayed antimalarial activity, with IC_{50} values ranging from 9.8 to 25 μM. Nigrosporin B and 9α-hydroxyhalorosellinia A exhibited activity against *M. tuberculosis* (MIC 41 and 39 μg/ml, respectively). In addition, from new isolated compounds, only fusarnaphthoquinones A exhibited cytotoxic activity against oral human carcinoma cells (KB, IC_{50} 130 μM), human breast cancer cells (MCF-7, IC_{50} 22 μM), without to be toxic against Vero cells lines (Trisuwan et al. 2010).

Isochaetochromins B_1, B_2, D_1 and oxychaetochromin B are bis-(naphtho-γ-pyrones) which were discovered as inhibitors of HIV-1 integrase from *Fusarium* sp. This strain (MF6252, ATCC74396) was isolated from leaf litter collected under *Podocarpus dacryoides* in New Zealand. They inhibit the coupled and strand transfer activity of HIV-1 integrase with IC_{50} values ranging of 1–3 and 4–12 μM, respectively. Although, isochaetochromins B1 and B2 have exhibited viral protection in a 24 h HIV-1 viral infectivity assay (IC_{50} of 3 and 2 μM), they were toxic in a MTT assay in a 7-day infectivity at 50 μM (Singh et al., 2003c).

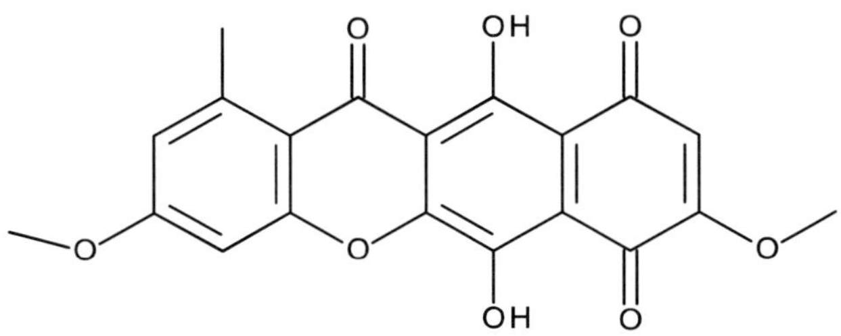

Bikaverin

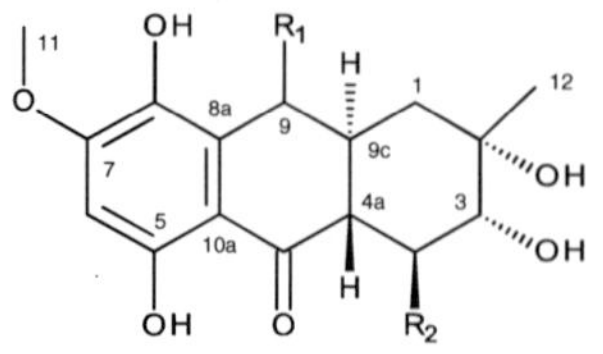

Fusaranthraquinone R1 = β-OH, R2 = H
9-α-Hydroxyhalorosellinia R1 = α-OH, R2 = OH

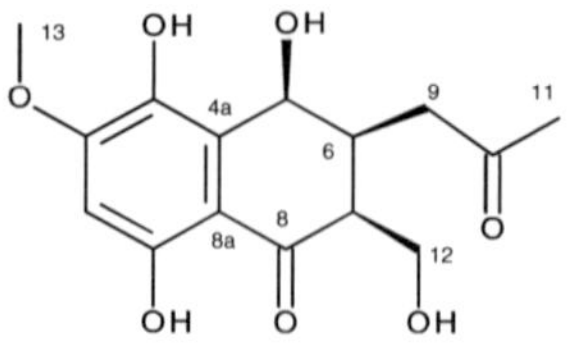

Fusamaphtoquinone A

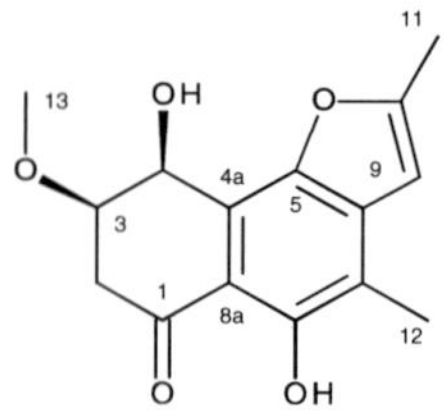

Fusamaphtoquinone B

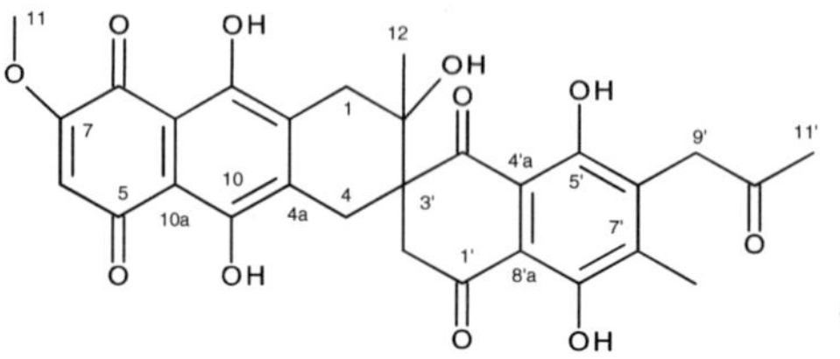

Fusamaphtoquinone C

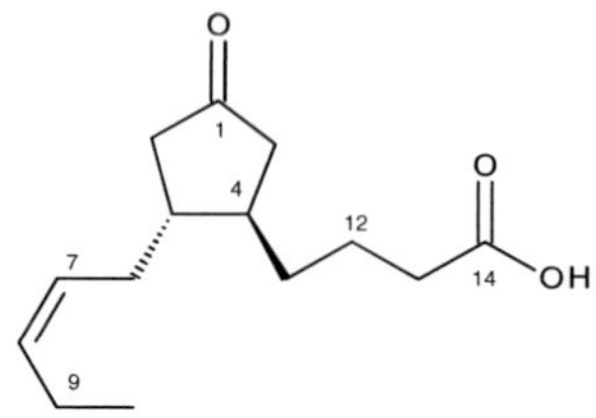

Fusarone

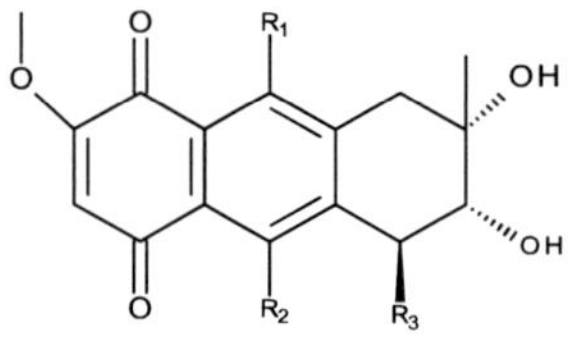

Bostrycin R1 = R2 = R3 =OH
Nigrosporin B R1 = R3 = H, R2 =OH

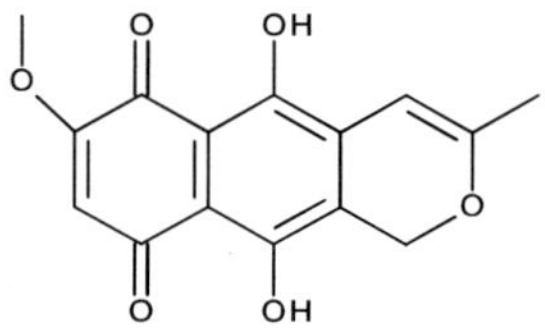

Anhydrofusarabin

An EtOAc extract, obtained from endophytic fungus *Fusarium oxysporum* (EPH2RAA) of the root tissue of *Ephedra fasciculate* (Mormon tea; Ephedraceae), afforded beauvericin, (-)-oxysporidinone, and two *N*-methyl-2-pyridones; (-)-4,6’-anhydrooxysporidinone and (-)-6-deoxyoxysporidinone. Other *F. oxysporum* (CECIS), isolated from the stem tissue of *Cylindropuntia echinocarpus* (silver cholla; Cactaceae), afforded rhodolamprometrin, bikaverin, and 6-deoxybikaverin (Zhan et al., 2007).

Fumonisins are important *Fusarium* mycotoxins which had their chemical structure first elucidated by Bezuidenhout et al. (1988). They characterized by a 19-20 carbon aminopolyhydroxyalkyl chain which is diesterified with propane-1,2,3-tricarboxylic acid. Until now, twenty-eight fumonisins have been isolated and they are divided in four series known as A, B, C and P. The A-series are acetylated on the amino group; the B-series have a free amine; the C-series is structurally similar to the B series; however, the C-1 terminal methyl group is missing. Hydroxylated fumonisin C_1 has an additional hydroxy group at C-3 (Seo et al., 1999).

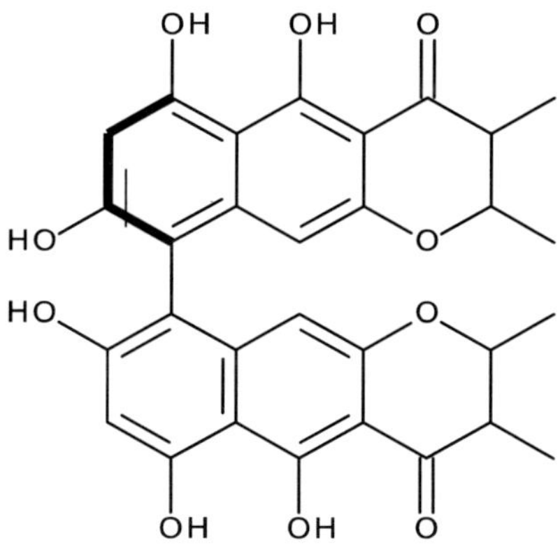

Isochaetochromin B_1

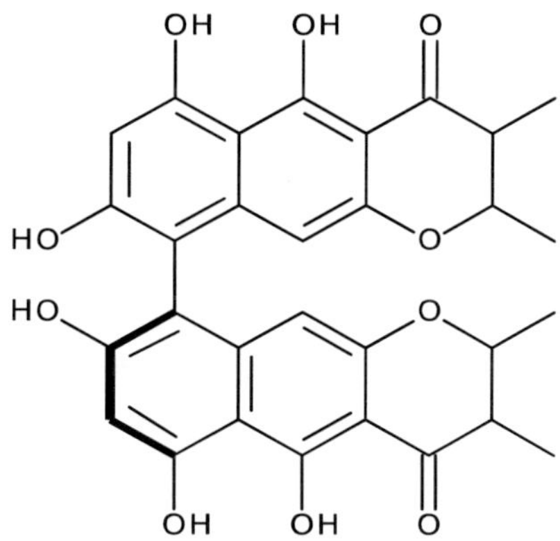

Isochaetochromin B_2

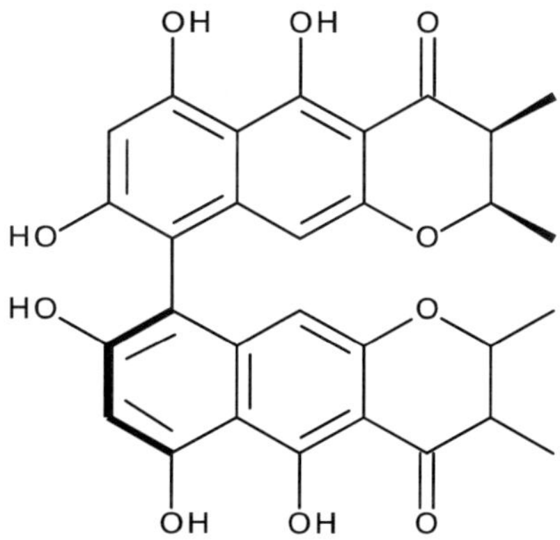

Isochaetochromin D_1

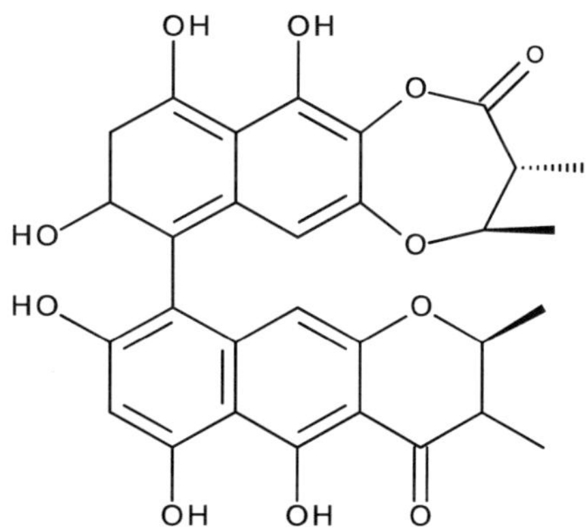

Isochaetochromin B

Fumonisin B1 (FB1), the most abundant fumonisin, occurs naturally in contaminated foods and feeds (Dutton, 1996). FB1 toxicity is specific to tissue, to treated animal and is associated with the high prevalence of human esophageal cancer in southern Africa and China (1980; Marasas, 1982, Lin et al., 1980; Yang). In a long-term carcinogenesis study, FB1 induced hepatocellular carcinomas in BDIX rats at a dietary level of 0.005% (Gelderblom et al., 1991) and, in a short-term carcinogenesis study, FB1 induced enzyme-altered foci in the BDIX at a dietary level of 0.1% (Gelderblom et al., 1992). Fumonisin B1 has backbones of the polyhydric alcohols that resembles to sphingoid bases such as sphingosine, a derivated from L-alanine. It suggests that the mechanism of action of FB1 is mainly via disruption of sphingolipid metabolism. Sphingolipids are believed to play critical roles in a number of cellular functions including cell-cell communication, cell growth, differentiation and transformation of cells (Merrill 1991; Wang et al., 1991). It inhibits the activity of sphingosine *N*-acyltransferase that results in accumulation of sphingoid bases in Swiss 3T3 fibroblasts and rat hepatocytes (Schroeder et al., 1994; Wang et al., 1991). Wistar rats treated by FB1 at 0.5 mg/kg b.w./day for 7 days caused DNA lesions in the kidney of animals and altered the sphingolipid metabolism that may play a role in kidney DNA damage (Domijan et al., 2007).

FB1 also inhibits the dephosphorylation protein serine/threonine phosphatases (IC_{50} 80 mM-3000 mM), being PP5 the most sensitive (IC_{50} of 80 mM) (Fukuda et al., 1996); epidermal growth factor (EGF)-induced DNA synthesis in rat hepatocytes up to 90% at concentrations of 150 to 300 mM (Gelderblom et al., 1995) and it stimulates DNA synthesis in Swiss 3T3 fibroblasts at concentrations of 10 to 50 mM (Schroeder et al., 1994). Although FB1 is not mutagenic in the *Salmonella thyphimurium* assay (10-114 µg/plate), intraperitoneal administration of 25 and 100 mg/kg in male mice induced an increase of micronuclei in mouse bone marrow polychromatic erythrocytes, being the low dose more effective and not dose-dependency (Aranda et al., 2000).

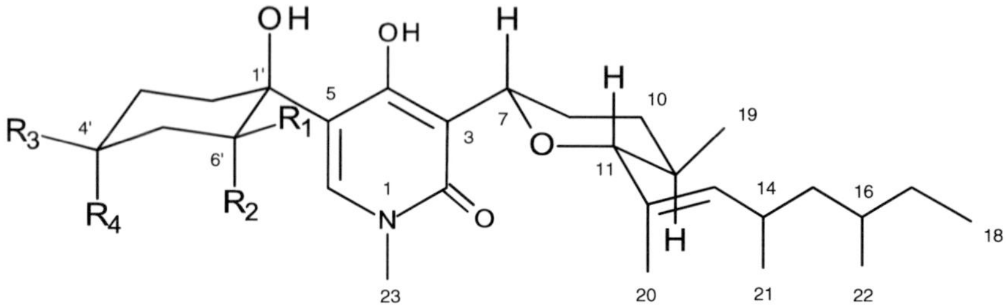

Name	R1	R2	R3	R4
(-)-Oxysporidinone	-OH	-H	=O	=O
6-*Epi*-oxysporidinone	-H	-OH	=O	=O
Dimethyl ketal of oxysporidinone	-OH	-H	$-OCH_3$	$-OCH_3$

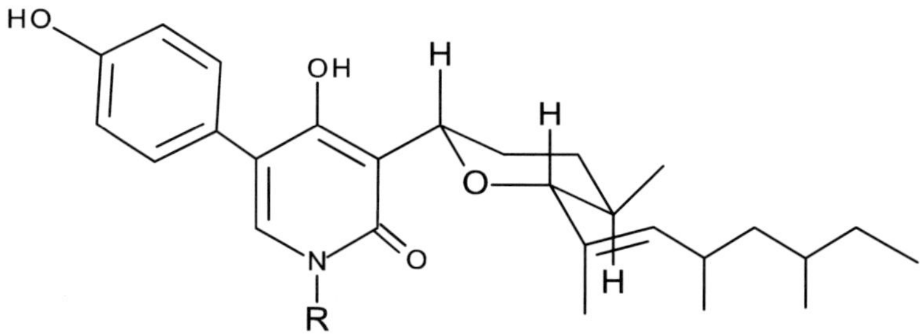

(-)-Sambutoxin $R = CH_3$

N-Demethylsambutoxin $R = H$

FB1 and FB2 are toxic against three cell lines: dog kidney (MDCK, IC_{50} 10 µM and 10 µM), rat liver hepatoma (H4TG, IC_{50} 15 µM and 8 µM) and low toxicity in swiss mouse fibroblasts (NIH3T3, IC_{50} 200 µM) (Abbas et al., 1995). FB1 concentrations encountered in naturally contaminated foodstuffs and feed might lead to oxidative damage by lipid peroxidation (at 0.14 µM, 0.1 g/ml), DNA base modifications and stimulate the carcinogenic process without inhibit protein (IC_{50} 30 µM) and DNA syntheses (IC_{50} 30 µM) in cell culture (Abado-Becognee et al., 1998).

The fusarins are a group of yellow pigments produced by *Fusarium* species that contaminate maize products. Isotope labelling experiments ([1,2-$^{13}C_2$] acetate) into cultures of *F. moniliforme*, (strain MRC 826), which produces mainly fusarin A, suggested a mixed biosynthetic pathway between polyketide derived from an acetyl-CoA starter unit and 6 malonyl-CoA units, and a C_4 unit most probably derived from oxaloacetate of the krebs tricarboxylic acid cycle (Steyn and Vleggaar, 1985).

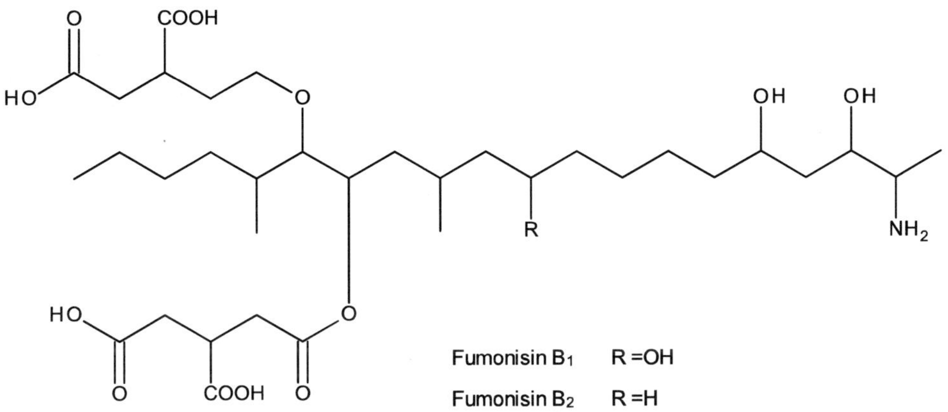

Fumonisin B_1 R =OH

Fumonisin B_2 R =H

The mutagenic mycotoxin fusarin C, a polyenic chromophore with a substituted 2-pyrrolidone moiety, in addition to nonmutagenic fusarins A and D, were isolated from a culture of *F. moniliforme* (MRC 826) (Gelderblom et al., 1984). Fusarin C is a unstable compound that due to spontaneous changes in the substituted 2-pyrrolidone moiety renders two isomeric forms with nonmutagenic properties (Gelderblom et al., 1984a,c). Mutagenic activity seems to be due to the epoxide group in 2-pyrrolidone moiety (Gelderblom et al., 1984c).

Lucilactaene, a γ-lactam compound with a polyunsaturated side chain structurally similar to fusarin A, and two epolactaenes, NG-391 and NG-393 which are geometric stereoisomers at the C-8 position, were isolated from *Fusarium* sp. (RK97-94). Lucilactaene at 5 μg/ml increased DNA contents, suggesting inhibition of the cell cycle progression in human non small cell lung cancer cells and activated the promoter of the p53 gene (Kakeya et al., 2001).

Sansalvamide A is a depsipeptide discovered by Belofsky et al. (1999) from a *Fusarium* (CNL 292), obtained from the seagrass (*Halodule wrightii*), collected in Bahamas. It is composed of four hydrophobic amino acids (Phe, 2 Leu, Val) and one hydrophobic hydroxy acid (*O*-Leu), with five chiral carbons all having *S*-stereochemistry. It exhibited antitumor activity against NCI's 60 cell line panel (mean IC_{50} of 27.4 μg/ml) and colon cancer cell line (IC_{50} 3.5 μg/ml) and melanoma cell-line (IC_{50} 5.9 μg/ml) (Belofsky et al., 1999). The anticancer activity can be explained by the binding of sansalvamide A to the *N*-middle domain of the heat shock protein 90 (Hsp90) and blocking the binding of proteins thought to Hsp90 C terminus domain (Vasko et al., 2010). Sansalvamide A also inhibits the virus-encoded type-1 topoisomerase, an enzyme from *Molluscus contagiosum* poxvirus, important for biological processes such as DNA replication, repair, and transcription. It inhibits DNA binding by the isolated catalytic domain and thereby the covalent complex formation (Hwang et al., 1999).

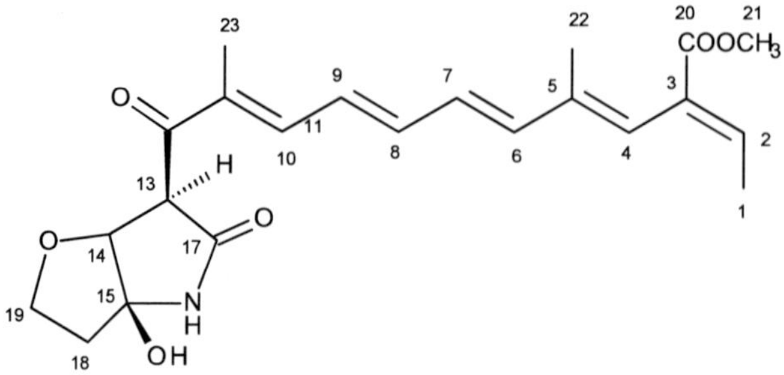

Lucilactaene

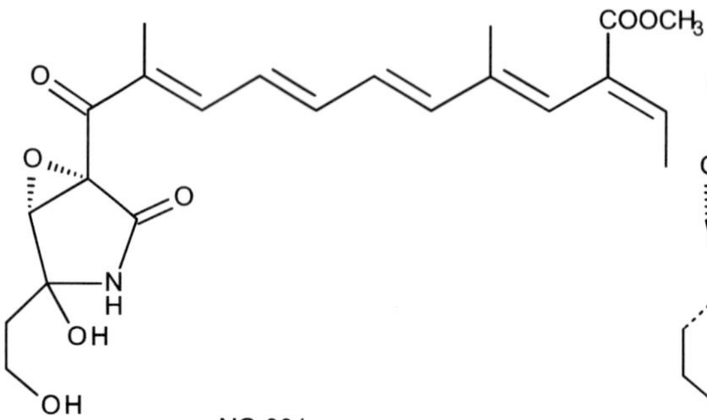

NG-391

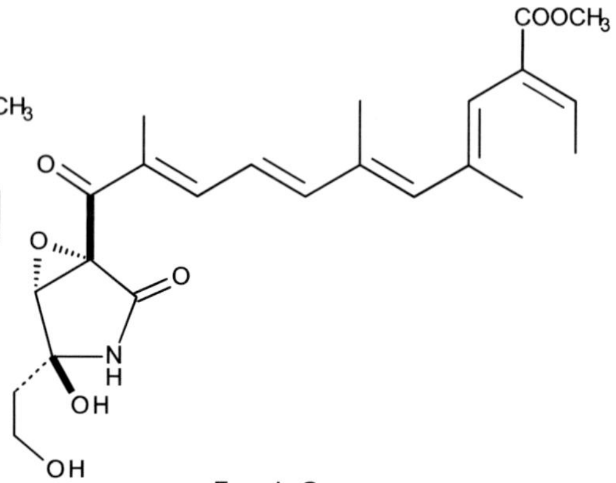

Fusarin C

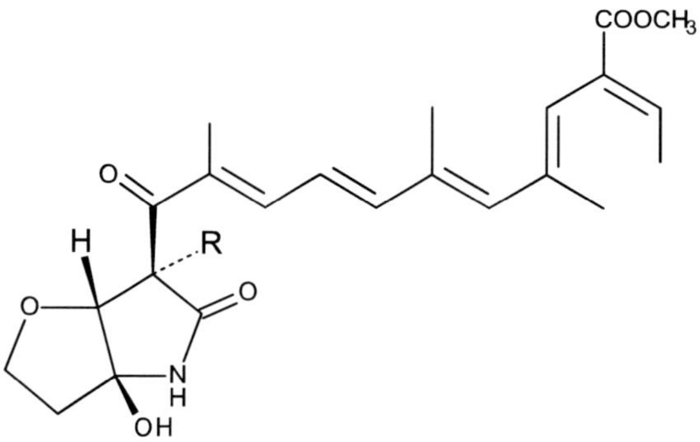

Fusarin A R = H

Fusarin D R = OH

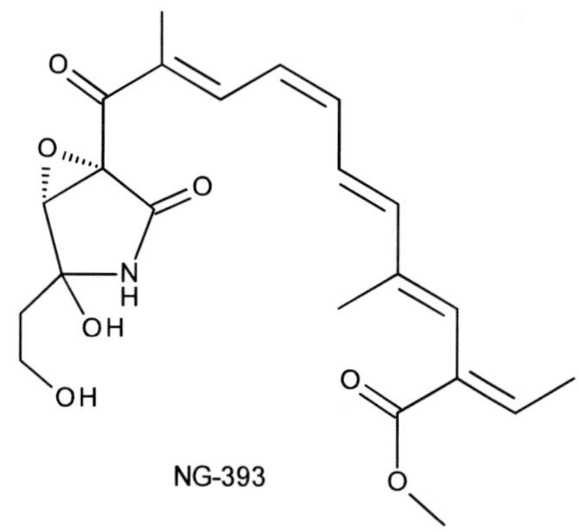

NG-393

N-methylsansalvamide, isolated from endophytic *Fusarium* (strain CNL-619) from Caribbean green alg

a (*Avrainvillea* sp.), exhibits weakness cytotoxicity in NCI's 60 cell line panel (GI_{50} 8.3 μM) in comparison to sansalvamide A (GI_{50} 3.6 μM) (Cueto et al., 2000). Neo-*n*-methylsansalvamide ((S)-2-hydroxy-4-methylpentanoic acid, *N*-methyl-l-leucine, l-valine, l-leucine and l-phenylalanine) was subsequently isolated from *F. solani* KCCM90040, a strain obtained from potato in Korea. From the three compounds, neo-*N*-methylsansalvamide requires the most high concentrations to inhibit lung cancer, ovarian cancer, skin melanoma and uterine sarcoma cell lines (GI_{50} 10.0-14.0 M) (Song et al., 2011).

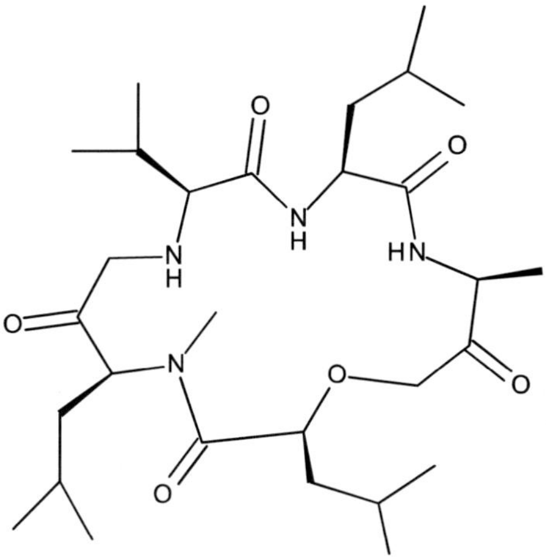

neo-*N*-Methylsansalvamide

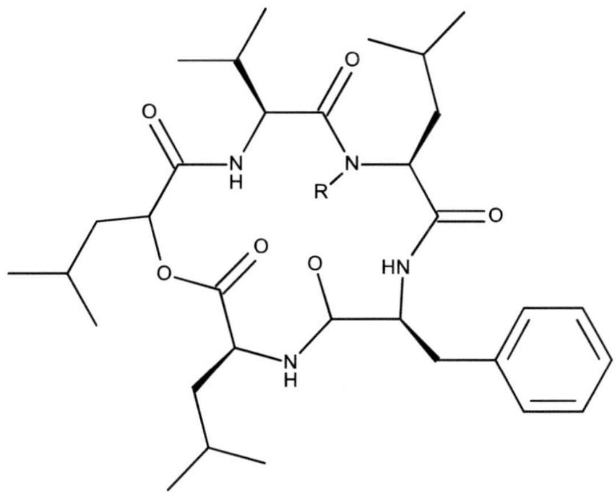

Sansalvamide A R = H

N-methylsansalvamide A R = CH_3

Enniatins (ENs) are cyclohexadepsipeptides that contain three D-hydroxy acid residues (hydroxyisovaleric acids) and three L-amino acid residues (Val, Ile, and Leu) alternately linked with ester and amide bonds forming an 18-membered ring (Tomoda et al., 1992; Visconti et al., 1992; Strongman et al., 1988). They have a wide range of activities including phytotoxic, (Burmeister and Plattner, 1987), antimicrobial (Tomoda et al., 1992), insecticidal (Strongman et al., 1988) and are inhibitors of cholesterol acyltransferase (Tomoda et al., 1992). ENs form ionophores with hydrophobic groups on the outside and polar groups in the core and can transport mono- and divalent cations either in sandwiched complexes or by creating channels in biological membranes what could explanation their broad spectrum of activity (Feudijo et al., 2010; Jestoi, 2008; Ovchinnikov et al., 1974).

Enniatin B (EN B), the most prevalent among the 28 homologues identified (Firáková et al., 2007), is a cyclic hexadepsipeptide consisting of three D-2-hydroxyisovaleric acid residues linked alternatively to three *N*-methyl-L-valines (Blais et al., 1992). EN B binds to calmodulin and inhibits the activity of 3',5'-cyclic nucleotide phosphodiesterase (IC_{50}> 5 μM) (Mereish et al., 1990), inhibits the growth of *S. aureus* and vancomycin resistant enterococci VRE788 (Jiang et al., 2002), has cytotoxic activity against V79 cells (IC_{50} 4 μM), although it has not been mutagenicity up to toxic levels (100 μM) in Ames assays (Behm et al., 2009; Jestoi, 2008; Ivanova et al., 2006).

Enniatins A, A_1, and B_1 show antifungal activity against *C. albicans* (IC_{50} 2 μg/ml, MIC 3.13-6.25 μg/ml), *Cryptococcus neoformans* (IC_{50} 3.5-9.0 μg/ml, MIC 12.5-25 μg/ml) and *Mycobacterium intracellulare* (IC_{50} 5.0-15.0 μg/ml, MIC 50 μg/ml) (Jayasinghe et al., 2006) while ENB_1 inhibits *Botrytis cinerea* spore germination at 75 μg/ml (Pohanka et al., 2004). Enniatins B, B_1 and D, purified from culture filtrate of the isolate, *Fusarium* sp. Y-53, competitively inhibit a multidrug efflux pumps whose overexpression confers multidrug resistance in *S. cerevisiae* at non-toxic concentration (5 μg/ml, 7.8 μM). According to Hiraga et al. (2005) the inhibition mechanism of ENs is different from their function as ionophores.

ENs interact with Pdr5p (pleiotropic drug resistance 5 protein) and specifically inhibit drug efflux by Pdr5p at non-toxic concentrations.

From the series of ENs (A, A_1, A_2, B, B_1, B_4 and J_3), EN A_1 was the most active between ENs A, B1 and B4 against the human epithelial colorectal adenocarcinoma, the human colon carcinoma and the human liver carcinoma with IC_{50} ranging from 9.1 to 12.3 μM. ENs at nanomolar concentrations stimulates cell proliferation, and they might have tumor-promoting functions, for example, via food intake. In contrast, elevated (micromolar) concentrations, they exert profound p53- dependent cytostatic and p53-independent cytotoxic activities especially against human cancer cells, suggesting a potential quality of ENs as an anticancer drug (Dornetshuber et al., 2007).

Fusafunginum is a mixture of ENs (A, A_1, B, B_1) produced by *F. lateritium* (strain 437) which have alternated D-α-hydroxyvaleric acid and L-*N*-methylamino acid residues. It is active against Gram-positive bacteria (Levy et al., 1995) and has anti-inflammatory property. It is used for the treatment of nasal and throat infections in the upper respiratory tract in a 1% nasal inhalation solution (Kroslák, 2002; Lohmann, 1988) and marketed by laboratory Servier in Europe under the trade names Locabiotal®, Bioparox® and Locabiosol®. The lack of correlation between *in vitro* and *in vivo* toxicity is presumably the result of low bioavailability by oral administration, which may be caused by impaired uptake from the gastrointestinal tract due to low compound water solubility (Blais et al 1992) and interaction with efflux pumps (Ivanova et al, 2010) or by elimination from the systemic circulation due to metabolism reactions. However, the pharmacokinetic properties and biotransformation products of ENs have not yet been characterized.

Beauvericin (BEA) is structurally related to the ENs and has been encountered in a number of fungal strains including *Beauveria bassiana* (Hamill et al., 1969), *Paecilomyces fumoso-roseus* (Bernardini et al., 1975), *Polyporus sulphureus* (Deol et al., 1978), *F. semitectum* and *F. moniliforme var. subglutinans* (Gupta et al., 1991). It is an insecticidal compound (Grove and Pople, 1980), toxic to *A. salina* (Hamill et al., 1969), to human hepatocellular carcinoma, to human cell lines (Ivanova et al., 2006; Logrieco et al., 1997; 1996). BEA at 50 μM induces apoptosis by triggering an increase in the cytoplasmic calcium (Dombrink-Kurtzman, 2003) and intercalates with double stranded DNA what can interfere with DNA replication (Pocsfalvi et al., 1997). Induction of apoptosis by BEA involves pro- and antiapoptotic Bcl-2 family proteins, mitochondrial membrane potential, mitochondrial cytochrome c and caspase 3 (Jow et al., 2004). BEA inhibits migration of the metastatic prostate cancer (PC-3M) and breast cancer (MDA-MB-231) cells and has antiangiogenic activity in endothelial cells at sublethal concentration of 3 μM (IC_{50} 7.5 μM) (Zhan et al., 2007). Although BEA and ENs have been cytotoxic, especially to cancer cells; inhibit bone resorption; induce increase of intracellular calcium levels and activate programmed cell death by apoptosis, they inhibit drug efflux pumps and are non-mutagenic, which suggest their potential as drug candidates to fight disseminated cancer (Feudjio et al., 2010).

Apicidin [cyclo-(N-O-Methyl-L-Trp-L-Ile-D-Pip-L-2-amino-8-oxo-decanoyl) belongs to the class of cyclic tetrapeptides produced by isolates of *F. semitectum* (Jin et al., 2008) and *Fusarium pallidoroseum* (Singh et al., 1996).

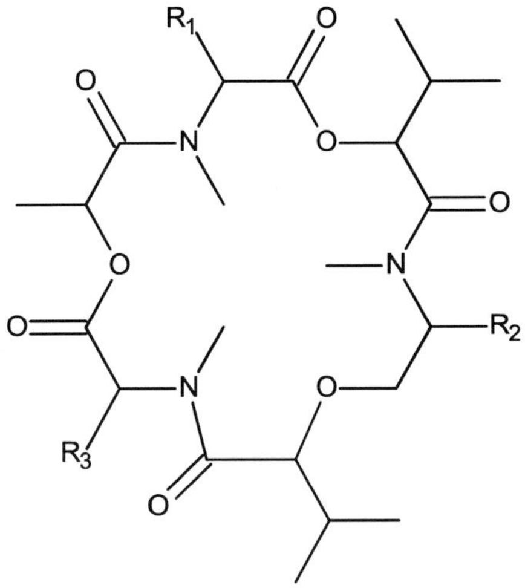

Name	R_1	R_2	R_3
Enniatin A	3	3	3
Enniatin A_2	3	3	2
Enniatin A_1	1	3	3
Enniatin B	1	1	1
Enniatin B_1	1	1	3
Enniatin B_4	2	1	1
Enniatin D	1	1	2
Enniatin E	1	2(3)	3 (2)
Enniatin F	2	3	3
Enniatin J_3	5	3	1
Beauvericin	4	4	4

1

2

3

4

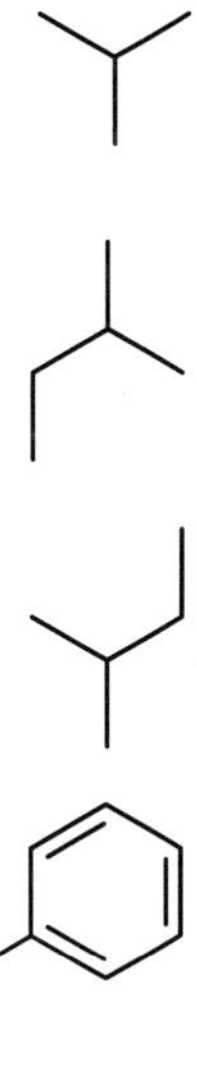

5 CH_3

Apicidin showed *in vivo* efficacy against *Plasmodium berghei* malaria in mice (<10 mg/kg i.p.) and it is a reversible inhibitor (IC_{50} 1-2 nM) of apicomplexan histone deacetylase (HDAC) (Darkin-Rattray et al., 1996). HDACs are nuclear isozymes that regulate gene transcription via acetylation and deacetylation of lysine residues of histones. Blockade of the deacetylation process causes hyperacetylation of histones, affects the transcriptional control in eukaryotic cells and this leads to untimely cell death. Apicidin decreases the growth of *P. falciparum* cells (IC_{50} value of 70nM) and inhibits HAD enzyme, resulting in increase acetylation of histones and collapse of the transcriptional cascade of parasite (Chaal et al., 2010).

Fusarium sp. (KCTC 16677), isolated of soybeans from Korea, has caused toxicity and death in rats feeding with a diet containing 0.05% of apicidin in ground wheat. This isolated produced a substantial amount of apicidin (680 μg/g) as a hemorrhagic factor. Apicidin was also toxic in *A. salina* assays (LD_{50} 40 μg/ml), cytotoxic for mouse leukemia cells (P388, IC_{50} 2.1 μg/ml), human leukemia (K562, IC_{50} 16 μg/ml) and breast carcinoma cells (MOF-7, IC_{50} 25 μg/ml) (Park et al, 1999). The overexpression of HAD enzyme has been observed in many primary human cancer tissues. Apicidin is HDAC inhibitor that induces apoptosis in K562 cells via the activation of the mitochondrial pathway dependent caspase cascades (Cheong et al, 2003). It inhibits (IC_{50} 0.41 μM) the proliferation of human endometrial cancer cells (Ishikawa cells) and induces apoptosis by blocking HDAC 3. Apicidin at 5 mg/kg by intraperitoneal administration to the mice for 21 days, induced 53% inhibition of tumor growth in the Ishikawa cell xenografts (Ahn et al., 2010). Kim et al. (2004) found that apicidin displays anti-angiogenic activity in chorioallantoic membrane assay in human vascular endothelial cells and is able of inhibiting the formation of new blood vessels *in vitro* at 1 μg/ disks. These micotoxin reduces HPV16-E6 and -E7 transcripts and protein levels in SiHa cervical cancer cells at concentration of 2 μg/ml (Łuczak & Jagodzinski, 2008) and has potential as drug for the treatment of autoimmune diseases by reducing the phospholipase C gamma-1 transcript and protein contents in Hut-78 T lymphoma cells at concentration ranging of 2 to 8 μg/ml (Debicki et al., 2009).

Congeners of the apicidin, apicidins B and C that contain proline and valine, were isolated from *F. pallidoroseum* (ATCC 74289) (Sigh et al., 2001). Apicidin B and apicidin C inhibit the binding of [^{3}H]-apicidin A to *Eimeria tenella* parasite HDAC with IC_{50} values of 10 and 6 nM, respectively. Apicidin B show selectivity for parasite HDAC over the host chick liver HDAC (IC_{50} 50 nM) and activity against *Besnoitia jellisoni* (MIC 12.8 nM), *E. tenella* (MIC 411 nM) and *P. falciparum* (MIC 189 nM). However, apicidin C was slightly more active than apicidin B (MIC 0.8, 101, and 69 nM, respectively) and apicidin (6.4, 100, and 201 nM) against the same microorganisms (Sigh et al., 2001).

Further, others congeners with the (*S*)-2-amino-8-oxodecanoic acid (Aoda) residues, named as apicidins D_1, D_2, and D_3 were isolated from the same fungus (Sigh et al., 2002). Between apidicins D, only apicidin D1 was active against binding of [^{3}H]-apicidin A to *E. tenella* parasite HDAC (IC_{50} 4 nM; MIC value of 13 nM) and against *P. falciparum* (IC_{50} 34 nM) (Sigh et al., 2002).

Apicidin D_1 was somewhat more potent across all the assays but the R-hydroxy keto turned out to be least selective of the series (apidicin, apidicins A and B, apidicins D_2 and D_3) presumably due to its ability to better chelate the bivalent ions which are required for HDAC function. The cyclic tetrapeptide with C-8 Aoda ((*S*)-2-amino-8-oxodecanoic acid) keto group and aromatic amino acid, mainly tryptophan, next to the residue Aoda is the minimum pharmacophore required for activity, and this has been confirmed by the lack of activities of the keto-reduced congeners (Sigh et al., 2002).

A mixed of the peptide-polyketide compounds, cyclodepsipeptide HA 23 and the antifungal secondary metabolite W 493-B, were isolated from *Fusarium* sp. (CANU-HA 23) (Feng et al., 2002). HA 23 is a unique depsipeptide that contains one 14-carbon polyketide unit, one substituted tyrosine and pipecolinic acid. This metabolite contains a polyketide moiety and three amino acid residues, whereas depsipeptides (and cyclic peptides) typically incorporate either four or more than five residues. Only W 493-B showed antifungal activity against *Cladosporium resinae* (Feng et al., 2002).

The fusarium toxin equisetin (EQ), a derivative of *N*-methyl-2,4-pyrrollidone, was first isolated in 1974 from the white mold *F. equiseti* (NRRL 5537). EQ belongs to acyl tetramic acid class by virtue of its *N*-methylserine-derived heterocycle which in turn is fused to a quaternary stereogenic center of a *trans* octalin skeleton substituted (Burke et al., 2000). The tetramic acid moiety (hydrophilic) is an amino acid-derived heterocycle and the bicyclic (hydrophobic) is originated from a polyketide-derived decalin.

Equisetin is active against Gram-positive bacteria (*B. subtilis*, *Mycobacterium phlei*, *S. aureus*) and the Gram-negative bacteria (*Neisseria perflava*) at concentrations ranging of 0.5~4.0 μg/ml. The LD_{50} in mice is 63 mg/kg body weight. It inhibits human immunodeficiency virus type 1 (HIV-1) integrase (IC_{50} 7.5-15 μM), an enzyme that leads to insertion of viral DNA into the genome of the host cell (Singh et al., 1998). EQ also inhibits in the 3' end processing, strand transfer and disintegration (IC_{50} 15-25 μM) catalysed by stable complexes (Hazuda et al., 1999). This metabolite inhibits the growth of seedlings of various dicotyledonous and monocotyledonous plants (2.5-10 μg/ml) and causes necrotic lesions on roots, cotyledons, or coleoptiles of seedlings at 10 μg/ml (Wheeler et al., 1999). EQ prevents translocation of variant surface glycoprotein into *T. brucei* microsomes (95% at

50 μM) and kills bloodstream *T. brucei in vitro* (IC_{50} 3.3 μM) without to be toxic against mammalian cells (HeLa) at 20 μM (Patham et al., 2009).

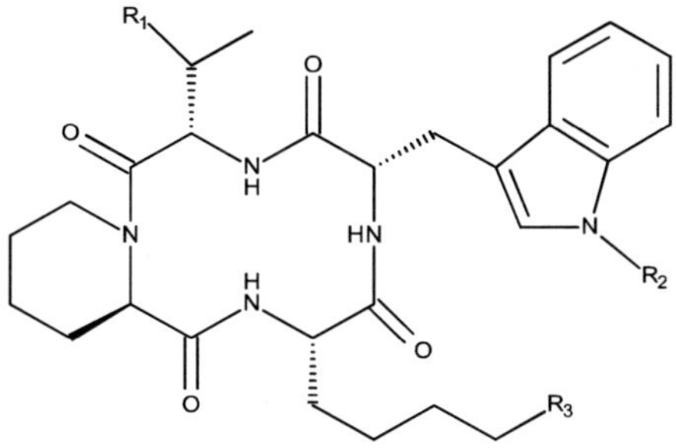

Name	R_1	R_2	R_3
Apicidin	$-CH_2CH_3$	$-OCH_3$	$-CH_2COCH_2CH_3$
Apicidin A	$-CH_2CH_3$	-H	$-CH_2COCH_2CH_3$
Apicidin C	$-CH_3$	$-OCH_3$	$-CH_2COCH_2CH_3$
Apicidin D_1	$-CH_2CH_3$	$-OCH_3$	$-CH_2COCH(OH)CH_3$
Apicidin D_2	$-CH_2CH_3$	$-OCH_3$	$-CH_2CH(S\text{-}OH)CH_2CH_3$
Apicidin D_3	$-CH_2CH_3$	$-OCH_3$	$-CH_2CH_2CH(OH)CH_3$

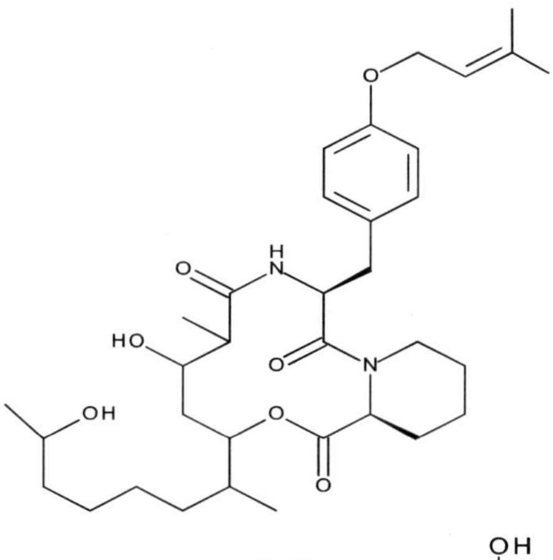

HA 23

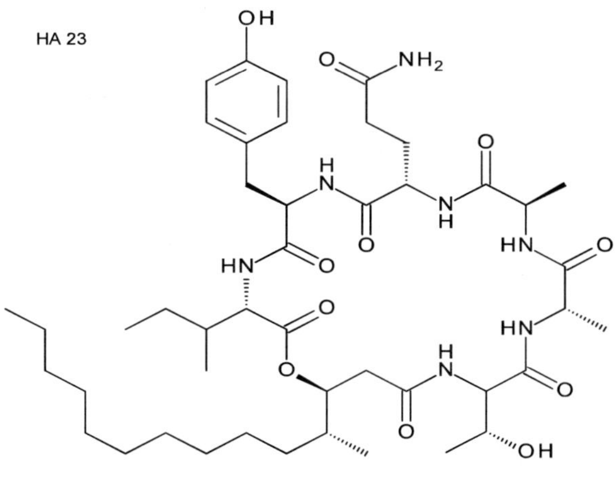

W 493-B

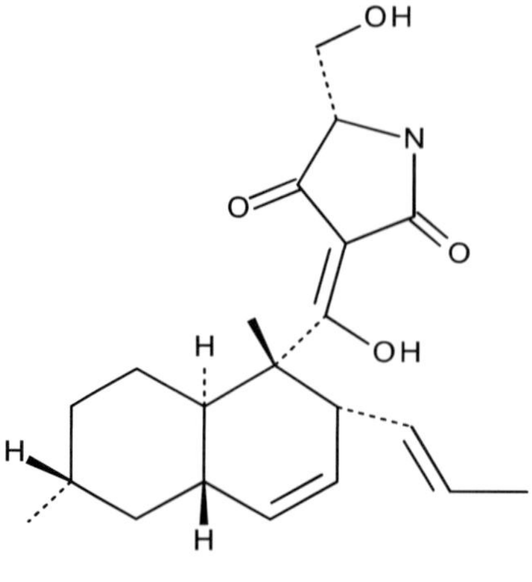

Equisetin

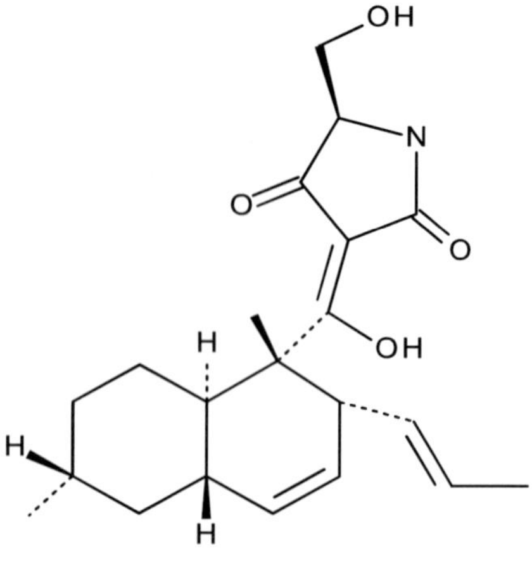

Epi-equisetin

Fusarium aquaeductuum, a strain (SANK1 1089) collected in Japan, furnished aquastatin A, a glycosylated depside with two aromatic rings and one galactose unit. Aquastatin A showed to be a negligible inhibitor of calcium channel in taenia (61% at 10^{-6}g/ml), of the calmodulin-dependent and independent phosphodiesterases (IC_{50} 58 μM and 100 μM, respectively) and of phospholipase A_2 (49% even at 148 μM, 100 μg/ml). It did not suppress the release of arachidonic acid at all 74 μM (50 μg/ml), did not show antimicrobial activity when tested at 1 mg/ml and inhibitory activity in a gastric secretory study at 100 mg/kg. However, aquastatin A inhibited ^{3}H-nitrendipine binding to porcine heart microsomes (IC_{50} 7.16 μM, 4.84 μg/ml), mammalian adenosine Na^+/K^+-ATPase (IC_{50} 7.1 μM) and H^+/K^+-ATPase (IC_{50} 6.2 μM) (Hamano et al., 1993) and protein tyrosine phosphatases 1B (IC_{50} 0.19 μM, competitive inhibitor), a target type 2 diabetes and obesity (Seo et al., 2009). Aquastatin A inhibited the enoyl-ACP reductases of *S. aureus* and *Streptococcus pneumoniae* (IC_{50} 3.2 and 9.2 μM, respectively) and prevented the growth of *S. aureus* and methicillin-resistant *S. aureus* (MIC 16-32 μg/ml). Dihydroxypentadecyl benzoic acid moiety in the molecule seems to be responsible for the anti-enoyl-ACP reductase and antibacterial activities (Kwon et al., 2009; 2011).

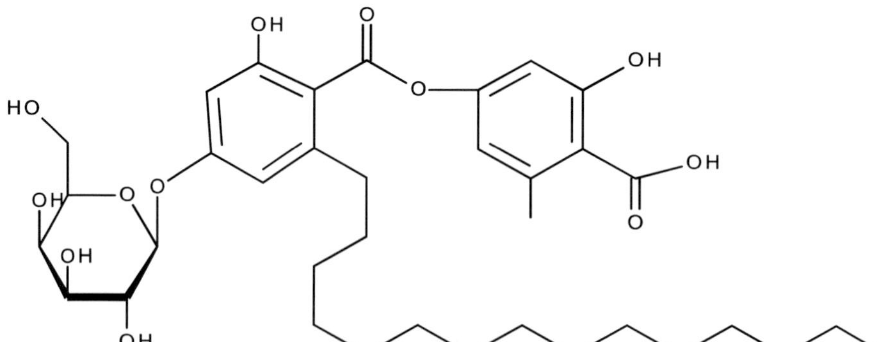

Aquastatin A

The fungus *Fusarium* sp. FN080326 was isolated from a soil sample collected in Korea. The EtOAc-soluble portion of the fungal extract afforded fusarisetin A, a compound that has an *N*-methyl group, a conjugated hexahydrofuro[2,3-c]pyrrole unit, a primary alcohol group, a pentacyclic ring system comprising decalin (6/6) and an oxygen-containing isoindol tricyclic moiety (5/5/5). Fusarisetin A inhibits 3D matrigel culture-induced colony formation in MDA-MB-231 cells at a concentration of 30 μg/ml and inhibits the cell migration and invasion at concentrations of 3 and 10 μg/ml, respectively, without to be toxic to MDA-MB-231 cells (Jan et al., 2011).

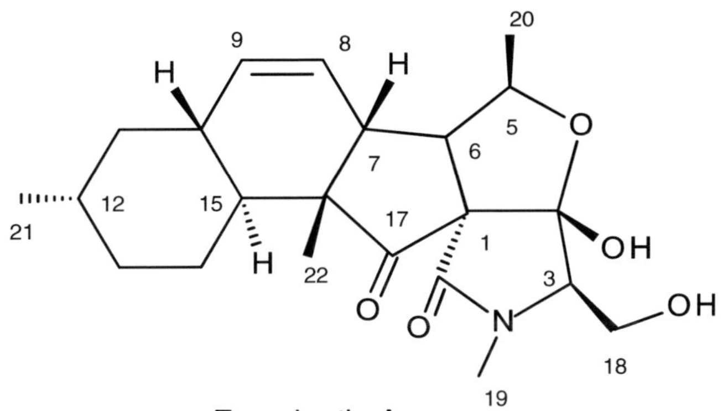

Fusarisetin A

Visoltricin (methyl ester of 3-[1-methyl-4-(3-methyl-2-butenyl)-imidazol-5yl]-2-propenoic acid), is an imidazole derivative produced by *Fusarium* spp. as *Fusarium tricinctum*. It is toxic to *A. salina* (LD_{50} = 8.5 x 10^{-7} M), inhibits the growth of six human tumour cell lines (out of 60 lines tested) at concentrations lower than 10^{-5}M and has miotic activity on rabbit eyes (Visconti and Solfrizzo, 1995). This last activity can be explained by its reversible anticholinesterase property towards human serum cholinesterase EC 3.1.1.8 (IC_{50} 1.9 x 10^{-4}M) (Visconti and Solfrizzo, 1994).

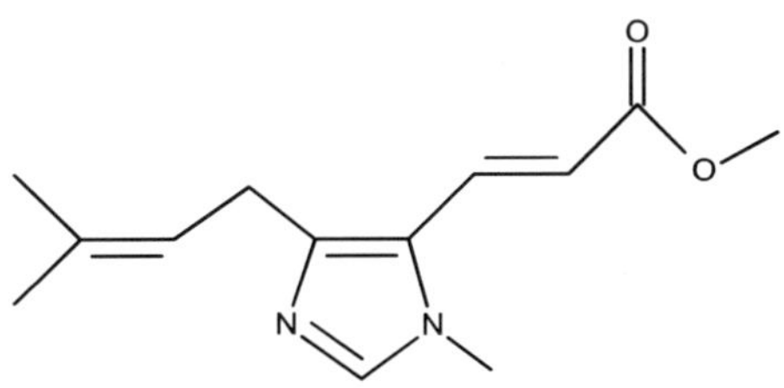

Visoltricin

Fungerin, 3-[1-methyl-5-(3-methyl-2-butenyl)-1H-imidazol- 4-yl]-2(E)-propenoic acid methylester, was isolated from a strain of *Fusarium sp.* It presented antifungal activity against *Penicillium chrysogenum, Colletotrichum langenarium, Alternaria mali* and *Pyricularia oryzae* with MIC values of 12.5, 25, 50 and 50 μg/ml, respectively (Kato et al., 1996). Fungerin also arrests the cell cycle at the M phase through inhibition of microtubule polymerization and induces apoptosis in Jurkat cells (Koizumi et al., 2004).

Piperazine derivatives, fusaperazine A and fusaperazine B, were isolated from *Fusarium chlamydosporum* OUPS-N124, an endophytic fungus obtained of the marine alga *Carpopeltis affinis*.

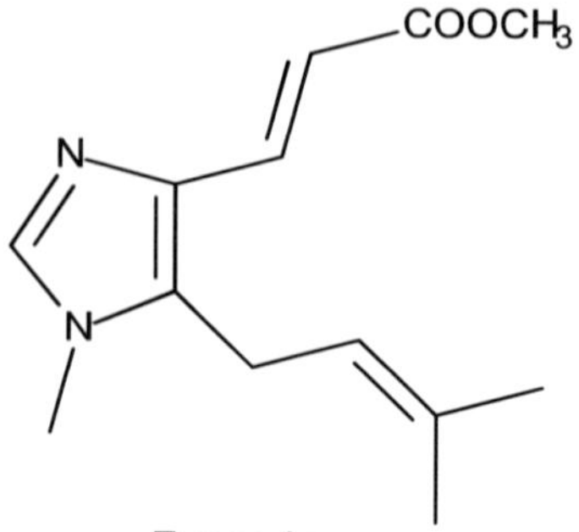

Fungerin

Others known compounds such as Sch 54794, Sch 54796 (Chu et al., 1993), 3-[(4-hydroxyphenyl)-methyl]-1,4-dimethyl-3,6-bis(methylthio)-2,5-piperazinedione (Hanson et al., 1981) and cis-bis(methylthio)silvatin (Kawahara et al., 1987, Wang et al., 1998) were also isolated. Fusaperazine A, Sch 54794 and cis-bis(methylthio)silvatin exhibited cytotoxic activities (ED_{50} 22.8, 21.5 and 7.7μg/ml, respectively) against P388 lymphocytic leukemia cells (Usami et al., 2002). Sch 54794 exhibited inhibitory activity in the platelet activating factor assay (Chu et al., 1993) and cis-bis(methylthio)silvatin displayed weak toxicity against NS-1 cells (LD_{99} 0.15 μM) (Capon et al., 2007).

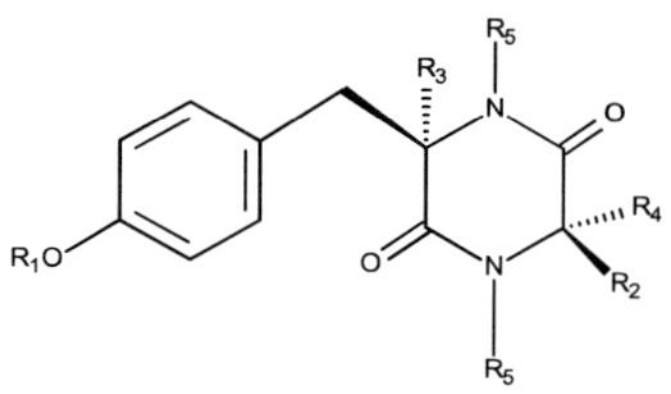

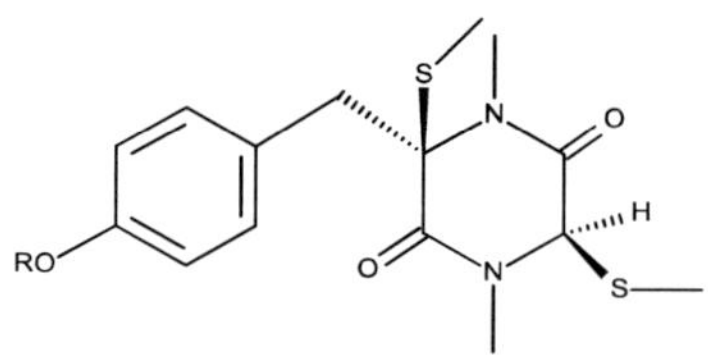

Name	R_1	R_2	R_3	R_4	R_5
Fusaperazine A	-H	-H	$-SCH_3$	$-SCH_3$	-H
Fusaperazine B	isoprenyl	$-SCH_3$	$-OCH_3$	-H	-H
Sch 54794	isoprenyl	-H	$-SCH_3$	$-SCH_3$	-H
Sch 54796	isoprenyl	$-SCH_3$	$-SCH_3$	-H	-H

3-[(4-Hydroxyphenyl)methyl1]-1,4-dimethyl-3,6-bis(methylthio)-2,5-piperazinedione R = H

cis-Bis(methylthio)silvatin R = isoprenyl

isoprenyl

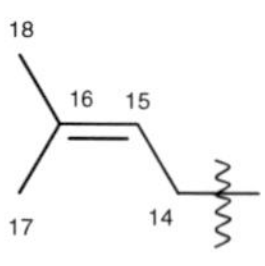

More structural simple pyridine can be represented by acuminatopyrone. It is produced by *F. chlamydospmum* (incidence in 55% of strains compared to 25% of *F. tricinctum*) and *F. tricinctum* at concentration levels ranging from 36 to 1800 mg/kg and from 3 to 37mg/kg, respectively (Visconti and Solfrizzo, 1994).

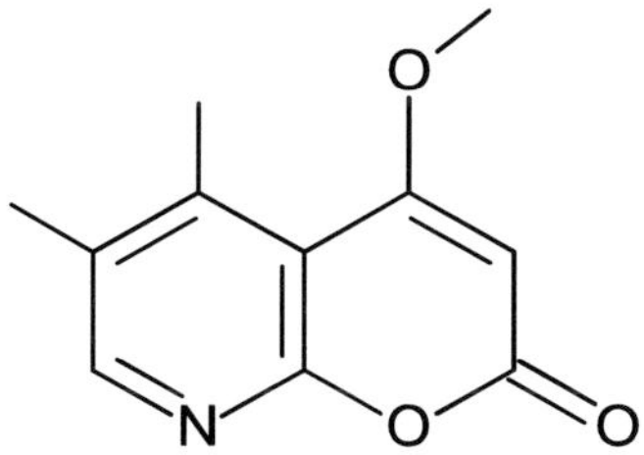

Acuminatopyrone

Fusaric acid (5-butylpicolinic acid) is a pyridine derivative compound that has been found in *Fusarium oxysporum*. It has hypotensive effect into rabbits, rats, cats or dogs and its mechanism of action likely to be the inhibition of dopamine-*β*-hidroxylase. The LD_{50} of fusaric acid for mice was 100 mg/kg by intravenous administration and 80 mg/kg by intraperitoneal administration. The death is explained by due to its hypotensive effect and not to other factors (Hidaka et al., 1969). A clinical trial with fusaric acid in manic patients decreased, in cerebrospinal fluid, 3-methoxy-4-hydroxy-phenylglycol, the major metabolite of norepinephrine, and increased in homovanillic acid, the major metabolite of dopamine. These changes are consistent with central dopamine-*β*-hydroxylase inhibition. However, patients with more severe mania became worse, i.e. more psychotic (Sack et al., 1974). Intraperitoneal administration of picolinic acid into C57BL/6 mice induced high levels of macrophage-mediated cytostatic activity without influencing natural killer cell activity (Ruffmann et al, 1984). Fusaric acid was also isolated from *F. oxysporum* EF119, an endophytic fungus from red pepper (*Capsicum annuum* L.) and suppressed the cell growth of various plant pathogenic oomycetes, fungi and bacteria with MIC values ranging from 11 to 100 μg/ml (Son et al., 2008).

Fusaric acid and some related analogs demonstrated to be phytotoxic in cotton cotyledonary leaf bioassay which nicotinic acid was considerably less phytotoxic than picolinic acid and fusaric acid methyl ester. The carboxylic acid group is essential for phytotoxicity and increasing the alkyl chain length at the 5-position on picolinic acid from two up to five carbons atoms enhance the activity (Stipanovic et al., 2011). Others analogues as cadmium (II) and copper (II) complexes of fusaric acid exhibited potent inhibitory activity against the *Mycobacterium bovis* BCG (MIC 4 μg/ml] and against the *M. tuberculosis* H37Rv (MIC 10 μg/ml), respectively (Pan et al, 2011).

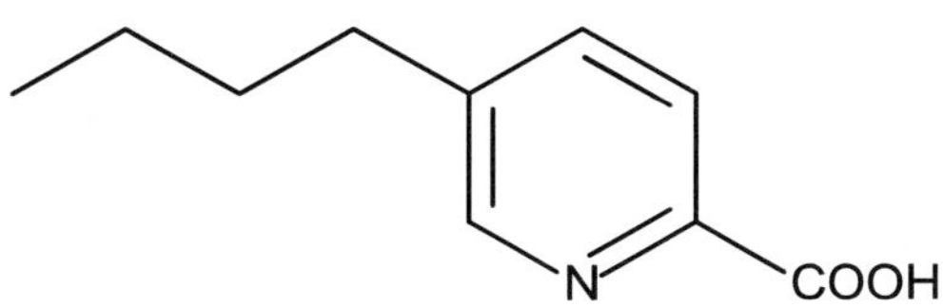

Fusaric acid

Parnafungins, natural products containing isoxazolidinone and xanthone moieties, have been isolated from *Fusarium larvarum* complex; mycoparisites associated with scale insects, aphids, lichens, *Septobasidium clelandii* (basidiomycete fungus), *Trichonectria rectipila* and

Cladobotryum pinarense (mycoparasitic fungi) (Jiang et al., 2008; Parisch et al., 2008; Tisson et al., 2005; Rossman et al., 1999; Booth, 1981). Parnafungins A and B, isolated as a mixture (Jiang et al., 2008), are inhibitors of the fungal polyadenylate polymerase, the enzyme responsible for the extension of the poly (A) tail of an mRNA transcript (Parisch et al., 2008). They were isolated together with parnafungins C and D from a *F. larvarum* strain F-155,597. The mixture of parnafungins A and B (MIC 0.008-2.5 μg/ml) and pure parnafungins C and D (MIC 0.08-2.5 and 0.06-5 μg/ml, respectively) have potent antifungal activity against wild-type *C. albicans* (Overy et al., 2009).

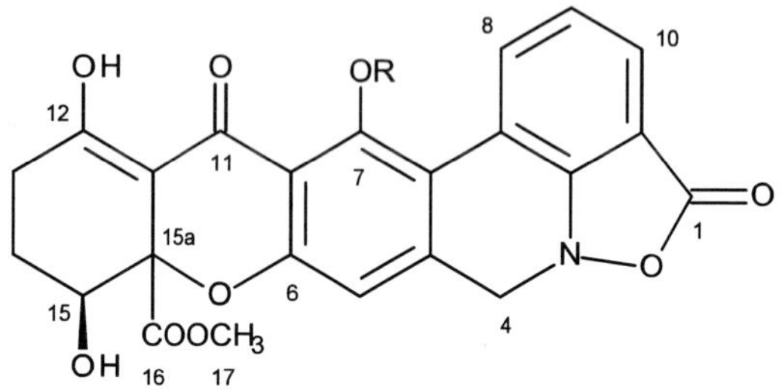

Parnafungin A R = H

Parnafungin C R = OCH_3

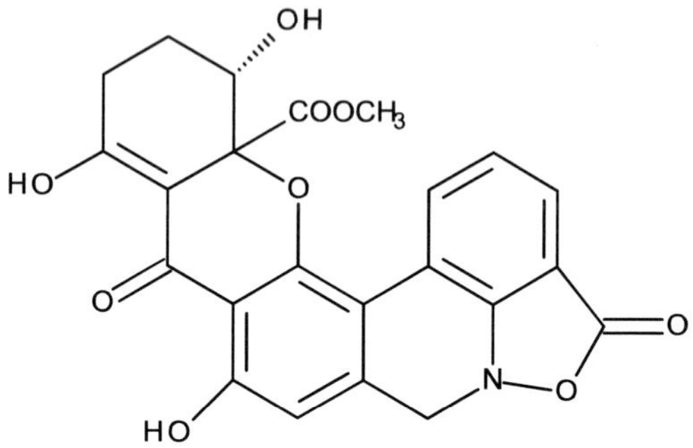

Parnafungin B

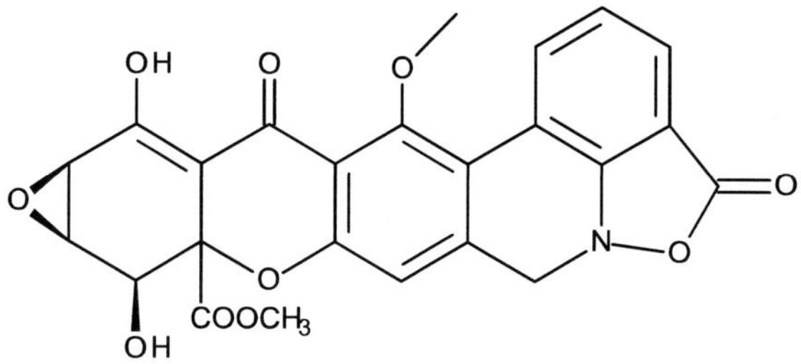

Parnafungin D

A group of secondary metabolites with 4-hydroxy-2-pyridone chromophore has been commonly occurring in fungi. They have the central 4-hydroxy-2-pyridinone linked to two additional ring systems at the C-3 and C-5 positions. H^{18} biosynthetic investigations demonstrated that the pyridinone rings are formed by condensation of phenylalanine and polyketide precursors, initially resulting in tetramic acid intermediates, which subsequently undergo rearrangement and ring expansion (Wright et al., 1977; Breinholt et al., 1997).

McBrien et al. (1996) identified from *Fusarium* sp WC-49758 isolated from flowers of sourwood (*Oxydendron arboreum*), an *N*-hydroxypyridone named fusaricide. The metabolite is an HIV regulator of virion expression. It is active in the R17 coat protein/R17 RNA binding assay (75 μg/ml) but is not specific for regulation of virion expression HIV protein. Fusaricide also show antifungal activity against *C. albicans* and *Penicillium chrysogenum* (MIC's 16 and 8 μg/ ml, respectively) and has antitumor activity against a murine tumor cell line (Madison lung carcinoma 109) at concentrations of 1 μg/ml (McBrien et al., 1996).

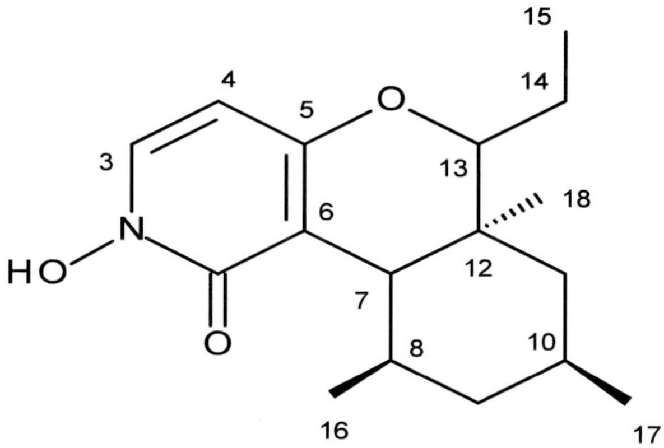

Fusaricide

Sambutoxin (3-{5-methyl-6-[(E)-1,3,5-trimethylhept-1-enyl]tetrahydropyranyl}-5-(*p*-hydroxyphenyl)-1-methyl-4-hydroxy-2(1H)–pyridone) was isolated from wheat cultures of *Fusarium sambucinum* PZF-4 as a hemorrhagic mycotoxin. It was also produced by *F. oxysporum* isolates obtained from rotted potato tubers. Rats fed with diets supplemented with 0.05 and 0.1% sambutoxin presented hemorrhages in the stomach and intestines, body weight loss and finally death. The toxin was also toxic to chick embryos (50% lethal dose, 29.6 μg/egg) (Kim et al., 1995).

N-methyl-4-hydroxy-2-pyridinone analogues such as 6-*epi*-oxysporidinone, the dimethyl ketal of oxysporidinone and *N*-demethylsambutoxin, along with (-)-oxysporidinone and (-)-sambutoxin, were obtained from ethyl acetate extract of *F. oxysporum* (N17B) (Jayasinghe et al., 2006). Oxysporidinone, a 3,5-disubstituted *N*-methyl-4-hydroxy-2-pyridone, was previously isolated from fermentations of *Fusarium oxysporum* (CBS 330.95) (Breinholt et al., 1997).

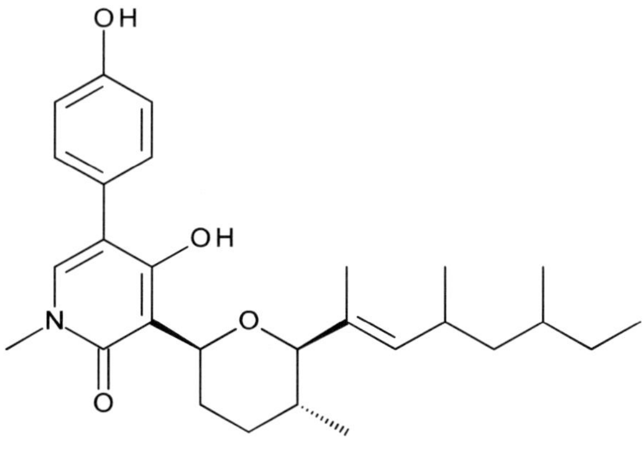

Sambutoxin

It has antifungal activity against *Aspergillus niger* (MIC 10 μg/ml), *Botrytis cinerea* (MIC 1 μg/ml), *Alternaria alternata* (MIC 50 μg/ml), *Venturia inequalis* (MIC 10 μg/ml) (Breinholt et al., 1997), *A. fumigatus* (IC_{50} 2 μg/ml), *C. neoformans* (IC_{50} 35 μg/ml) (Jayasinghe et al.,

2006), albeit it has not been active against bacteria (*B. subtilis* and *Pseudomonas aeruginosa*) and yeasts (*C. albicans* and *S. cervisiae*) when tested at 1 mg/ml. 6-*Epi*-oxysporidinone had only marginal activity against *A. fumigatus* (IC_{50} 35 μg/ml). The cyclohexyl ring of 3,5-disubstituted *N*-methyl-4-hydroxy-2-pyridones is critical for the anti-*A. fumigatus* activity. The removal of the carbonyl or aromatization of the cyclohexyl ring led to complete loss of the antifungal activity and the change of configuration of the 6'-hydroxyl group resulted in reduction of activity (Jayasinghe, et al., 2006).

Fusapyrone and deoxyfusapyrone are two 3-substituted-4-hydroxy-6-alkyl-2-pyrones isolated from *F. semitectum* (strain ITEM 1623) grown on autoclaved wheat kernels. They have a functionalized aliphatic chain and a 4-deoxy-*β*-xylohexopyranosyl *C*-glycosyl moiety bound to the C-6 and C-3 of the 2-pyrone ring, respectively. Antibiotic properties of these compounds is related to their amphiphilic nature because of hydrophilic sugar residue opposed to a long hydrophobic chain. Even though their structural similarities, fusapyrone has been shown to be a potent antifungal toward *Botrytis cinerea*, *Aspergillus parasiticus* and *Penicilliun brevi-compactum* (MIC 0.78-6.25 μg/ml) with low toxic in *A. salina* assay (LC_{50} >200 μM), whereas deoxyfusapyrone was toxic in *A. salina* assay (LC_{50} 37.1 μM) besides to demonstrate antifungal activity (MIC 0.78 μg/ml) (Altomare et al., 2004).

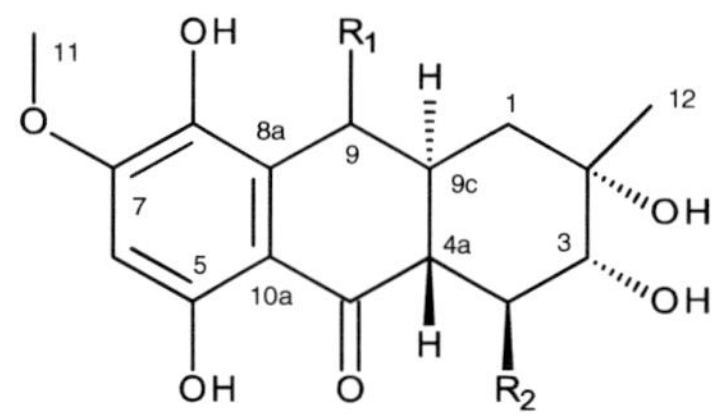

Fusaranthraquinone R_1 = β-OH, R_2 = H
9-α-Hydroxyhalorosellinia R_1 = α-OH, R_2 = OH

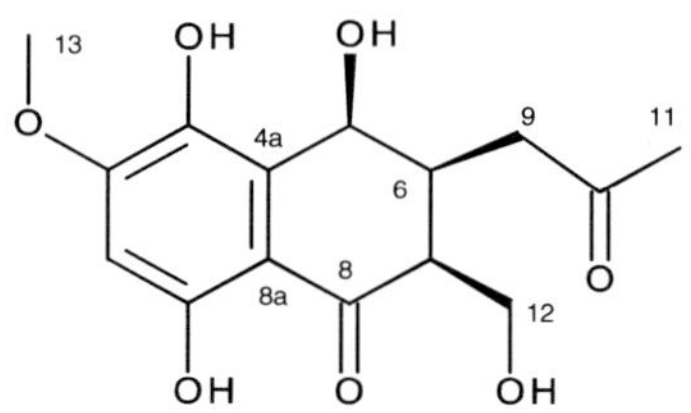

Fusarnaphtoquinone A

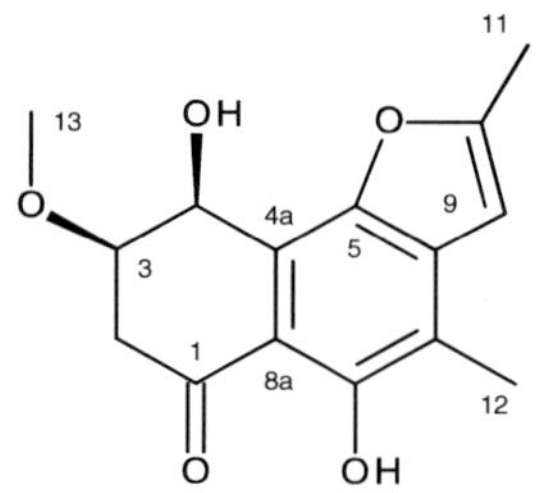

Fusarnaphtoquinone B

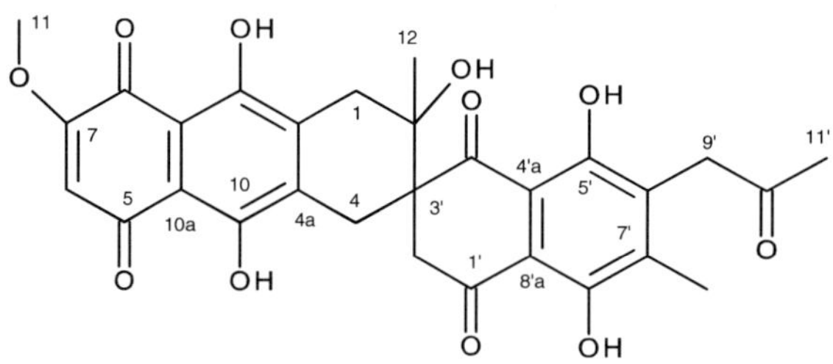

Fusarnaphtoquinone C

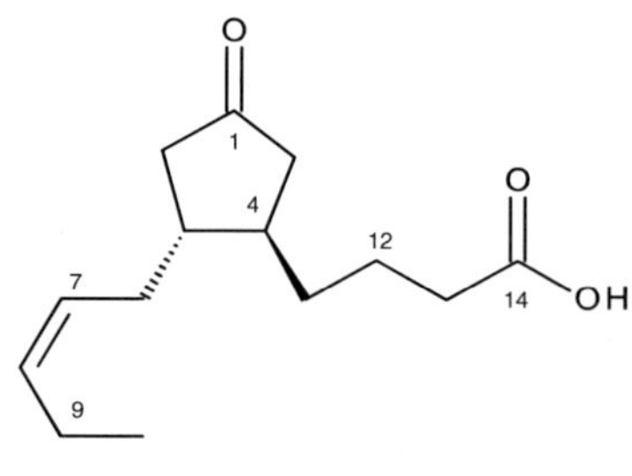

Fusarone

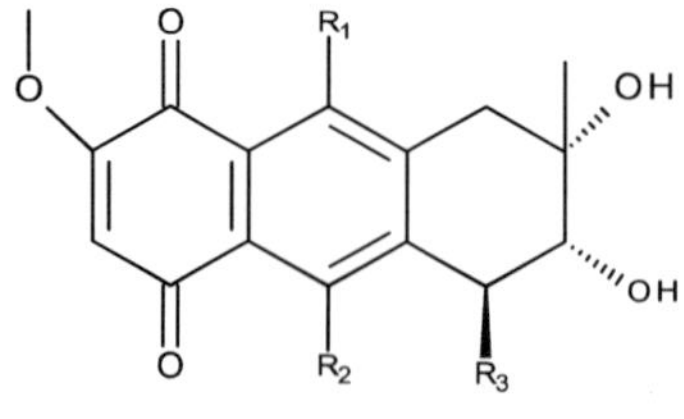

Bostrycin R_1 = R_2 = R_3 =OH
Nigrosporin B R_1 = R_3 = H, R_2 =OH

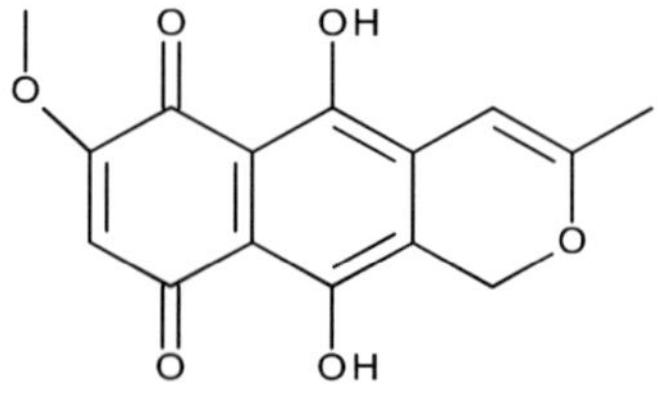

Anhydrofusarabin

Further, Hiramatsu et al. (2006) isolated fusapyrone, deoxyfusayrone and neofusapyrone from *Fusarium* sp. (FH-146) obtained from driftwood collected Japan. All compounds demonstrated antifungal activity against *Aspergillus clavatus* (MIC 3.12-6.25 μg/ml) but only fusapyrone and deoxyfusapyrone were weakly active against *P. aeruginosa* (MIC 50 and 12.5 μg/ml, respectively).

Balanol (4-[2-[[[Hexahydro-4-[(4-hydroxybenzoyl)oxy]-1H-azepin-3-yl]amino] carbonyl]-6-hydroxybenzoyl]-3,5-dihydroxybenzoic acid) consists of two principal structural domains, a chiral hexahydroazepine-containing core and a benzophenone fragment. Based on NMR spectra, optical rotations and total synthesis, Boros et al. (1994) concluded that ophiocordin, first isolated from *Cordyceps ophioglossoides* (Kneifel et al., 1977), was identical to balanol, isolated from *Verticillium balanoides* (Kulanthaivel et al., 1993).

Further, *Fusarium merismoides* (NR6356) was discovered as the source of the rat brain protein kinase C (PKC) inhibitor, azepinostatin (IC_{50} 70 nM), which do not affect others enzymes utilizing ATP such as hexokinases. It was recognized that azepinostatin had the same structure than balanol (Ohshima et al., 1994). Balanol is a potent inhibitor (IC_{50} 4-9 nM) of human protein kinase (PK) C in assays against human PKC enzymes α, βI, βII, γ, δ, ε and η (Boros et al., 1994). This metabolite has some inhibition selectivity in enzymes of the GRK subfamilies because its IC_{50} values for G protein-coupled receptor kinase 2 (GRK2-GRK7) ranged of 42 to 490 nM (Tesmer et al., 2010).

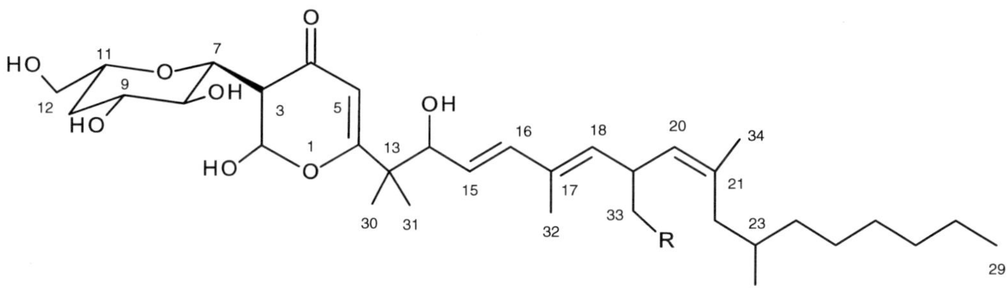

Fusapyrone R = OH

Deoxyfusapyrone R = H

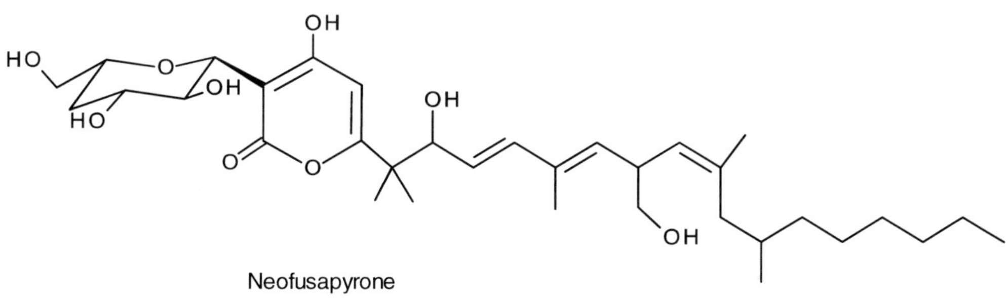

Neofusapyrone

The biosynthetic proposal for balanol is the formation of benzophenone by polyketide pathway, while nonribosomal peptide generates the 3-amino-hexahydroazepine-4-ol with the *N*-terminal *p*-hydroxybenzoate protecting group. The hexahydroazepine originates from intramolecular cyclization of 3-hydroxylysine by amide formation between the terminal amino group of the side chain and the main chain carboxylate, followed by reduction. Coupling of the two parallel products is then achieved by esterification of the hexahydroazepine-4-ol intermediate (Molnár et al., 2010).

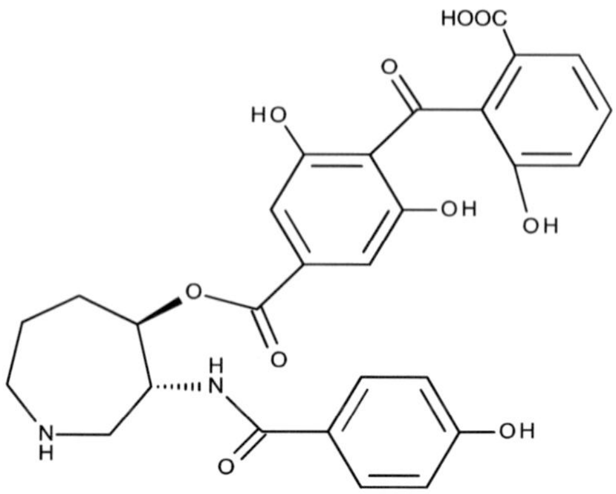

Balanol

The mycotoxin moniliformin was discovered by Cole et al. (1973) while screening for toxic products of a North American isolate of *Fusarium moniliforme* Sheldon (*Fusarium verticillioides* [Sacc.] Nirenberg) cultured on corn, which was actually misidentified and should have been identified as *F. proliferatum* (Rabie et al., 1982). Moniliformin has been isolated as sodium (1, M = Na) and potassium (1, M = K) salts of 1-hydroxycyclobut-l-ene-3,4-dione (Springer et al., 1974). It is produced in large quantities (approx. 10 g/kg of growth medium) by some fungal isolates under laboratory conditions (Kriek et al., 1977).

Moniliformin has been found to be negative in the Ames test for mutagenicity (Wehner et al., 1978). In addition, a single oral LD_{50} for moniliformin in white mice is 47.6 mg/kg of body weight and even a dosage about three times the single LD_{50}, mouse survived (Burmeister et al., 1980). Below this level, mice can tolerate remarkably large amounts of moniliformin without apparent ill effects.

Moniliformin and pyruvate are structurally similar and was proposed that toxicity of moniliformin is caused by the selective inhibition of pyruvate dehydrogenase enzyme (suicide inactivator at 304.4 μM) (Gathercole et al., 1986). Chronic exposure to moniliformin in animals results in the development of myocardial hypertrophy but it does not show a cardiostimulatory effect in cumulative concentration-response model using isolated guinea pig atria (Reams et al., 1996). In addition, the compound has been connected with endemic chronic deformative osteoarthropathy and cardiomyopathy, the Kashin-Beck disease (KBD). The infection of maize in China with the moniliformin-producing *F. fujikuroi* and *F. subglutinans* is common (Yu et al., 1995). The major pathophysiological changes in KBD patients are chondrocyte necrosis in the hypertrophied layer of the cartilage, with cell cluster formation nearby (Wang et al., 2006). KBD patients have also heart lesions that resemble the lesions that can be observed as a result of acute moniliformin intoxication in animal experiments (Zhao et al., 1993). Moniliformin inhibits the syntheses of aggrecan and type II collagen, suggesting the pro-catabolism effect on the metabolism in articular cartilage that can lead to degradation (Zhang et al., 2010).

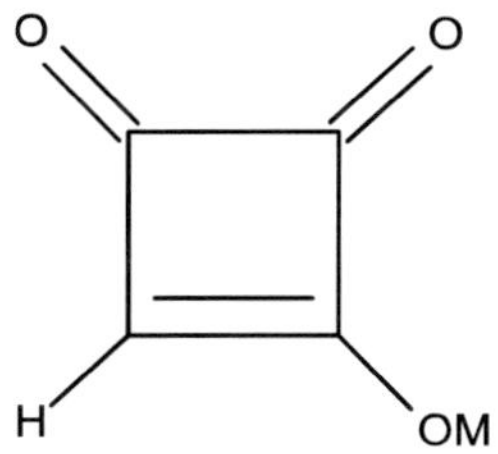

M = Na or K

Moniliformin

Secondary metabolites with several kinds of scaffolds have been isolated for more than 70 years from *Fusarium* species. Most of natural products produced by this genus have presented some biological activity that can be related to the key for survival, ecological adaptations and diversification of fungus. These metabolites have shown to confer an advantage on the plant in which the fungus grow symbiotically or are virulence factors that cause significant disease on agricultural crop. The production of mycotoxins depends on genetic variation, pattern of inheritance of this secondary gene cluster during evolution, geographic area of origin and environmental microbiota. Culture media or habits (nutritional regulation) and the presence of inducing substances (regulatory factors) can regulate the expression of cluster genes which regulate the secondary metabolite production. The pharmaceutical utility of these compounds is still in progress since regulation involving multiple proteins and complexes that respond to various environmental and host stimuli (Fox and Howlett, 2008; Yu and Keller, 2005).

5. Synthesis of Fusarium Metabolites

Fusarium metabolites are auspicious substances, since very complex structures were isolated from this source and presented biologic activity. Despite the fact of complex substances be produced by fermentations, where the biologic processes work at mild conditions like pH, temperature and salinity, the downstream treatments are very expensive, mainly in case of constraints for very pure substances. Quantity, quality and cost are factors that walk in opposite directions.

Several metabolites have been isolated from *fusarium* genera since 30´s, as example, Ashley et al. (1937) named Aurofusarin and Rubrofusarin the colored compounds isolated from *Fusarium* species. For last decades, with the development of more powerful analytical techniques, like Nuclear Magnetic Resonance (NMR) and High Resolution Mass Spectrometry (HRMS), the number of compounds isolated from *Fusarium* genera increased substantially, and some of them were assayed to verify their biologic activity. Among the classes of products isolated from *Fusarium*, we can emphasize the azanaphthoquinone (Bostrycoidin), benzoxanthentrione (Bikaverin), butyrolactam (Fusarin C), cyclodepsipeptide (Enniatins), sphingoid base (fumonisin), furanocoumarin (Bergapten), gibberelene (Gibberellic Acid), naphthopyrone (Rubrofusarin), naphthoquinone (Juglone), pyridine derivative (Fusaric Acid), resorcyl lactone (Zearalanone), terpene (Culmorin), trichothecene (T-2-toxin).

Synthetic routes are intellectual approaches to discovery inexpensive and efficient mechanism to obtain compounds where its usual production is economically not feasible. In terms of academic search, some routes are only to prove concepts.

Cameron et al. (1982) presented a work relating the study of the regiochemistry for carbocyclic quinones by photochemistry 1,2 addition of 1, 1-dimethoxyethene with azanaphtoquinones (3, 4, 5). These studies based them to develop the synthesis of Bostrycoidin (1) and 8-O-Methylbostrycoidin (2).

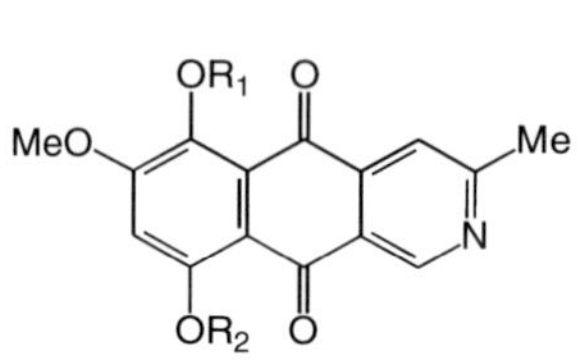

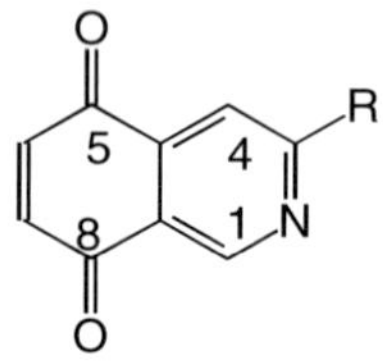

(**1**) R_1 = H; R_2 = H
(**2**) R_1 = H; R_2 = Me

(**4**) R = H
(**5**) R = Me

Bikaverin (6), a benzoxanthentrione was synthesized by Hauser et al. (1988) as regiospecific studies of the chromones condensations. The authors describe the synthon 7 as main block for formation of the A and B rings and after condensation with chromone 8, in the preparation of Bikaverin.

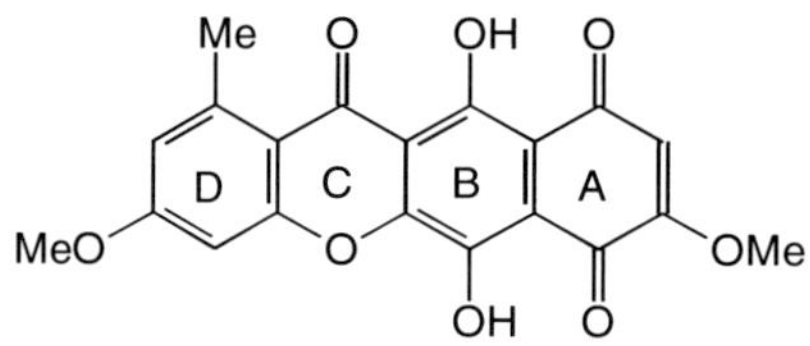

(**6**)

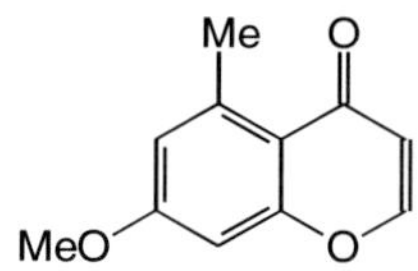

(**8**)

However Alain Bekaert et al. (1992), presented other route based on condensation of Mannich´s base 10 from 2-Hydroxy-5,7,8-trimethoxy-1,4-naphtoquinone 9 with orcinol monomethylether.

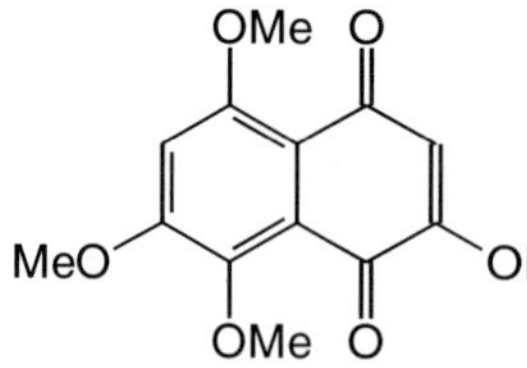

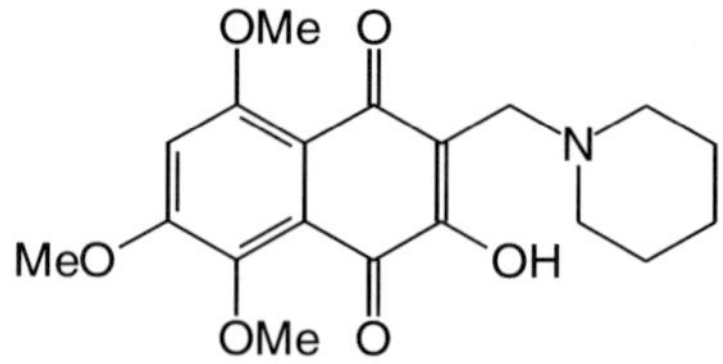

(**9**) (**10**)

Fusarin C (11) belongs to lactamic ring group of natural compounds like epolactaene (12), oteromycin (13) and talaroconvolutin A (14).

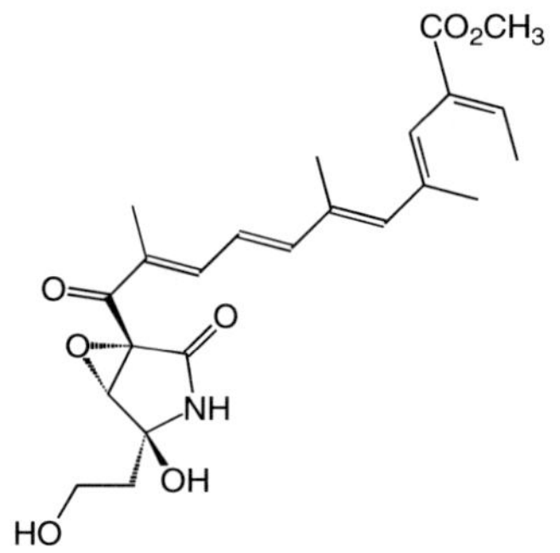

(11)

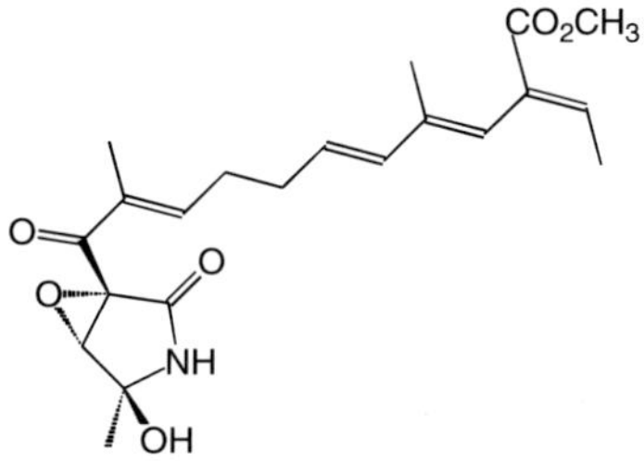

(12)

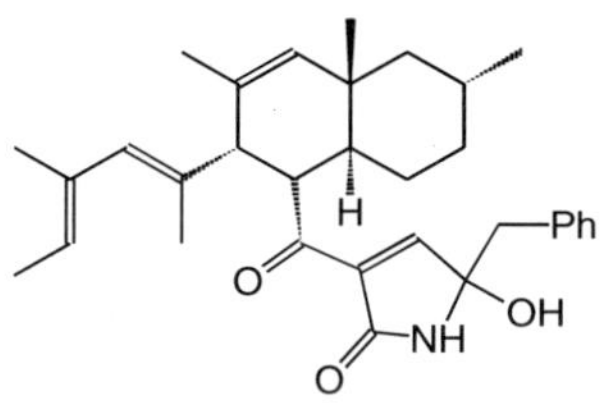

(13)

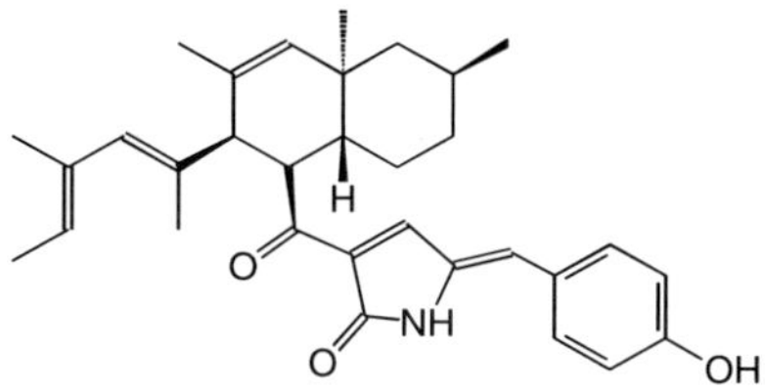

(14)

Snider and Neubert (2004) presented a retrosynthetic route to synthesize the lactamic core of Fusarin C (15) from synthon 20, according to scheme 1. Thus, the main conception of the lactamic core 15 synthesis was to produce the aldehyde 20. It was possible by Moffat oxidation process of an alcohol.

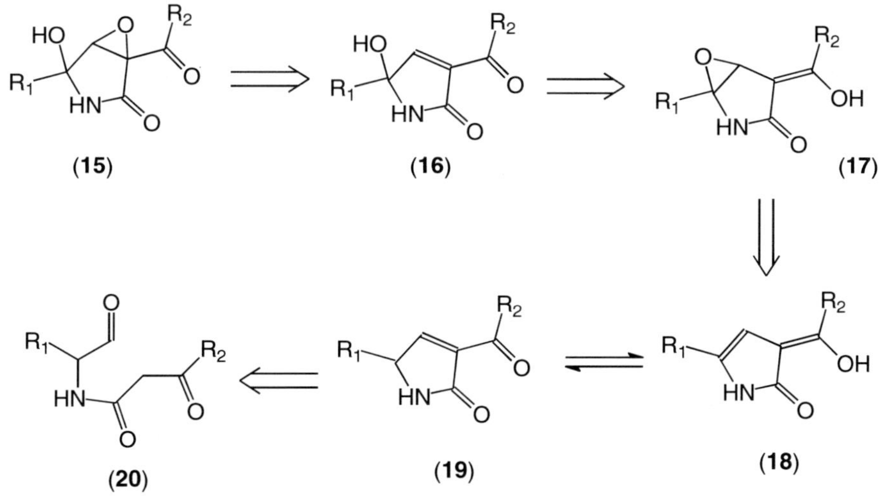

Scheme 1. Retrosynthesis in the formation of lactamic ring core of Fusarin C.

This methodology was used to produce the models 21, 22, 23 and oteromycin model 24.

Enniatins are cyclic depsipeptides (peptide where the amide bond was replaced by ester bond). The different isomers are classified by A (25), B (26), C (27), etc, according to chemical groups linked to core. Beauvericin (28) is other kind of Enniatins. It is important to note that these chemical modifications occur at amino acid residue.

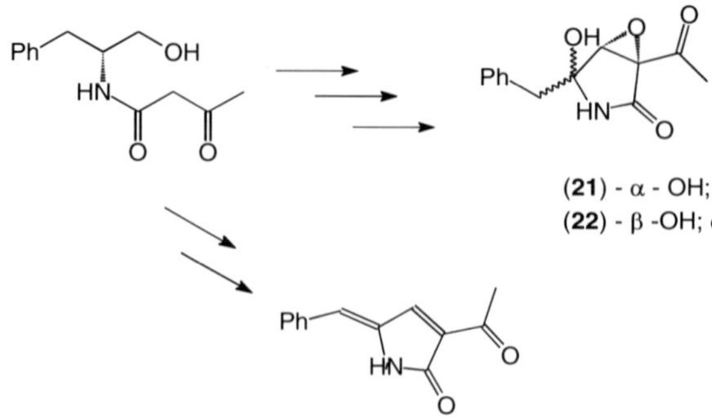

(21) - α - OH; β -$PhCH_2$
(22) - β -OH; α - $PhCH_2$

(23)

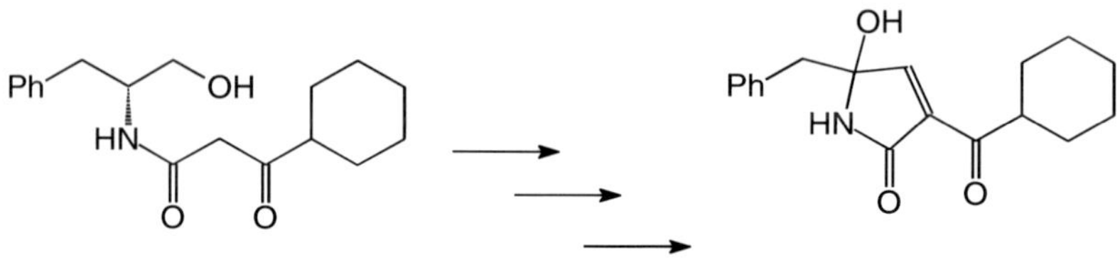

(24)

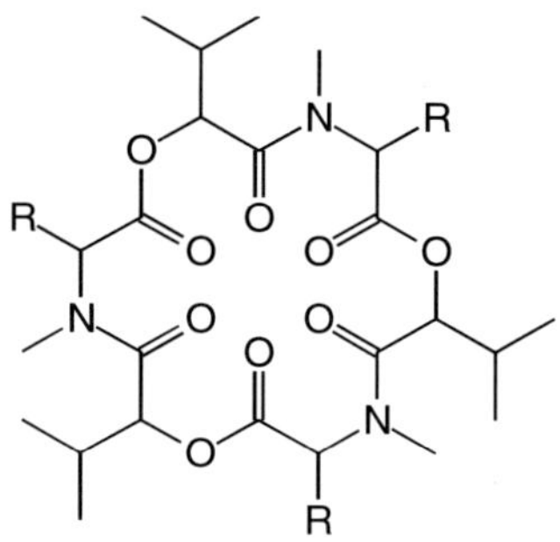

Enniatin A (25)- R = $CH(CH_3)CH_2CH_3$

Enniatin B (26) - R = $CH(CH_3)_2$

Enniatin C (27) - R = $CH_2CH(CH_3)_2$

Beauvericin (28) - R = CH_2Ph

Shemyakin et al. (1963) presented the configuration of Enniatin B by means of synthesis, comparing the physicochemical and biological data between naturally occurring antibiotic and synthetic compounds. It was presented that the cyclotetradepsipeptide, although constituted for the same fragments of the alkaline hydrolysis from enniatin B, was not one. They started from (L)-N-Methyl-valine and □ hidroxi-isovaleryl acid to prepare tetra, hexa and octadepsipeptide. Thus it was possible to show that enniantin B is a cyclohexa-depsipetide.

Tomoda and Doi (2008) presented a protocol to synthesize a group of beauveriolides by combinatorial chemistry in solid-phase using a 2-chlorotrityl linker followed by solution-phase cyclization. They built 104 compounds that were evaluated as Antiatherosclerotic Agents.

Fumonisins are structurally similar to the sphingoid moiety of the sphingolipids, as example, sphingosine (29) and Fumonisin B (30).

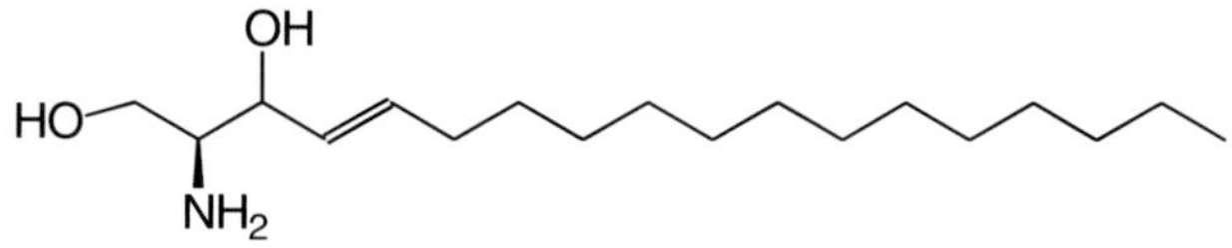

(**29**)

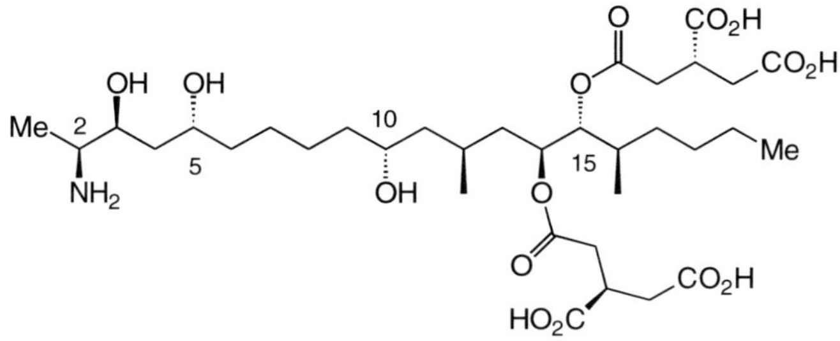

(**30**)

The main challenger to synthesize the compound 30 is to protect the absolute chiral configuration at centers 2, 3, 5, 10, 12, 14, 15 and 16, respectively. Pereira et al. (2009) built Fumonisin B from two parts, the fragment C1-C9 and the fragment C10-C20. The chirality of these sectors was produced by means of chiral assistant of the reagents and chirality transfer of the previous chiral synthons. Similar concept to build the Fumonisine B1 hexaacetate derivative (31) was developed by Gurjar et al. (1998) where the chiral centers started from D-glucose.

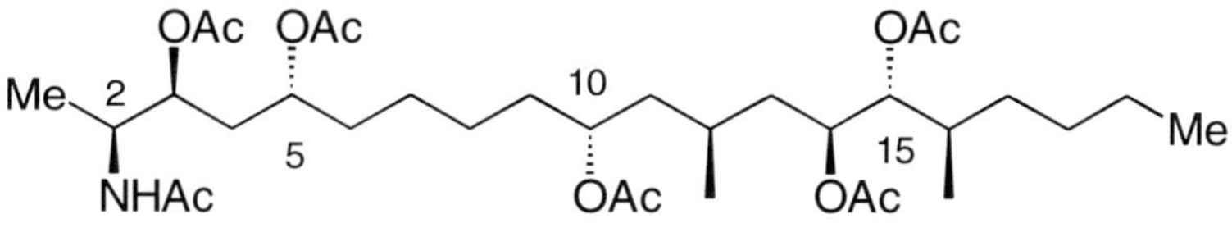

(**31**)

Psoralenes (32), a serie of coumarins with the addition of furan ring, are represented by Bergapten (33) and Xanthotoxin (34).

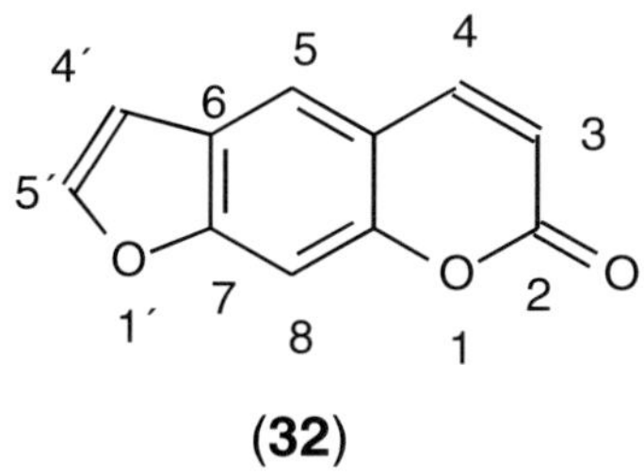

(**32**)

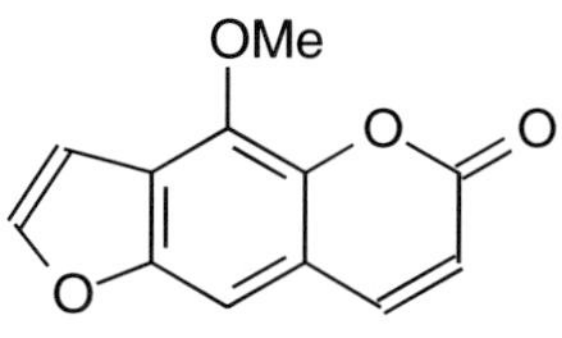

(**35**)

By means of [4+2] cycloaddition strategy, Mal et al. (1998) elaborated the synthesis of 34 and its isomer 35. They started the synthesis from commercial reagent acetylacetaldehyde dimethyl acetal with methyl chloroacetate in the presence of NaOMe to produce the intermediate compounds 36. Annulations of this compounds with 37 (derived from ciclopentadiene, ref), produced 34 and 35.

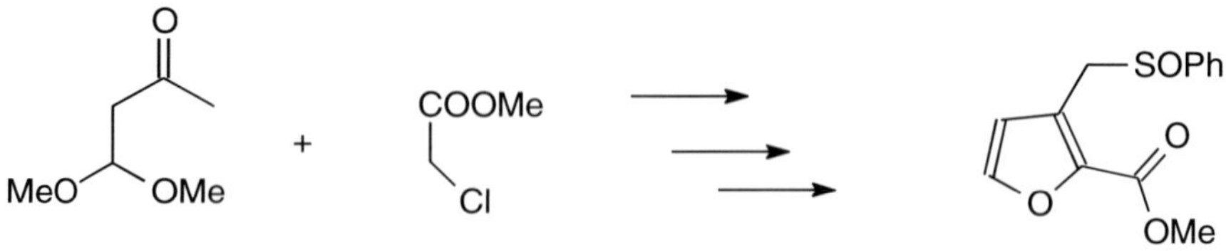

(36)

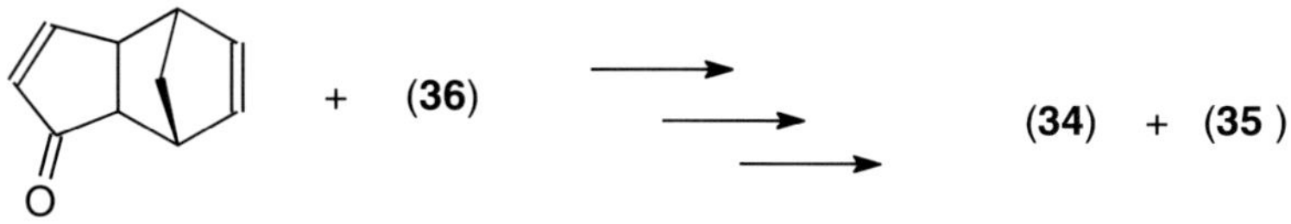

(37)

Gibberellic acid is a component of the family of Gibberellins (GAs), a series of plants´ hormones like GA1 (38), GA3 (39), *ent*-Gibberellane (40) and *ent* Kaurene (41).

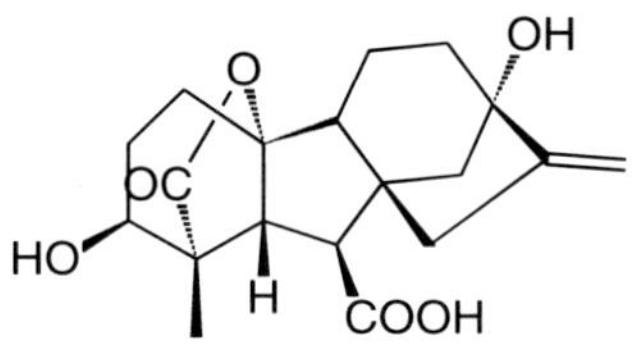

(38)

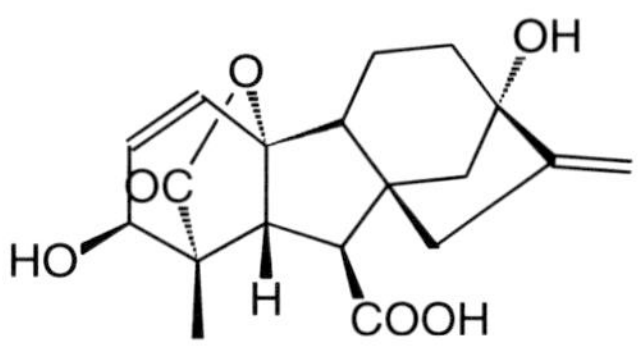

(39)

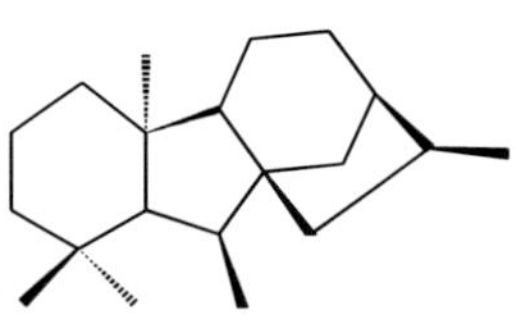

(40)

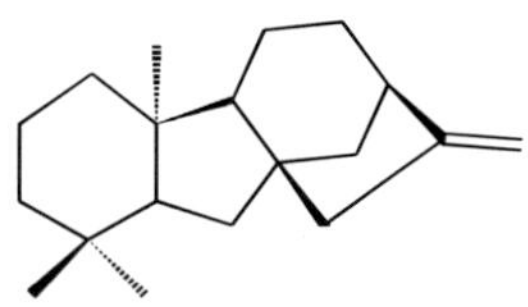

(41)

Core et al. (1978) started in 70´s developing strategies to synthesize chiral fused rings of Gibberellic Acid by Diels-Alder coupling. The retrosynthetic analysis, in Scheme 2, shows the main synthons for preparations of the 38. Later they developed new methodologies for preparations of the 44 synthon, a crucial molecule for synthesis of Gibberellic acid (ref).

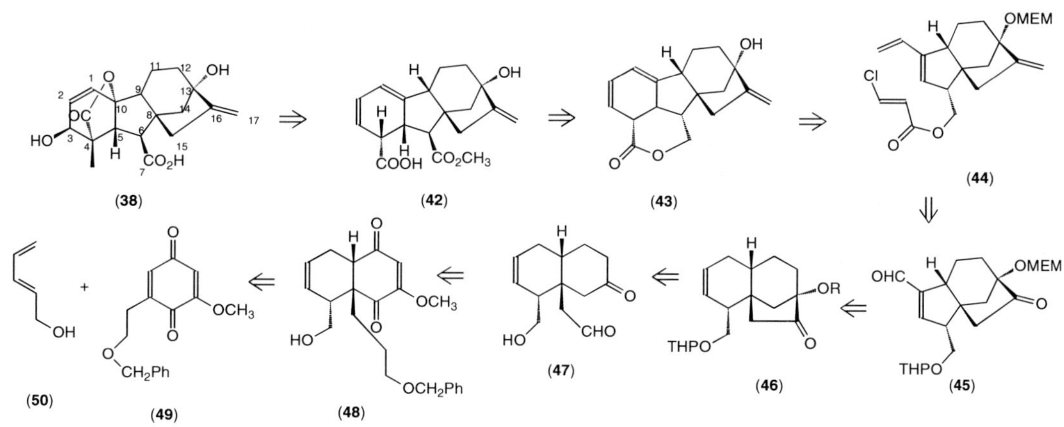

Scheme 2. Retrosynthesis in the formation of Gibberellic Acid.

Rubrofusarin (51), a colored compound isolated from *Fusarium* genus, belongs to naphthopyrone group. Shibata et al.(1967) synthesized it and its monomethyl ether derivatives (52, 53 and 54) by Claisen condensation of resorcylic acid and acetoacetate. This synthetic strategy was efficient to prepare derivatives and to study the reactivity of the substituents on fused rings.

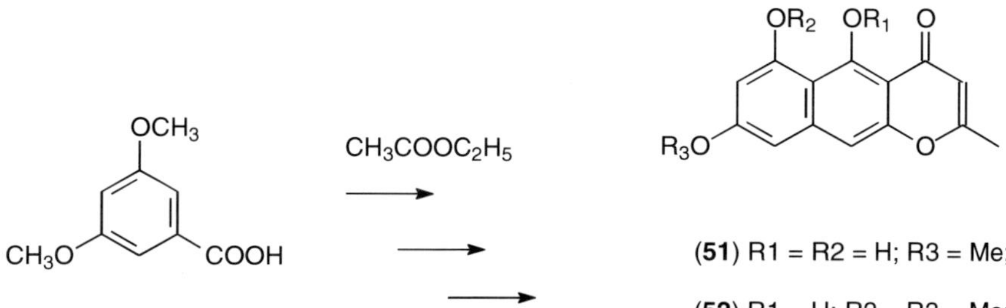

(51) R1 = R2 = H; R3 = Me;

(52) R1 = H; R2 = R3 = Me;

(53) R1 = R2 = R3 = H;

(54) R1 = R2 = R3 = Me

Juglone (55) and Javanicin (56) represent compounds of the naphthoquinones classes.

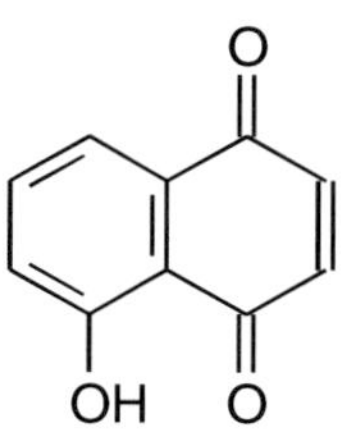

(55)

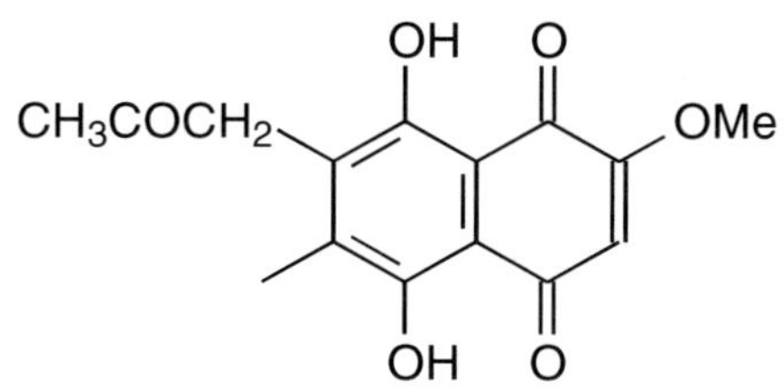

(56)

In 1959, MacLeod and Thomson presented a paper where a series of Juglones derivatives were synthesized and the mechanisms of the reactions were elucidated (McLeod and

Thomson, 1959). Bergeron et al. (1993) presented the total synthesis of Javanicin from Juglones derivatives. In this case the author showed that the difference of reactivity among the aromatic and quinone ring was fundamental to direct the reaction to end product.

Renslo and Danheiser (1998) studied the synthesis of substituted pyridines via regiocontrolled [4+2] Diels-Alder cycloadditions. They realized a protocol to synthetize a series of pyridines derived from (57) with a set of dienes. These studies were sufficient to develop the conditions to give a regiocontrolled reaction to synthesize Fusaric Acid (58).

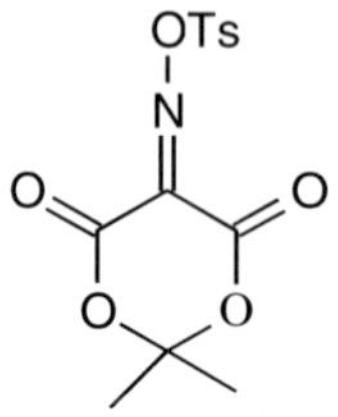

(57)

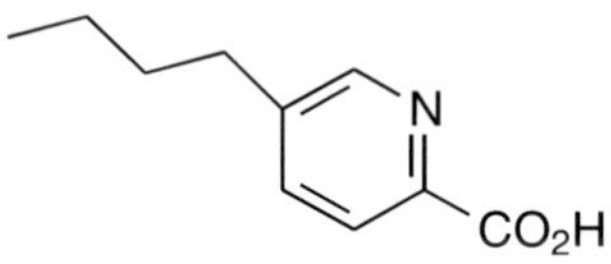

(58)

The macrocyclic lactones Zearalenone (59) and Zearalanone (60) are structures constituted by 14 atoms from lactone ring fused to aromatic resorcyl group. They have a chiral methyl group linked to carbon C3, therefore the simple coupling of aryl moiety and aliphatic part of the precursors to form 59 or derivatives is not a trivial process.

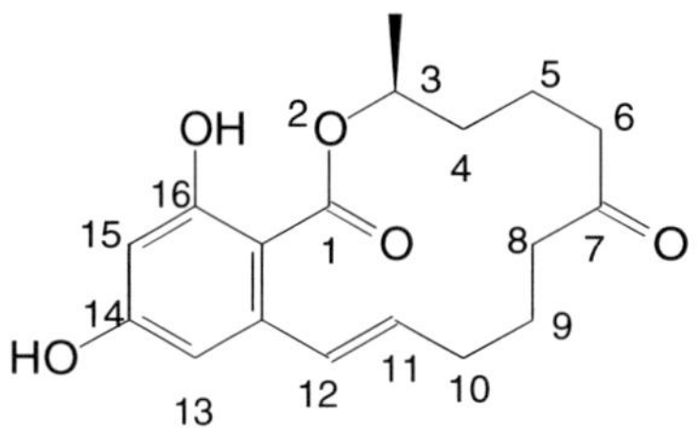

(59)

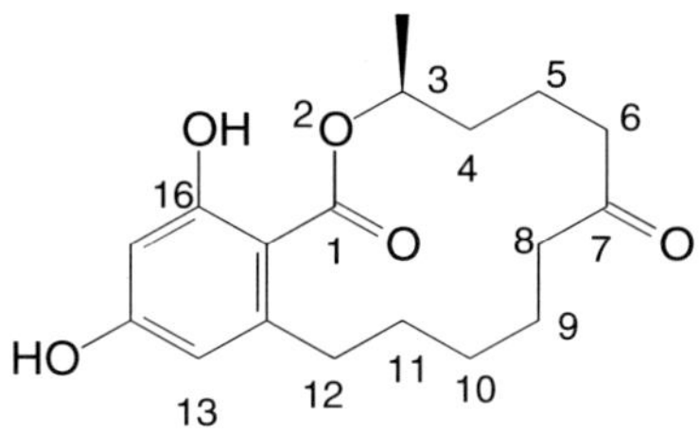

(60)

Bracher and Krauss (2001) synthesized the racemic 60 by coupling of the racemic aliphatic portion with arene triflate. They started from racemic 61, whereas Hitchcock and Pattenden (1992) produced the chiral (59), based on macrocyclisation, where the aliphatic portion started from chiral 62.

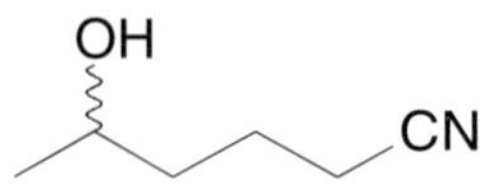

(61)

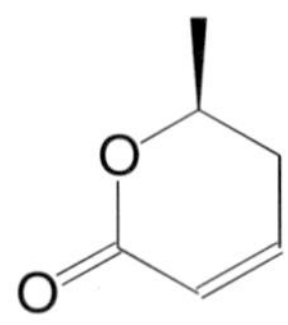

(62)

Racemic Culmorin (64), a tricyclic sesquiterpene, was synthesized by Takasu et al (1999). They produced 64 by intramolecular double Michael addition as the key step from 63.

Although Culmorin had been synthesized later by others groups, the Takasu's methodology was significant because it was possible to produce Culmorin in a single operation, protective and deprotective processes could be omitted and stereo- and regioselectivities could be controlled.

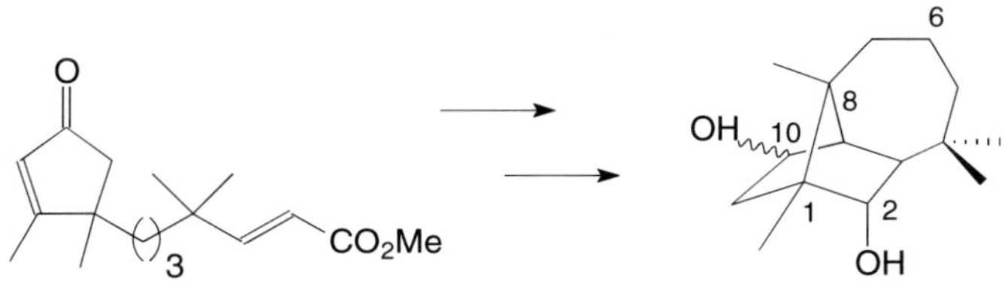

(63) **(64)**

Trichothecenes, a large class of sesquiterpenic compounds that present chemical structure like 65, are described by isomers forms like, Verrucarol, Anguidine, Scirpentriol, Deoxyverrucarol, HT-2 Toxin and T2- Toxin among other.

Schuda and Potlock (1984) presented a synthetic methodology to prepare Deoxyverrucarol (69) from Verrucarol (66). In this case, the fundamental constrain was to remove specifically the group OH at C4 without affect OH at C15. The tetra hydroxyl derivative of T-2 Tetraol (R1 = R2 = R3 = R5 = OH; R4 = H), was synthesized by Colvin et al. (1990) using five strategic sequences: a) construction of the AB fused ring by Diels-Alder cycloaddition; b) formation of ring C by aldol reaction, c) creation of the diol functionality of C ring; d) ring A enone formation and e) reduction of this enone to alcohol.

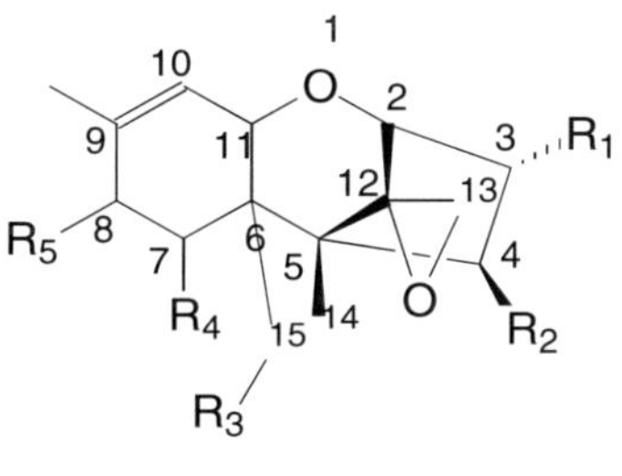

(65)

Verrucarol (**66**) R1 = H; R2 = R3 = OH; R4 = R5 = H

Anguidine (**67**) R1 = OH; R2 = R3 = OAc; R4 = R5 = H

Scirpentriol (**68**) R1 = R2 = R3 = OH; R4 = R5 = H

Deoxyverrucarol (**69**) R1 = R2 = H; R3 = OH; R4 = R5 = H

HT-2 Toxin (**70**) R1 = R2 = OH; R3 = OAc; R4 = H; R5 = 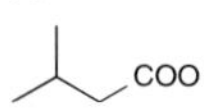

T-2 Toxin (**71**) R1 = OH; R2 = OAc; R3 = OAc; R4 = H; R5 = 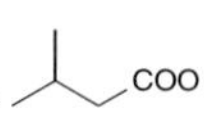

Considerations - The information contained in this chapter attempted to introduce the reader a glimpse of the vast horizon of organic synthesis, their challenges and complexity.

Although extremely complex, molecules are synthesized so trivially in biological environments. There is a great challenge for researchers: try to reproduce these structures by means of efficient protocols, aimed at new routes, higher yields and purer products.

Conclusion

Over the past 100 years, the fungi of *Fusarium* genus has been extensively studied by research groups around the world. Several papers report about the taxonomy of this genus, by the way is very complex. Others papers describe the importance of the genus in agriculture due to huge economic losses that are caused by diseases that affect plants. This review highlights the potential of *Fusarium* to produce primary and secondary metabolites with biotechnological applications. Although *Fusarium* species are considered highly toxic, they are used as food in some countries, such as meat substitute. They produce several extracellular enzymes, especially with applications in the food industry and, known secondary metabolites with vast therapeutic potential. The natural products discussed in this review represent a subset of all the natural metabolites that have been isolated, characterized and described in published literature to date. There are a huge number of other molecules obtained by *Fusarium* that have not been included here. Metabolites from *Fusarium* species there are a great potential for use to treat diseases in humans as well as use in cosmetics and food industry and investigations about their toxicity are required.

References

Abado-Becognee, K; Mobio, T.A.; Ennamany, R.; Fleurat-Lessard, F., Shier, W.T., Badria, F., Creppy, E.E. (1998). Cytotoxicity of fumonisin B1: implication of lipid peroxidation and inhibition of protein and DNA syntheses. *Archives of toxicology*, 72, 233-236.

Abbas, H.K.; Mirocha, C.J.; Shier, W.T. (1992). Isolation, identification and biological-activity of Chlamydosporol from Fusarium culmorum HM-8. *Mycopathologia*, 118, 115-123.

Abbas, H.K.; Tanaka, T.; Shier, W.T. (1995). Biological activities of synthetic analogues of Alternaria alternata toxin (AAL-toxin) and fumonisin in plant and mammalian cell cultures. *Phytochemistry*, 40, 1681-1689.

Ahn, M.Y., Chung, H.Y.; Choi, W.S.; Lee, B.M.; Yoon, S.; Kim, H.S. (2010). Anti-tumor effect of apicidin on Ishikawa human endometrial cancer cells both in vitro and in vivo by blocking histone deacetylase 3 and 4. *International Journal of Oncology* 36, 125-131.

Alkorta, I.; Garbisu, C.; Llama, M.J.; Serra, J.L. (1998). Industrial applications of pectic enzymes: a review. *Process Biochemistry*, 33, 21–28.

Altomare, C.; Pengue, R.; Favilla, M.; Evidente, A.; Visconti, A. (2004). Structure-activity relationships of derivatives of fusapyrone, an antifungal metabolite of Fusarium semitectum. *Journal of Agricultural and Food Chemistry*, 52, 2997-3001.

Anderson, C.; Solomons, G.L. (1984). Primary metabolism and biomass production from Fusarium. In: Moss, M.O.; Smith, J.E. (eds) *The applied mycology of Fusarium.* Cambridge University Press, Cambridge, 231–250.

Aranda, M.; Pérez-Alzola, P.; Ellahueñe, M.F.; Sepúlveda, C. (2000). Assessment of *in vitro* mutagenicity in Salmonella and in vivo genotoxicity in mice of the mycotoxin fumonisin B_1. *Mutagenesis,* 15, 469-471.

Arnstein, H.R.V.; Cook, A.H.; Lacey, M.S. (1946a). An antibiotic pigment from Fusarium javanicum. *Nature*, 157, 333,

Arnstein, H.R.V.; Cook, A.H.; Lacey, M.S. (1946b). Production of antibiotics by fungi. Part II. Production by Fusarium javanicum and other Fusaria. British *Journal of Experimental Pathology*, 27, 349-355.

Ashe, B.M.; Fletcher D.S. (1989). *Anti-inflammatory and antidegenerative compounds isolated from L-681,512.* US 4,874,755, October 17.

Ashley, J.N.; Hobbs, B.C.; Raistrick, H. (1937). LV. Studies in the biochemistry of micro-organisms LIII. The crystalline colouring matters of Fusarium culmorum (W. G. Smith) *sacc*. and related forms. *Biochemical Journal*, 31, 385–397.

Ávalos, J.; Cerdá-Olmedo, E. (1987). Carotenoid mutants of Gibberella fujikuroi. *Current genetics*, 11, 505-511.

Ávalos, J.; Cerdá-Olmedo, E.; Reyes, F.; Barrero, A. F. (2007). Gibberellins and other metabolites of Fusarium fujikuroi and related fungi. *Current Organic Chemistry,* 11, 721-737.

Backhouse, D.; Burgess W. L.; Summerell B.A. (2001). Biogeography of Fusarium, p. 122-137. In B. A. Summerell, J. ; Leslie.; Backhouse, D.; Bryden, W. L.; Burgess L.W. *Fusarium: Paul e Nelson Memorial Symposium. aps press, St. Paul, Minnesota.*

Balan, J.; Fuska, J.; Kuhr, I.; Kuhrová, V. (1970). Bikaverin, an antibiotic from Gibberella fujikuroi, effective against Leishmania brasiliensis. *Folia microbiologica*, 15, 479-484.

Barrero, A.F.; Oltra, J.E.; Herrador, M.M.; Cabrera, E.; Sánchez, J. F.; Quílez, J. F.; Rojas, F. J.; Reyes, J. F. (1993). Gibepyrones: a-Pyrones from Gibberella fujikuroi. *Tetrahedron,* 49, 141-150.

Rojas M.C.; Urrutia, O.; Cruz, C.; Gaskin, P.; Tudzynski, B.; Hedden, P. (2004). Kaurenolides and fujenoic acids are side products of the gibberellin P450-1 monooxygenase in Gibberella fujikuroi. *Phytochemistry,* 65, :821-830.

Barton, D.H.R.; Werstiuk, N.H. (1968). Sesquiterpenoids. Part XIV. The constitution and stereochemistry of culmorin. *Journal of the Chemical Society. C: Organic*. 148-155.

Barton, D.H.R; Werstiuk, N.H. (1967). The constitution and stereochemistry of culmorin. *Chemical communications*,1, 30-31.

Behm, C.; Degen, G.H.; Follmann, W. (2009). The Fusarium toxin enniatin B exerts no genotoxic activity, but pronounced cytotoxicity in vitro. *Molecular Nutrition & Food Research,* 53, 423-430.

Bekaert, A.; Andrieux, J.; Plat, M. (1992). New total synthesis of Bikaverin. *Tetrahedron Letters*, 33, 2805-2806.

Belofsky, G.N.; Jensen, P.R.; Fenical W. (1999) Sansalvamide: a new cytotoxic cyclic depsipeptide produced by a marine fungus of the genus Fusarium. *Tetrahedron Letters*, 40, 2913-2916.

Berek, L.; Petri, I.B.; Mesterhazy, A.; Teren, J.; Molnar, J. (2001). Effects of mycotoxins on human immune functions in vitro. *Toxicology in vitro*, 15, 25-30.

Bergeron, D.; Caron, B.; Brassard, P. (1993). An Expeditious Synthesis of Javanicin. *Journal of Organic Chemistry*, 58, 509-611.

Bernardini, M., Carilli, A., Pacioni, G.; Santurbano, B. (1975). Isolation of beauvericin from Paecilomyces fumosoroseus. *Phytochemistry* 14, 1865.

Bezuidenhout, G.C.; Gelderblom, W.C.A.; Gorst-Allam, C.P.; Horak, R.M.; Marasas, W. F.O.; Spiteller, G.; Vleggaar, R. (1988). Structure elucidation of the fumonisins, mycotoxins from Fusarium moniliforme. *Journal of the Chemical Society Chemical Communications,* 743-745.

Birch, A.J.; Richards, R.W.; Smith, H. (1958). The biosynthesis of gibberellic acid. *Proceedings of the Chemical Society* 1, 192-193.

Birch, A.J.; Richards, R.W.; Smith, H.; Harris, A.; Whalley, W.B. (1959). Studies in relation to biosynthesis: XII. Rosenonolactone and gibberellic acid. *Tetrahedron* 7: 241-251.

Blais, L.A.; Apsimon, J.W.; Blackwell, B.A.; Greenhalgh, R., Miller, J.D. (1992). Isolation and characterization of enniatins from *Fusarium-avenaceum* DAOM-196490. *Canadian Journal of Chemistry*, 70, 1281-1287.

Bode, H.B.; Muller, R. (2005). The impact of bacterial genomics on natural product research. *Angewandte Chemie International edition*, 44, 6828.

Booth, C (1981). Nectria aurantiicola. *CMI Descriptions of pathogenic fungi and bacteria* 714, 1-2.

Booth, C. (1971). The genus Fusarium. *Commonwealth Mycological Institute, Kew, Surrey, United Kingdom.*

Boros, S.M.; Hamilton, B.K.; Kulanthaivel P. (1994). Comparison of balanol from Verticillium balanoides and ophiocordin from Cordyceps ophioglossoides. *Journal of antibiotics*, 47, 1010-1016.

Bracher, F.; Krauss, J. (2001). Total synthesis of (+/-)-Zearalanone. *Monatshefte für Chemie* 132, 805-811.

Branco, A.; Pinto, A.C.; Braz-Filho, R.; Silva, E.F.;. Grynberg, N.F.; Echevarria, A. (2008). Rubrofusarina, um policetídeo natural inibidor da topoisomerase II-α humana. *Brazilian Journal of Pharmacognosy*, 18, 703-708.

Breinholt, J.; Ludvigsen, S.; Rassing, B.R.; Rosendahl, C.N. (1997). Oxysporidinone: a novel, antifungal *N*-methyl-4-hydroxy-2-pyridone from Fusarium oxysporum. *Journal of Natural Products*, 60, 33-35.

Brewer, D.; Arsenault, G.P. (1973). Productionof bikaverin by Fusarium oxysporum and its identity with lycopersin *Journal of Antibiotics XXVI*, 778-781.

Brill, G.M.; Kati, W.M.; Montgomery, D.; Karwowski, J.P.; Humphrey, P.E.; Jackson, M.; Clement, J.J.; Kadam, S.; Chen, R.H.; McAlpine, J.B. (1996). Novel triterpene sulfates from Fusarium compactum using a rhinovirus 3C protease inhibitor screen. *The Journal of Antibiotics*, 49, 541-546.

Bu'Lock, J. D.; Wright, C.E.; Mooney, J.E. (1986). Use of a protoplast fusion test to establish the status of mycotoxin genes in an edible Fusarium. *Biotechnology Letters*, 8 621-624.

Bukowski, R.; Vaughn, C.; Bottomley, R., Chen, T. (1982). Phase-II study of anguidine in gastrointestinal malignancies - A southwest oncology group-study. *Cancer Treatment Reports*, 66, 381-383.

Burgess, L.W. (1981). General ecology, 225-235. In Nelson, P.E.; Toussoun, T.A.; Cook, R.J. (eds.) *Fusarium: Diseases, Biology and Taxonomy*. Pennsylvania State University Press, University Park, Pennsylvania.

Burgess, L.W.; Nelson, P.E.; Toussoun, T.A.; Forbes, G.A. (1988). Distribution of Fusarium species in sections Roseum, Arthrosporiella, Gibbosum, and Discolor recovered from grassland, pasture, and pine nursery soils of eastern Australia. *Mycologia,* 80, 815-824.

Burgess, L.W.; Summerell, B.A.; Bullock, S.; Gott, K.P.; Backhouse, D. (1994). Laboratory manual for Fusarium research, 3ed. *University of Sydney Royal Botanic Gardens,* Sydney, Australia.

Burkert, J.F.M.; Maugeri, F.; Rodrigues, M.I. (2004). Optimization of extracellular lipase production by *Geotrichum* sp. using factorial design. *Bioresource Technology*, 91, 77-84.

Burmeister, H.R.; Grove, M.D.; Kwolek, W.F. (1980). Moniliformin and butenolide: effect on mice of high-level, long-term oral intake. *Applied and Environmental Microbiology*, 1142-1144.

Burmeister, H.R.; Plattner, R.D. (1987). Enniatin production by Fusarium tricinctum and its effect on germinating wheat seeds. *Physiology and Biochemistry,* 77, 1483-1487.

Burmeistert, H.R.; Vesonder, R.F. (1990). Production of a novel steroid sulfate metabolite [4,4,24- trimethylcholesta-8,14,24(28)-trien-2α,3β,11α,12α-tetrol 12-acetate, 3-sulfate] by *Fusarium* species and its biological activity. *Applied And Environmental Microbiology*, 3209-3212.

Bushnell G.W.; LI Y.L., Poulton G.A. (1984). Pyrones. X.[1] Lateropyrone, a new antibiotic from the fungus *Fusarium lateritium* Nees. *Canadian Journal of Chemistry*, 62, 2101-2106.

Calvo, A.M.; Wilson, R.A.; Bok, J.W.; Keller, N.P. (2002). Relationship between secondary metabolism and fungal development. *Microbiology and Molecular Biology Reviews*, Sept. 2002, 447–459.

Cameron, D.W.; Deutscher, K.R.; Feurill, G.I. (1982). Nucleophilic Alkenes. IX. Addition of 1,1-Dimethoxyethene to azanaphtoquinones: Synthesis of Bostrycoidin and 8-O-Methylbostrycoidin. *Australian Journal of Chemistry,* 35, 1439-1450.

Campos F.F.; Johann, S.; Alves, T.M.A.; Rosa L.H.; Caligiorne R.B.; Cisalpino, P.S.; Rosa, C.A.; Zani, C.L. (2011). Antifungal activity of trichothecenes from *Fusarium* sp. against clinical isolates of Paracoccidioides brasiliensis. *Mycoses,* 54,122-129.

Cannell, R. J. P. (1998). *How to approach the isolation of a natural product, in natural products isolation*. 1ed. Humana Press, New Jersey, 1–51.

Capon, R.J.; Stewart, M.; Ratnayake, R.; Lacey, E.; Gill, J.H. (2007). Citromycetins and bilains a–c: new aromatic polyketides and diketopiperazines from Australian marine-derived and terrestrial *Penicillium* spp. *Journal of Natural Products*, 70, 1746-1752.

Celius, T.; Haugen, T.B.; Grotmol, T.; Walther, B.T. (1999). A sensitive zonagenetic assay for rapid in vitro assessment of oestrogenic potency of xenobiotics and mycotoxins. *Environmental Health Perspectives*, 107, 63-68.

Chaal, B.K.; Gupta, A.P.; Wastuwidyaningtyas, B.D.; Luah, Y.H.; Bozdech, Z. (2010). Histone deacetylases play a major role in the transcriptional regulation of the Plasmodium falciparum life cycle. PLoS Pathogens, 6, 1-13.

Chelkowski, J.; Zajkowski, P.; Visconti, A. (1992). Bikaverin production of Fusarium species. *Mycotoxin Res* 8, 73-76.

Chen, R.H.; Tennant, S.; Frost, D.; O'Beirne, M.J.; Karwowski, J.P.; Humphrey, P.E.; Malmberg, L.H.; Choi, W., Brandt, K.D., West, P.; Kadam, S.K.; Clement, J.J.; McAlpine, J.B. (1996). Discovery of saricandin, a novel papulacandin, from a Fusarium species. *The Journal of Antibiotics*, 49, 596-598.

Cheong, J.W.; Chong, S.Y.; Kim, J.Y.; Eom, J.I.; Jeung, H.K.; Maeng, H.Y.; Lee, S.T.; Min, Y.H. (2003). Induction of apoptosis by apicidin, a histone deacetylase inhibitor, via the activation of mitochondria-dependent caspase cascades in human Bcr–Abl-positive leukemia cells. *Clinical Cancer Research*, 9, 5018-5027.

Chilton, W.S. (1968). Isolation and structure of norjavanicin. *Journal of Organic Chemistry,* 33, 4299-4300.

Chu, M.; Mierzwa, R.; Truumees, I.; Gentile, F.; Patel, M.; Gullo, V.; Chan, T.M.; Mohindar, S.P. (1993). Two novel diketopiperazines isolated from the fungus Tolypocladium sp. *Tetrahedron Letters*, 34, 7537-7540.

Ciegler, A.; Bennett, J.W. (1980). Mycotoxins and mycotoxicoses. *Bioscience* 30, 512-515.

Claridge, C.A.; Bradner, W.T.; Schmitz, H. (1978). Antitumor activity of 15-acetoxyscirpen-3,4-diol. *The Journal of Antibiotics*, XXXI, 485-486.

Cole, R.J.; Kirksey, J.W.; Cutler, H.G.; Doupnik, B.L.; Peckham, J.C. (1973). Toxin from Fusarium moniliforme: effects on plants and animals. *Science*, 179, 1324-1326.

Colvin, E. W.; Egan, M.J.; Kerr, F.W. (1990). Synthesis of the Trichothecene mycotoxin, 1-2 Tetraol. Journal of the Chemical Society, *Chemical Communications,* 1200-1202, 1990.

Cook, A.H.; Cox, S.F.; Farmer, T.H.; Lacey, M.S. (1947). Production of antibiotics by Fusaria. *Nature,* 160, 31.

Corey, E.J.; Danheiser, R.L.; Chandrasekaran, S.; Siret, P.; Keck, G.E.; Gras, J.L. (1978). Stereospecific total synthesis of Gibberellic Acid. A Key Tricyclic Intermediate. *Journal of the American Chemical Society*, 100, 8031-8034.

Cornforth, J.W.; Ryback, G.; Robinson, P.M.; Park, D. (1971). Isolation and characterization of a fungal vacuolation factor (bikaverin). *Journal of the Chemical Society,* 16, 2786-2788.

Cragg, G. M.; Newmann, D.J.; Snader, K.M. (1997). Natural products in drug discovery and development. *Journal of Natural Products*, 60, 52–60.

Cross, B.E.; Galt, R.H.B.; Hanson, J.R. (1963b). New metabolites of Gibberella fujikuroi. Part III. The structure of 7-hydroxykaurenolide. *Journal of the Chemical Society*, 2944-2961.

Cross, B.E.; Galt, R.H.B.; Hanson, J.R.; Curtis, P.J.; Grove, J.F.; Morrison, A.; (1963a). New metabolites of Gibberella fujikuroi. Part II. The isolation of fourteen new metabolites. *Journal of the Chemical Society*, 2937-2943.

Cueto, M.; Jensen, P.R.; Fenical, W. (2000). *N*-Methylsansalvamide, a cytotoxic cyclic depsipeptide from a marine fungus of the genus Fusarium. *Phytochemistry*, 55,223-226.

de Wolf, E.D.; Madden, L.V.; Lipps, P.E. (2003). Risk assessment models for wheat *Fusarium* head blight epidemics based on within-season weather data. *Phytopathology,* 93, 428-435.

Darkin-Rattray, S.J.; Gurnett, A.M.; Myers, R.M.; Dulski, P.M.; Crumley, T.M.; Alloco, J.; Cannova, C.; Meinke, P.T.; Colletti, S.L.; Bednarek, M.A.; Singh, S.; B.; Goetz, M.A.; Dombrowski, A.W.; Polishook, J.D.; Schmatz, D.M. Proc. Natl. Acad. Sci. (U. S. A.) 1996 (in press). In: Singh, S.B.; Zink, D.L.; Polishook, J.D.; Dombrososki, A.W.; Darkin-Rattray, S.J.; Schmatz, D.M.; Goetz, M.A. (1996). Apicidins: novel cyclic tetrapeptides as coccidiostats and antimalarial agents from *Fusarium pallidoroseum. Tetrahedron Letters*, 37, 8077-8080.

Debicki, S.; Jagodzinski P.P. (2009). Apicidin decreases phospholipase C gamma-1 transcript and protein in Hut-78 T lymphoma cells. *Biomedicine & Pharmacotherapy*, 63, 543-547.

Deol, B.S.; Ridley, D.D.; Singh, P.A. (1978). Isolation of cyclodepsipeptides from plant pathogenic fungi Australian. *Journal of Chemistry* 31, 1397-1399.

DeSimone, P.A.; Greco, F.A.; Lessner, H.F.; Bartolucci, A. (1986). Phase-II evaluation of anguidine (NSC-141537) in 5-day courses in colorectal adenocarcinoma- A Southeastern Cancer Study-Group Trial. American *Journal of Clinical Oncology*, 9, 187-188.

Desjardins, A.E.; Plattner, R.D.; Nelson, P.E. (1997) Production of fumonisinB1 and moniliformin by *Gibberella fujikuroi* from rice from various geographic areas. *Applied and Environmental Microbiology*, 63, 1838-1842.

Dombrink-Kurtzman, M.A. (2003). Fumonisin and beauvericin induce apoptosis in turkey peripheral blood lymphocytes. *Mycopathologia*, 156, 357-364.

Domijan, A.M.; Zeljezic, D.; Milic, M.; Peraica, M. (2007). Fumonisin B1: Oxidative status and DNA damage in rats. *Toxicology* 232 163-169.

Dornetshuber, R.; Heffeter, P.; Kamyar, M.R.; Peterbauer, T.; Berger, W.; Lemmens-Gruber, R. (2007). Enniatin exerts p53-dependent cytostatic and p53-independent cytotoxic activities against human cancer cells. *Chemical Research in Toxicology*, 20, 465-473.

Dosik, G.M.; Barlogie, B.; Johnston, D.A.;Murphy,W.K.; Drewinko, B. (1978). Lethal and cytokinetic effects of anguidine on a human colon cancer cell line. *Cancer Research*, 38, 3304-3309.

Dowd, P.F., Burmeistert, H.R., Vesonder, R.F. (1992). Antiinsectan properties of a novel *Fusarium*-derived sterol sulfate. *Entomologia Experimentalis et Applicata* 63, 95-100.

Dutton, M.F. (1996). Fumonisins, mycotoxins of increasing importance: their nature and their effects. *Pharmacology & Therapeutics*, 70, 137-161.

Edelman, J.; Fewella, A.; Solomongs, L. (1983). Mycoprotein – A new food. *Nutrition Abstracts and Reviews*, 53, 471-480.

Feng, Y.; Blunt, J.W.; Cole, A.L.J.; Munro, M.H.G. (2002). A novel cyclodepsipeptide, HA23, from a Fusarium sp. *Organic Letters*, 4, 2095-2096.

Feudjio, F. T.; Dornetshube, R.; Lemmens, M.; Hoffmann, O.; Lemmens-Gruber, R.; Berger, W. (2010). Beauvericin and enniatin: emerging toxins and/or remedies? *World Mycotoxin Journal*, 3, 415-430.

Feudjio, F.T.; Dornetshuber, R.; Lemmens, M.; Hoffmann, O.; Lemmens-Gruber, R.; Berger, W. (2010). Beauvericin and enniatin: emerging toxins and/or remedies? *World Mycotoxin Journal*, 3, 415-430.

Firáková, S.; Proksa, B.; Sturdíková, M. (2007). Biosynthesis and biological activity of enniatins. Die Pharmazie - *An International Journal of Pharmaceutical Sciences,* 62, 563-568.

Fitzpatrick, D.W.; Picken, C.A.; Murphy, L.C.; Buhr, M.M. (1989). Measurement of the relative binding affinity of zearalenone, α-zearalenol and β-zearalenol for uterine and oviduct estrogen receptors in swine, rats and chickens: An indicator of estrogenic potencies. *Comparative Biochemistry and Physiology Part C: Comparative Pharmacology*, 94, 691-694.

Flannigan, B. (1991). Mycotoxins. In: D'Mello, J.P.F.; Duffus, C.M. and Duffus, J.H. (eds.) Toxic substances in crop plants. *The Royal Society of Chemistry, Cambridge, UK*, pp. 226-257.

Fox, E.M; Howlett, B.J. (2008). Secondary metabolism: regulation and role in fungal biology. *Current Opinion in Microbiology*, 11, 481-487.

Fukuda H., Shima, H.; Vesonder, R.F.; Tokuda, H.; Nishino, H.; Katoh, S.; Tamura S.;¶ Sugimura, T.; Nagao, M. (1996). Inhibition of protein serine/threonine phosphatases by fumonisin B1, a mycotoxin. *Biochemical and Biophysical Research Communications* 220, 160-165.

Fuminori Hiramatsu, Takeshi Miyajima, Tetsuya Murayama, Koetsu Takahashi, Takuya Koseki, Yoshihito Shiono. Isolation and Structure Elucidation of Neofusapyrone from a Marine-derived Fusarium species, and Structural Revision of Fusapyrone and Deoxyfusapyrone. *J. Antibiot.* 59(11): 704–709, 2006

Fuska, J.; Proska, B.; Fuskova, A. (1975). New potential cytotoxic and antitumor substances. I. In vitro effect of bikaverin and its derivatives on cells of certain tumors. *Neoplasma* 22, 335-338.

Garrett, S.D. (1981). *Soil fungi and soil fertility: An introduction to soil mycology.* 2 ed. Oxford, Pergamon Press.

Gathercole, P.S.; Thiel, P.G.; Hofmeyr, J.H.S. (1986) Inhibition of pyruvate dehydrogenase complex by moniliformin. *Biochemical Journal*, 233, 719-723.

Gelderblom, W.C.A.; Kriek, N.P.J.; Marasas, W.F.O.; Thiel, P.G. (1991). Toxicity and carcinogenicity of the Fusarium moniliforme metabolite, fumonisin B1, in rats. *Carcinogenesis* 12, 1247-1251.

Gelderblom, W.C.A.; Semple, E.; Marasas, W.F.O.; Farber, E. (1992) The cancer-initiating potential of the fumonisin B mycotoxins:*Carcinogenesis*. 13, 433-437.

Gelderblom, W.C.A.; Snyman, S.D.; van der Westhuizen, L.; Marasas, W.F.O. (1995) Mitoinhibitory effect of fumonisin B1 on rat hepatocytes in primary culture. *Carcinogenesis.* 16, 625-631.

Gelderblom, W.C.A.; Thiel, P.G.; Marasas, W.F.O.; van der Merwe, K.J. (1984a). Natural occurrence of fusarin C, a mutagen produced by Fusarium moniliforme, in corn. *Journal of Agricultural and Food Chemistry*, 32, 1064-1067.

Gelderblom, W.C.A.; Thiel, P.G.; Marasas, W.F.O.; van der Merwe, K.J.; vanRooyen, P.H.; Vleggar, R.; Wessels, P.L. (1984b). Structure elucidation of fusarin C, a mutagen produced by Fusarium moniliforme. *Journal of the Chemical Society, Chemical Communications, issue 2*, 122-124.

Gelderblom, W.C.A.; Thiel, P.G.; van der Merwe, K.J. Metabolic activation and deactivation of fusarin C, a mutagen produced *by Fusarium moniliforme*. *Biochemical Pharmacology*, 33, 1601-1603.

Goliński, P.; Wnuk, S.; Chełkowski, J.; Visconti, A.; Schollenberger M. (1986). Antibiotic Y: biosynthesis by Fusarium avenaceum (Corda ex Fries) Sacc., isolation, and some physicochemical and biological properties. *Applied and Environmental Microbiology*, 743-745.

Goodwin, J.W.; Bottomley, R.H.; Vaughn, C.B.; Frank, J.; Pugh, R.P. (1983). Phase-II Evaluation of anguidine in central nervous-system tumors - a southwest oncology group-study. *Cancer Treatment Reports*, 67, 285-286.

Gordon, W.L. (1960). The taxonomy and habitats of *Fusarium* species from tropical and temperate regions. *Canadian Journal of Botany*, 38, 643-58.

Gottelieb, D & Etten, J.L.V. (1964). Biochemical changes during the growth of fungi. I. Nitrogen compounds and carbohydrate changes in Penicillium atrovenetum. *Journal of Bacteriology*, 88, 114-121.

Greuter, W.; McNeill, F.R.; Barrie, H.M.; Burdet, V.; Demoulin, T. S.; Filgueiras, D.H.; Nicolson, P.C.; Silva, J.E.; Skog, P.; Trehane, N.J.; Turland, D.L.; Hawksworth, E. (2000). *International code of botanical nomenclature (saint louis code) adopted by the sixteenth international botanical congress st. louis, missouri, july – august 1999.* Koeltz Scientific Books, Königstein, Germany.

Grove, J.F.; Pople, M. (1980). The insecticidal activity of beauvericin and the enniatin complex. *Mycopathologia*, 70, 103-105.

Gupta, S.; Krasnoff, S.B.; Underwood, N.L.; Renwick, J.A.A.; Roberts, D.W. (1991). Isolation of beauvericin as an insect toxin from *Fusarium semitectum* and Fusarium moniliforme var. subglutinans. *Mycopathologia*, 115, 185-189

Gupta, V.K.; Gaur, R.; Gautam, N.; Kumar, P.; Yadava, I.J.; Darmwal, N.S. (2009). Optimization of xylanase production by Fusarium solani f7. *Journal of Food Technology*, 4, 20-29.

Gurjar, M.K.; Rajendran, V.; Rao, V. (1998). Chiron approach towards a potent toxin Fumonisin B1 backbone: synthesis of its hexaacetate derivatives. *Tetrahedron Letters*, 39, 3803-3806.

Haig, A.M. Jr. *Chemical warfare in southeast Asia and Afghanistan*. Washington, DC: US Government Printing Office; March 22, 1982. Report to the Congress.

Hamano, K.; Kinoshita-Okami, M.; Minagawa, K.; Haruyama, H.; Kinoshita, T., Hosoya, T.; Furuya, K.; Kinoshita, K., Tabata, K.A. (1993). Aquastatin a, an inhibitor of mammalian adenosine triphosphatases from *Fusarium aquaeductuum*. taxonomy, fermentation, isolation, structure determination and biological properties. *Journal of Antibiotics*, 46, 1648-1657.

Hamill, R.L.; Higgens,C.E.; Boaz,H.E.; Gorman,M. (1969).The structure of beauvericin, a new depsipeptide antibiotic toxic to Artemia salina. *Tetrahedron Letters* 49, 4255-4258.

Hanson, J.R. Nyfeler, R.. Journal of Chemical Society Perkin Trans. 1, 247 1 (1976). In: Kasitu, G.C., ApSimon J.W., Blackwell B.A., Fielder D.A., Greenhalgh R., Miller J.D. (1992). Isolation and chracterization of culmorin derivatives produced by Fusarium culmorum CMI 14764. *Canadian journal of chemistry,* 70, 1308-1316.

Hauser, F.M.; Hewawasam, P.; Baghdanov, V.M. (1988). Regiospecific preparation of the benz[b] xanthen-12-one ring system: total synthesis of Bikaverin. *Journal of Organic Chemistry*, 53, 223-226.

Hazuda, D,; Blau, C. U.; Felock, P.; Hastings, J., Pramanik, B.; Wolfe, A.; Bushman, F.; Farnet, C.; Goetz, M.; Williams, M.; Silverman, K.; Lingham, R.; Singh, S. (1999). Isolation and characterization of novel human immunodeficiency virus integrase inhibitors from fungal metabolites. *Antiviral Chemistry & Chemotherapy* 10, 63-70.

Hedden, P.; MacMillan, J.; Grinsted, M.J. (1973). Fungal products. Part VIII. New kaurenolides from Gibberella fujikuroi. *J. Chem. Soc. Perkin Trans*, 1, 2773-2778.

Hedden, P.; Phillips, A.L.; Rojas, M.C.; Carrera, E.; Tudzynski, B. (2002). Gibberellin biosynthesis in plants and fungi: a case of convergent evolution? *Journal of Plant Growth Regulation*, 20, 319-331.

Hidaka, H.; Nagatsu T.; Takeya K.; Takeuchi T.; Suda H.; Kojiri K.; M. Matsuzaki; Umezawa H. (1969). Fusaric acid, a hypotensive agent produced by fungi. *Journal of Antibiotics*, 22, 228-230.

Hiraga, K.; Yamamoto, S.; Fukuda, H.; Hamanaka, N.; Oda, K. (2005). Enniatin has a new function as an inhibitor of Pdr5p, one of the ABC transporters in Saccharomyces cerevisiae. *Biochemical and Biophysical Research Communications* 328, 1119-1125.

Hitchcock, S.A.; Pattenden, G. (1992). Total synthesis of the Mycotoxin (-)-Zearalenone based on macrocyclisation using a cinnamyl radical intermediate. Journal of the Chemical Society. *Perkin Trans.* 1, 1323-1328.

Hochlowski, J.E., Whittern, D.N.; Buko, A.; Alder, L.; McAlpine, J.B. (1995). Fusacandins A and B; Novel antifungal antibiotics of the papulacandin class from fusarium sambucinum II. Isolation and structural elucidation. *The Journal of Antibiotics,* 48, 614-618.

Hodge, E.B.; Hidy, P.H.; Wehrmeister H.L.; Haute, T. (1966). *Estrogenic compounds and animal growth promoters.* U.S. Patent 3373025, Mar. 12.

Hodge, E.G.; Hidy, P.H.; Wehrmeister, H.J. (1966). *Estrogenic compounds and animal growth promoters.* U.S. Patent 3239345.

Hwang, Y.; Rowley, D.; Rhodes, D.; Gertsch, J.; Fenical, W.; Bushman, F. (1999). Mechanism of inhibition of a poxvirus topoisomerase by the marine natural product sansalvamide A. *Molecular Pharmacology.* 55, 1049-1053.

Ivanova, L.; Skjerve, E.; Eriksen, G.S.; Uhlig, S. (2006). Cytotoxicity of enniatins A, A1, B, B1, B2 and B3 from Fusarium avenaceum. *Toxicon*, 47, 868-876.

Ivanova, L; Uhlig, S.; Eriksen, G.S.; Johannessen, L.E. (2010). Enniatin B-1 is a substrate of intestinal P-glycoprotein, multidrug resistance-associated protein 2 and breast cancer resistance protein. *World Mycotoxin Journal*; 3, 271-281.

Jackson, M.; Frost, D.J.; Karwowski, J.P.; Humphrey, P.E.; Dahod, S.K.; Choi, W.S.; Brandt, K.; Malmberg, L.H.; Rasmussen, R.R.; Scherr, M.H., Flamm, R.K., Kadamand, S.; McAlpine, J.B. (1995). Fusacandins A and B; novel antifungal antibiotics of the papulacandin class from Fusarium sambucinum. I. Identity of the producing organism, fermentation and biological activity. *The Journal of Antibiotics*, 48, 608-613.

Jang, J.H.; Asami, Y., Jang, J.P.; Kim, S.O.; Moon. D.O.; Shin, K.S.; Hashizume, D.; Muroi, M.; Saito, T.; Oh, H.; Kim, B.Y.; Osada, H.; Ahn, J.S. (2011). Fusarisetin A, an acinar morphogenesis inhibitor from a soil fungus, *Fusarium* sp. FN080326. *Journal of the American Chemical Society*, 133, 6865-6867.

Jayasinghe, L.; Abbas, H.K.; Jacob, M.R., Herath, W.H.M.W.; Nanayakkara, D.N.P. (2006). *N*-methyl-4-hydroxy-2-pyridinone analogues from Fusarium oxysporum. *Journal of Natural Products*, 69, 439-442.

Jestoi, M. (2008). Emerging fusarium mycotoxins fusaprofilin, beauvericin, enniatins, and moniliformin – a review. *Critical Reviews in Food Science and Nutrition,* 48, 21-49.

Jiang, B.; Xu, D. M.; Allocco, J.; Parish, C.; Davison, J.; Veillette, K.; Sillaots, S.; Hu, W. Q.; Rodriguez-Suarez, R.; Trosok, S.; Zhang, L.; Li, Y.; Rahkhoodaee, F.; Ransom, T.; Martel, N.; Wang, H.; Gauvin, D.; Wiltsie, J.; Wisniewski, D.; Salowe, S.; Kahn, J. N.; Hsu, M. J.; Giacobbe, R.; Abruzzo, G.; Flattery, A.; Gill, C.; Youngman, P.; Wilson, K.; Bills, G.; Platas, G.; Pelaez, F.; Diez, M. T.; Kauffman, S.; Becker, J.; Harris, G.; Liberator, P.; Roemer, T. (2008). PAP inhibitor with in vivo efficacy identified by Candida albicans genetic profiling of natural products. *Chemistry & Biology*, 15, 363-374.

Jiang, Z.; Barret, M.O.; Boyd, K.G.; Adams, D.R.; Boyd, A.S.F.; Burgess, J.G. (2002). JM47, a cyclic tetrapeptide HC-toxin analogue from a marine Fusarium species. *Phytochemistry* 60, 33-38.

Jin, J.; Baek, S.R.; Lee, K.R.; Lee, J.K., Yun, S.H.; Kang, S.C.; Lee, Y.W. (2008). Purification and phytotoxicity of apicidins produced by the Fusarium semitectum KCTC16676. *The Plant Pathology Journal*, 24, 417-422.

Jones, A.; Pharis, R.P. (1987). Production of gibberellins and bikaverin by cells of Gibberella fujikuroi immobilized in carrageenan. *Journal of Fermentation Technology,* 65, 717-722

Jow, G.-M.; Chou, C.-J.; Chen, B.-F.; Tsai, J.-H. (2004). Beauvericin induces cytotoxic effects in human acute lymphoblastic leukemia cells through cytochrome c release, caspase 3 activation: the causative role of calcium. *Cancer Letters*, 216, 165-173.

Kakeya, H.; Kageyama, S.; Nie, L.; Onose, R.; Okada, G.; Beppu, T.; Norbury, C.J.; Osada, H. (2001). Lucilactaene, a New Cell Cycle Inhibitor in p53-Transfected Cancer Cells, Produced by a Fusarium sp. *The Journal of Antibiotics,* 54, 850-854.

Kasitu, G.C., ApSimon J.W., Blackwell B.A., Fielder D.A., Greenhalgh R., Miller J.D. (1992). Isolation and chracterization of culmorin derivatives produced by *Fusarium culmorum* CMI 14764. *Canadian journal of chemistry*, 70, 1308-1316.

Kato, Y.; Koshino, H.; Uzawa, J.; Anzai, K. (1996). Fungerin, a new antifungal alkaloid from Fusarium sp. *Bioscience Biotechnology and Biochemistry*, 60, 2081-2083,

Kawahara, N.; Nozawa, K.; Nakajama S.; Kawai, K. (1987). Studies on fungal products. Part 13. Isolation and structures of dithiosilvatin and silvatione, novel dioxopiperazine derivatives from Aspergillus silvaticus. *Journal of Chemical Society, Perkin Trans.* 1: 2099-2101.

Kim, J.-C.; Lee, Y.-W.; Tamura, H.; Yoshizaw, T. (1995). Sambutoxin: a new mycotoxin isolated from Fusarium sambucinum. *Tetrahedron Letters*, 36, 1040-1050.

Kim, S.H.; Ahn, S.; Han, J.W.; Lee, H.W.; Lee, H.Y.; Lee, Y.W.; Kim, M.R.; Kim, K.W.; Kim, W.B.; Hong, S. (2004). Apicidin is a histone deacetylase inhibitor with anti-invasive and antiangiogenic potentials. *Biochemical and Biophysical Research Communications,* 315, 964-970.

Kimura, M.; Tokai, T.; Takahashi-Ando, N.; Ohsato, S.; Fujimura, M. (2007). Molecular and genetic studies of fusarium trichothecene biosynthesis: pathways, genes, and evolution. *Bioscience, Biotechnology and Biochemistry*, 71, 2105-2123.

Kjær, D.; Kjær, A.; Risbjerg, E. (1983). A synthesis of bikaverin (7,12-dihydro-6,11-dihydroxy-3,8-dimethoxy-1-methyl-10*H*-benzo[*b*]xanthene-7,10,12-trione) and some related benzoxanthones and quinones. *Journal of the Chemical Society, Perkin Transactions* 1, 2815-2820.

Kneifel, H.; Konig, W.A., Loeffler, W.; Muller, R. (1977). Ophiocordin, an antifungal antibiotic of Cordyceps ophioglossoides. *Archives of Microbiology,* 113, 121-130.

Koizumi, Y.; Arai, M.; Tomoda, H.; Omura, S. (2004). Fungerin, a fungal alkaloid, arrests the cell cycle in M phase by inhibition of microtubule polymerization. *Journal of Antibiotics*, 57, 415-420.

Kriek, N.P.; Marasas, W.F.; Steyn, P.S.; van Rensburg, S.J.; Steyn, M. (1977). Toxicity of a moniliformin-producing strain of *Fusarium moniliforme var. subglutinans* isolated from maize. *Food and Cosmetics Toxicology*, 15, 579-587.

Kroslák, M. (2002). Efficacy, and acceptability of fusafungine, a local treatment for both nose and throat infections, in adult patients with upper respiratory tract infections. *Current Medical Research and Opinion*, 18,194-200.

Kuiper, G.G.; Lemmen, J.G.; Carlsson, B.; Corton, J.C.; Safe, S.H.; van der Saag, P.T.; van der Burg, B.; Gustafsson, J.A. (1998). Interaction of estrogenic chemicals and phytoestrogens with estrogen receptor β. *Endocrinology,* 139, 4252-4263.

Kuiper-Goodman, T.; Scott, P.M.; Watanabe, H. (1987). Risk assessment of the mycotoxin zearalenone. *Regulatory Toxicology and Pharmacology*, 7, 253-306.

Kulanthaivel, P.; Hallock, Y.F.; Boros, C.; Hamilton, S.M.; Janzen, W.P.; Ballas, L.M.; Loomis, C.R.; Jiang, J.B.; Katz, B.; Steiner, J.R.; Clardy, J. (1993). Balanol: a novel and potent inhibitor of protein kinase C from the fungus Verticillium balanoides. *Journal of the American Chemical Society*, 115, 6452-6453.

Kumar, S.; Kikon, K.; Upadhyay, A.; Kanwar, S.S.; Gupta, R. (2005). Production, purification, and characterization of lipase from thermophilic and alkaliphilic *Bacillus coagulans* BTS-3. *Protein Expression and Purification*, 41, 38-44.

Kurobane, I.; Zaita, N.; Fukuda, A. (1986). New metabolites of *Fusarium marth* related to dihydrofusarubin. *The Journal of Antibiotics*, XXXIX, 205-214.

Kwon, H.W,; Yoon, J.H,; Kim, S.H,; Hong, S.B,; Cheon, Y,; Ko, S.J. (2007).Detection of extracellular enzyme activities in various Fusarium spp. *Mycobiology*, 35, 162-165.

Kwon, Y.J.; Fang, Y.; Xu, G.H.; Kim, W.G. (2009). Aquastatin A, a new inhibitor of enoyl-acyl carrier protein reductase from *Sporothrix* sp. FN611. *Biological Pharmaceutical Bulletin,* 32:2061-2064.

Kwon, Y.J.; Sohn, M.J.; Kim, W.G. (2011). Aquastatin C (I), a new glycoaromatic derivative from *Sporothrix* sp. FN611. *The Journal of Antibiotics*, 64, 213-215.

Lang, A (1956). Bolting and flowering in biennial Hyoscyamus niger, induced by gibberellic acid. *Plant Physiol 31*[Suppl]:XXXV.

Langseth, W.; Ghebremeskel, M.; Kosiak, B.; Kolsaker, P.; Miller, D. (2000). Production of culmorin compounds and other secondary metabolites by *Fusarium culmorum* and F. graminearum strains isolated from Norwegian cereals. *Mycopathologia*, 152, 23-34.

Le Guevel, R.; Pakdel, F. (2001). Assessment of oestrogenic potency of chemicals used as growth promoter by in vitro methods. *Human Reproduction*, 16, 1030-1036.

Li, Y.; Wang, Z.; Beier, R.C.; Shen, J.; De Smet, D.; De Saeger, S.; Zhang, S. (2011). T-2 Toxin, a trichothecene mycotoxin: review of toxicity, metabolism, and analytical methods. *Journal of Agricultural and Food Chemistry,* 59, 3441-3453.

Lin, M.; Lu, S.; Ji, C.; Wang, M.; Cheng, S.; Tian, G. (1980) in Genetics and Environmental Factors in Experimental and Human Cancer (Gelboin H. V., Ed.), pp. 139-148, *Japan Scientific Societies* Press, Tokyo.

Logrieco, A.; Moretti, A.; Fornelli, F.; Fogliano, V.; Ritieni, A.; Caiaffa, M.F.; Randazzo, G., Bottalico, A.; Macchia, L. (1996). Fusaproliferin production by *Fusarium subglutinans* and its toxicity to *Artemia salina*, SF-9 insect cells, andIARC/LCL 171 human B lymphocytes. *Applied and Environmental Microbiology*, 62, 3378- 3384.

Lohmann, A. (1988). Analytische, pharmakologische und mikrobiologische Untersuchungen des ionophoren antibiotikums fusafungin. *PhD thesis, Westfälische Wilhelms Universität, Münster, Germany.*

Łuczak, M.W.; Jagodzinski P.P. (2008). Apicidin down-regulates human papillomavirus type 16 E6 and E7 transcripts and proteins in SiHa cervical cancer cells. *Cancer Letters*, 272 53-60.

Macleod, J.W.; Thomson, R.H. (1952). Studies in the Juglone series. IV. The addition of aniline and toluene-p-thiol to 5-Substituted 1,4-Naphthoquinones. *Journal of Organic Chemistry*, 25, 36-42.

Maia, M.M.D,; Morais, M.M.C,; Morais, M.A,; Melo, E.H.M,; Filho, J.L.L. (1999). Production of extracellular lipase by the phytopathogenic fungus *Fusarium solani* fs1. *Revista de Microbiologia*, 30, 304-309.

Mal, D.; Bandhyopadhyay, M.; Datta, K.; Marry, K.V.S.N. (1998). Anionic [4+2] cycloaddition strategy to linear Furocoumarins: synthesis of 8-Methoxypsoralen and its lsoster. *Tetrahedron,* 54, 7525-7538.

Malonek, S.; Bömke, C.; Bornberg-Bauer, E.; Rojas, M.C.; Hedden, P.; Hopkins, P.; Tudzynski, B. (2005). Distribution of gibberellin biosynthetic genes and gibberellin production in the Gibberella fujikuroi species complex. *Phytochemistry,* 66, 1296-1311.

Marasas, W.F.O. (1982). Mycotoxicological investigations on corn produced in esophageal cancer areas in Transkei. *In Cancer of the Esophagus*; Pfeiffer, C. J., Ed.; CRC Press: Boca Raton, FL, pp 29-40.

Marin, D.E.; Taranu, I.; Burlacu, R.; Tudor, D.S. (2010). Effects of zearalenone and its derivatives on the innate immune response of swine. *Toxicon*, 56, 956-963.

McBrien, K.D.; Gao, Q.; Huang, S.; Klohr, S.E.; Wang, R.R.; Pirnik, D.M.; Neddermann, K.M.; Bursuker, I.; Kadow, K.F.; Leet, J.E. (1996). Fusaricide, a new cytotoxic *N*-hydroxypiridone from Fusarium sp. *Journal of Natural Products*, 59, 1151-1153.

McCormick, S.P.; Stanley, A.M.; Stover, N.A.; Alexander, N.J. (2011). Trichothecenes: from simple to complex mycotoxins. *Toxins,* 3, 802-814.

Meca, G.; Sospedra, I.; Soriano, J.M.; Ritieni, A.; Valero, M.A.; Mañes, J. (2009). Isolation, purification and antibacterial effects of fusaproliferin produced by Fusarium subglutinans in submerged culture. *Food and Chemical Toxicology*, 47, 2539-2543.

Medentsev, G.; Akimenko, V.K. (1998). Naphthoquinone metabolites of the fungi. *Phytochemistry*, 47, 935-959.

Mereish, K.A.; Solow, R.; Bunner, D.L.; Fajer, A.B. (1990). Interaction of cyclic peptides and depsipeptides with calmodulin. *Peptide Research* 3, 233-237.

Merrill, A.H.Jr.(1991). Cell regulation by sphingosine and more complex sphingolipids. *Jounal of Bioenergectics Biomembranes*, 23, 83-104.

Miksicek, R.J. (1994). Interaction of naturally occurring non-steroidal estrogens with expressed recombinant human estrogen receptor. *The Journal of Steroid Biochemistry and Molecular Biology,* 49, 153-160.

Miller, S.A. & Dwyer, J.T. (2001). Evaluating the safety and nutritional value of mycoprotein. *Food Technology*, 55, 42-47.

Minervini, F.; Dell'Aquila, M.E. (2008). Zearalenone and reproductive function in farm animals. *International Journal of Molecular Sciences*, 9, 2570-2584.

Mirocha, C.J. (1982). Hazards of scientific investigation: analysis of samples implicated in biological warfare. *Journal of Toxicology-Toxin Reviews*, 1, 199-203.

Molnár, I.; Gibson, D.M.; Krasnoff, S.B. (2010). Secondary metabolites from entomopathogenic Hypocrealean fungi. *Natural Product Reports*, 27, 1241-1275.

Nelson, P.E.; Juba, J.H.; Ross, P.F.; Rice, L.G. (1994). Fumonisin production by *Fusarium* species on solid substrates. *Journal of Aoac International*, 77, 522-525.

Nenkep, V.; Yun, K.; Zhang, D.; Choi, H.D.; Kang, J.S.; Son, B.W. (2010). Induced production of bromomethylchlamydosporols A and B from the marine-derived fungus Fusarium tricinctum. *Journal of Natural Products*, 73, 2061-2063.

Nirenberg, H. I. (1995). Morphological differentiation of *Fusarium sambucinum* Fückel sensu stricto, F. torulosum (Berk. & Curt.) Nirenberg and *F. venenatum* Nirenberg sp. nov. *Mycopathologia*, 129, 131-141.

Nwagu, T.N & Okolo, B.N. (2011). Extracellular amylase production of a thermotolerant *Fusarium* sp. isolated from eastern Nigerian soil Tochukwu. *Brazilian Archives of Biology and Technology*, 54, 649-658.

O'Donnell, K.; Cigelnik, E.; Casper, H.H. (1998). Molecular, phylogenetic, morphological, and mycotoxin data support reidentification of the Quorn mycoprotein fungus as Fusarium venenatum. *Fungal Genetics and Biology*, 23 57-67.

Ohshima, S.; Yanagisawa, M.; Katoh, A.; Fujii, T.; Sano, T.; Matsukuma, S.; Furumai, T.; Fujiu, M.; Watanabe, K.; Yokose, K., et al. (1994). *Fusarium merismoides* Corda NR 6356, the source of the protein kinase C inhibitor, azepinostatin. Taxonomy, yield improvement, fermentation and biological activity.*Journal of Antibiotics*, 47, 639-647.

Othmen, Z.O.; Golli, E.E.; Abid-Essefi, S.; Bacha, H. (2008). Cytotoxicity effects induced by Zearalenone metabolites, α Zearalenol and β Zearalenol, on cultured Vero cells. *Toxicology,* 252, 72-77.

Ovchinnikov, Y.A.; Ivanov, V.T.; Evstratov, A.V.; Mikhaleva, I.I.; Bystrov, V.F.; Portnova, S.L.; Balashova, T.A.; Meshcheryakova, E.N.; Tulchinsky, V.M. (1974). The enniatin ionophores. Conformation and ion binding properties. *International Journal of Peptide and Protein Research*, 6, 465-498.

Overy, D.; Calati, K.; Kahn, J.N.; Hsu, M.J.; Martín, J.; Collado, J; Roemer, T.; Harris, G., Parish, C.A. (2009). Isolation and structure elucidation of parnafungins C and D, isoxazolidinone-containing antifungal natural products. *Bioorganic & Medicinal Chemistry Letters*, 19, 1224-1227.

Pan, J.H.; Chen, Y.; Huang, Y.H.; Tao, Y.W.; Wang. J.; Li. Y.; Peng. Y.; Dong, T.; Lai, X.M.; Lin, Y.C. (2011). Antimycobacterial activity of fusaric acid from a mangrove endophyte and its metal complexes. *Archives of Pharmacal Research,* 34, 1177-1181.

Pandey, A.; Benjamin, S.; Soccol, C.R.; Nigam, P.; Kriger, N.; Soccol, V.T. (1999). The realm of microbial lipases in biotechnology. *Biotechnology and Applied Biochemistry,* 29, 119-131.

Parish, C.A.; Smith, S.K.; Calati, K.; Zink, D.; Wilson, K.; Roemer, T.; Jiang, B.; Xu, D.M.; Bills, G.; Platas, G.; Pelaez, F.; Diez, M. T.; Tsou, N.; McKeown, A.E.; Ball, R.G.; Powles, M.A.; Yeung, L.; Liberator, P.; Harris, G. (2008). Isolation and structure elucidation of parnafungins, antifungal natural products that inhibit mRNA polyadenylation. *Journal of the American Chemical Society,* 130, 7060-7066.

Park, J.S.; Lee, K.R.; Kim, J.C.; Lim, S.H.; Seo, J.A.; Lee, Y.W. (1999). A hemorrhagic factor (apicidin) produced by toxic fusarium isolates from soybean seeds. *Applied and Environmental Microbiology*, 65, 126-130.

Patham, B.; Duffy, J.; Lane, A.; Davis, R.C.; Wipf, P.; Fewell, S.W.; Brodsky, J.L.; Mensa-Wilmot, K. (2009). Post-translational import of protein into the endoplasmic reticulum of a trypanosome: an in vitro system for discovery of anti-trypanosomal chemical entities. *Biochemical Journal*, 419, 507-517.

Pedersen, P.B.; Miller, J.D. (1999). The fungal metabolite culmorin and -related compounds. *Natural Toxins*, 7, 305-309.

Peng, H.X.; Sivasithamparan, K.; Turner, D.W. (1999). Chlamidospore germination and *Fusarium* wilt of banana plantlets in suppressive and conducive soil are afforded by physical and chemical factors. *Soil Biology & Biochemistry,* 31, 1363-1374.

Pereira, C.L.; Chen, Y.H.; McDonald, F.E. (2009).Total synthesis of the sphingolipid biosynthesis inhibitor Fumonisin B1. *Journal of the American Chemical Society*, 131, 6066–6067.

Plasencia, J.; Mirocha, C.J. (1991). Isolation and characterization of zearalenone sulfate produced by Fusarium spp. Applied and Environmental Microbiology, 57, 146-150.

Pocsfalvi, G.; Di Langa, G.; Ferranti, P.; Ritieni, A.; Randazzo, G.; Malorni, A. (1997). Observation of non-covalent interactions between beauvericin and oligonucleotides using electrospray ionization mass spectrometry. *Rapid Communications in Mass Spectrometry,* 11, 265-272.

Pohanka, A.; Capieau, K.; Broberg, A.; Stenlid, J.; Stenström, E.; Kenne, L. (2004). Enniatins of *Fusarium* sp. strain F31 and their inhibition of Botrytis cinerea spore germination. *Journal of Natural Products*, 67, 851-857.

Porter, J.K.; Voss, K.A.; Chamberlain, W.J.; Bacon, C.W.; Norred, W.P. (1993). Neurotransmitters in rats fed fumonisin B1. *Proceedings of the Society for Experimental Biology and Medicine*, 202, 360-364.

Prado-Cabrero, A.; Schaub, P.; Díaz-Sánchez, V.; Estrada, A.F.; Al-Babili, S.; valos, J. (2009). Deviation of the neurosporaxanthin pathway towards β-carotene biosynthesis in Fusarium fujikuroi by a point mutation in the phytoene desaturase gene. *FEBS Journal*, 276, 4582-4597.

Prazeres, J.N,; Cruz, J.A.B,; Pastore, G.M. (2006). Characterization of alkaline lipase from *Fusarium oxysporum* and the effect of different surfactants and detergents on the enzyme activity, *Brazilian Journal of Microbiology*, 37:505-509.

Raaijmakers, JM.; Paulitz, TC.; Steinberg, C.; Alabouvette, C.; Moenne-Loccoz, Y. (2009). The rhizosphere: a playground and battlefield for soilborne pathogens and beneficial microorganisms. *Plant and Soil*, 321, 341-361.

Rabie, C.J.; Marasas, W.F.; Thiel, P.G.; Lübben, A.; Vleggaar, R. (1982). Moniliformin production and toxicity of different Fusarium species from Southern Africa. *Applied and Environmental Microbiology*, 43, 517-521.

Raimbault, M. (1998). General and microbiological aspects of solid substrate fermentation. *Electronic Journal of Biotechnology*, 1, 1-15.

Randazzo, G.; Fogliano, V.; Ritieni, A.; Mannina, L.; Rossi, E.; Scarallo, A.; Segre, A.L. (1993). Proliferin, a new sesterterpene from Fusarium proliferatum. *Tetrahedron,* 49, 10883-10896.

Reams, R.Y.; Thacker, H.L.; Harrington, D.D.; Novilla, M.N.; Wilson, B. (1996). Purified moniliformin does not affect the force or rate of contraction of isolated guinea pig atria. *Mycopathologia*, 133, 115-118.

Reddy, K.R.N.; Nurdijati, S.; Salleh, B. (2010). An overview of plant derived products on control of mycotoxigenic fungi and mycotoxins. *Asian Journal of Plant Science*, 9, 126-133.

Renner, M.K.; Jensen, P.R.; Fenical, W. (1998). Neomangicols: structures and absolute stereochemistries of unprecedented halogenated sesterterpenes from a marine fungus of the genus Fusarium. *Journal of Organic Chemistry*, 63, 8346-8354.

Renner, M.K.; Jensen, P.R.; Fenical, W. (2000). Mangicols: structures and biosynthesis of a new class of sesterterpene polyols from a marine fungus of the genus Fusarium. *Journal of Organic Chemistry* 65, 4843-4852.

Renslo, A.R.; Danheiser R.L. (1998). Synthesis of substituted pyridines via regiocontrolled [4 + 2] cycloadditions of oximinosulfonates. *Journal of Organic Chemistry,* 63, 7840-7850.

Richard, C. (1982). Examen de la microflore endoracinaire de la luzerne en fonction de l'âge, de l'état sanitaire et de l'emplacement dans la racine. *Phytoprotection*, 62, 67–78.

Richard, WB.; Yujin, L.D. (1989). *Anti-inflammatory and antidegenerative compounds isolated from L-681,512*. J. US 4871727, Oct. 3.

Richardson, K.E.; Hagler, W.M. Jr.; Haney, C.A.; Hamilton, P.B. (1985). Zearalenone and trichothecene production in soybeans by toxigenic Fusarium. *Journal of Food Protection*, 48, 240-243.

Richardson, K.E.; Hamilton, P.E. (1990). Comparative toxicity of scirpentriol and its acetylated derivatives. *Poultry Science*, 69, 397-402.

Rojas, M.C.; Urrutia, O.; Cruz, C.; Gaskin, P.; Tudzynski, B.; Hedden, P. (2004) Kaurenolides and fujenoic acids are side products of the gibberellin P450-1 monooxygenase in Gibberella fujikuroi. *Phytochemistry*, 65, 821-830.

Ross, P.F.; Ledet, A.E.; Owens, D.L.; Rice, L.G.; Nelson, H.A.; Osweiler, G.D.; Wilson, T.M. (1993). Experimental equine leukoencephalomalacia, toxic hepatitis, and encephalopathy caused by corn naturally contaminated with fumonisins. *Journal of Veterinary Diagnostic Investigation*, 5, 69-74.

Rossi, V.; Languasco, L.; Pattori, E.; Giosue, S. (2002). Dynamics of airborne *Fusarium* macroconidia in wheat fields naturally infected by head blight. *Journal of Plant Pathology & Microbiology*, 84, 53-64.

Rossman, A.Y.; Samuels, G.J.; Rogerson, C.T.; Lowen, R. (1999). Genera of Bionectriaceae, Hypocreaceae and Nectriaceae (Hypocreales, Ascomycetes). *Studies in Mycology*, 42, 1-248.

Rudd, J.C.; Horsley, R.D.; McKendry, A.L.; Elias, E.M. (2001). Host plant resistance genes for fusarium head blight: sources, mechanisms, and utility in conventional breeding systems. *Crop Science*, 41, 620-627.

Ruffmann, R.; Welker, R.D.; Saito, T.; Chirigos, M.A.; Varesio, L. (1984). In vivo activation of macrophages but not natural killer cells by picolinic acid (Pla). *Immunopharmacology and Immunotoxicology*, 6, 291-304.

Sack, R.L., Goodwin, F.K. (1974). Inhibition of dopamine-β-hydroxylase in manic patients. a clinical trial with fusaric acid. *Archives of General Psychiatry,* 31, 649-654.

Saenz de Rodriguez, C.A., Toro-Sola, M.A. (1982). Anabolic steroids in meat and premature telarche. *Lancet*, 1, 1300.

Sakaki, H.; Kaneno, H.; Sumiya, Y.; Tsushima, M.; Miki, W.; Kishimoto, N.; Fujita, T.; Matsumoto, S.; Komemushi, S.; Sawabe, A. (2002). A new carotenoid glycosyl ester isolated from a marine microorganism, *Fusarium* strain T-1. *Journal of Natural Products*, 65, 1683-1684.

Samuelsson,G. (1999). Drugs of natural origin: a textbook of pharmacognosy. 4 ed. *Swedish Pharmaceutical Press*, Stockholm, Sweden.

Sangalang, A.E.; Backhouse, D.; Burgess, L.W. (1995). Survival and growth in culture of four *Fusarium* species in relation to occurrence in soils from hot climatic regions. *Mycological Research*, 99, 529-533.

Santini, A.; Ritieni, A.; Fogliano, V.; Randazzo, G.; Mannina, A.; Logrieco, A.;

Benedetti, E. (1996). Structure and absolute stereochemistry of fusaproliferin, a toxic metabolite from Fusarium proliferatum. *Journal of Natural Products*, 59, 109-112.

Saremi, H.; Burgess, L. W.; Backhouse, D. (1999).Temperature effects on the relative abundance of Fusarium species in a model plant-soil ecosystem. *Soil Biology and Biochemistry Journal*, 31, 941-947.

Savard, M.E.; Miller, J.D.; Salleh, B.; Strange, R.N. (1990). Strange chlamydosporol, a new metabolite from Fusarium chlamydosporum. *Mycopathologia,* 110, 177-181,

Scherm, B.; Orrù, M.; Balmas, V.; Spanu, F.; Azara, E.; Delogu, G.; Hammond, T.M.; Keller, N.P.; Migheli, Q. (2011). Altered trichothecene biosynthesis in TRI6-silenced transformants of Fusarium culmorum influences the severity of crown and foot rot on durum wheat seedlings. *Molecular Plant Pathology*, 12, 759-771

Schoental, R. (1983). Precocious sexual development in Puerto Rico and oestrogenic mycotoxins (zearalenone). *Lancet* 1, 537.

Schollenberger, M.; Drochner, W.; Müller, H.M. (2007). Fusarium toxins of the scirpentriol subgroup: a review. *Mycopathologia*, 164, 101-118.

Schroeder, J.J.; Crane, H.M.; Xia, J.X.; Loitta, D.C.; Merrill, A.H. (1994). Disruption of sphingolipid metabolism and stimulation of DNA synthesis by fumonisin B1. A molecular mechanism for carcinogenesis associated with Fusarium moniliforme. *The Journal of Biological Chemistry*, 269, 3475-3481.

Schuda, P.F.; Potlock, S.J. (1984). Trichothecenes, 1: the synthesis of 4-deoxyverrucarol from verrucarol and diacetoxyscirpenol. *Journal of natural products,* 47, 514-519.

Seo, C.; Sohn, J.H.; Oh, H.; Kim, B.Y.; Ahn, J.S. (2009). Isolation of the protein tyrosine phosphatase 1B inhibitory metabolite from the marine-derived fungus Cosmospora sp. SF-5060. *Bioorganic & Medicinal Chemistry Letters,* 19, 6095-6097.

Sheehan, D.M.; Branham, W.S.; Medlock, K.L.; Shanmugasundaram, E.R. (1984). Estrogenic activity of zearalenone and zearalanol in the neonatal rat uterus. *Teratology,* 29, 383-392.

Shemyakin, M.M.; Ovchinniikov, Y.A.; Kiryushkin, A.A.; Ivanov, V.T. (1963). The Structure and Total Synthesis of Enniatin B. *Tetrahedron Letters,* 14, 885-890.

Shibata, S.; Morishita, E.; Arima, Y. (1967). Metabolic products of fungi. XXV. 2 synthesis of Rubrofusarin and Its derivatives. *Chemical & Pharmaceutical Bulletin,* 15, 1757-1764.

Shier, W.T.; Abbas, H.K. (1992). Occurrence of the mycotoxin chlamydosporol in Fusarium species. *Toxicon*, 30, 1295-1298.

Singh, S.B.; Ondeyka, J.G.; Schleif, W.A.; Felock, P.; Hazuda, D.J. (2003a). Chemistry and structure-activity relationship of HIV-1 integrase inhibitor integracide b and related natural products. *Journal of Natural Products*, 66, 1338-1344.

Singh, S.B.; Zink, D.L.; Bills, G.F.; Teran, A.; Silverman, K.C.; Lingham, R.B.; Felock, P.; Hazuda, D.J. (2003c). Four Novel bis-(naphtho-γ-pyrones) isolated from *Fusarium* species as inhibitors of HIV-1 integrase. *Bioorganic & Medicinal Chemistry Letters,* 13 713-717.

Singh, S.B.; Zink, D.L.; Dombrowski, A.W.; Polishook, J.D.; Ondeyka, J.G.; Hirshfield, J.; Felock, P.; Hazuda, D.J. (2003b). Integracides: tetracyclic triterpenoid inhibitors of HIV-

1 integrase produced by Fusarium sp. *Bioorganic & Medicinal Chemistry*, 11, 1577-1582.

Singh, S.B.; Zink, D.L.; Goetz, M.A.; Dombrowski, A.W.; Polishook, J.D.; Hazuda, D.J. (1998). Equisetin and a novel opposite stereochemical homolog phomasetin, two fungal metabolites as inhibitors of HIV-1 integrase. *Tetrahedron Letters,* 39, 2243-2246.

Singh, S.B.; Zink, D.L.; Liesch, J.M.; Dombrowski, A.W.; Darkin-Rattray, S.J.; Schmatz, D.M.; Goetz, M.A. (2001). Structure, histone deacetylase, and antiprotozoal activities of apicidins b and c, congeners of apicidin with proline and valine substitutions. *Organic Letters*, 3, 2815-2818.

Singh, S.B.; Zink, D.L.; Liesch, J.M.; Mosley, R.T.; Dombrowski, A.W.; Bills, G.F.; Darkin-Rattray, S.J.; Schmatz, D.M.; Goetz, M.A. (2002). Structure and chemistry of apicidins, a class of novel cyclic tetrapeptides without a terminal alpha-keto epoxide as inhibitors of histone deacetylase with potent antiprotozoal activities. *The Journal of Organic Chemistry*, 67, 815-825.

Singh, S.B.; Zink, D.L.; Polishook, J.D.; Dombrososki, A.W.; Darkin-Rattray, S.J.; Schmatz, D.M.; Goetz, M.A. (1996). Apicidins: novel cyclic tetrapeptides as coccidiostats and antimalarial agents from Fusarium pallidoroseum. *Tetrahedron Letters,* 37, 8077-8080.

Smith, S.N. & Snyder, W.C. (1971). Relationship of inoculum density and soil types to severity of Fusarium wilt of sweet potato. *Phytopathology*, 61, 1049–1051.

Smoragiewicz, W.; Cossette, B.; Boutard, A.; Krzystyniak K. (1993). Trichothecene mycotoxins in the dust of ventilation systems in office buildings. *International Archives of Occupational and Environmental Health,* 65, 113-117.

Snider, B.B.; Neubert, B.J. (2004). A Novel Biomimetic Route to the 3-Acyl-5-hydroxy-3-pyrrolin-2-one and 3-Acyl-3,4-epoxy-5-hydroxypyrrolidin-2-one ring systems. *Journal of Organic Chemistry*, 69, 8952-8955

Solfrizzo, M.; Visconti, A. (1994). Anticholinesterase activity of the fusarium metabolite visoltricin and its *N*-methyl derivative. *Toxicology in Vitro*, 8, 461-465.

Solomons, G.L. (1986). Microbial proteins and regulatory clearance for RHM myco-protein. *In Microbial Biomass Proteins, Elsevier*, New York, 19–26.

Son, S.W.; Kim, H.Y.; Choi, G.J.; Lim, H.K.; Jang, K.S.; Lee, S.O.; Lee, S.; Sung, N.D.; Kim, J.C. (2008). Bikaverin and fusaric acid from Fusarium oxysporum show antioomycete activity against Phytophthora infestans. *Journal of Applied Microbiology*, 104, 692-698.

Song, H.-H.; Lee, H.-S.; Lee, C. (2011). New cytotoxic cyclic pentadepsipeptide, neo-*N*-methylsansalvamide produced by *Fusarium solani* KCCM90040, isolated from potato. *Food Chemistry*, 126, 472-478.

Sørensen, J.L.; Phipps, R.K.; Nielsen, K.F.; Schroers, H.-J.; Frank, J.; Thrane, A.U. (2009). Analysis of *Fusarium avenaceum* metabolites produced during wet apple core rot. *Journal of Agricultural and Food Chemistry*, 57, 1632-1639.

Springer, J.P.; Clardy, J.; Cole, R.J.; Kirksey, J.W.; Hill, R.K.; Carlson, R.M; Isidor, J.L. (1974). Structure and synthesis of moniliformin, a novel cyclobutane microbial toxin. *Journal of the American Chemical Society*, 96, 2267-2268.

Sridevi, B & Charya, M.A.S. (2011). Isolation, identification and screening of potential cellulase-free xylanase producing fungi. *African Journal of Biotechnology,* 10, 4624-4630.

Steyn, P.S.; Vleggaar, R. (1985). Biosynthetic studies on the fusarins, metabolites of Fusarium moniliforme. Journal of the Chemical Society, *Chemical Communications,* 1189-1191.

Stipanovic, R.D.; Puckhaber, L.S.; Liu, J.; Bell, A.A. (2011). Phytotoxicity of fusaric acid and analogs to cotton. *Toxicon*, 57, 176-178.

Stob, M., Baldwin, R.S., Tuite, J., Andrews, F.N., Gillette, K.G. (1962). Isolation of an anabolic, uterotrophic compound from corn infected with Gibberella zeae. *Nature*, 196, 1318.

Strongman, D.B.; Strunz, G.M.; Giguere, P.; Yu, C.M.; Calhoun, L. (1988). Enniatins from *Fusarium avenaceum* isolated from balsam fir foliage and their toxicity to spruce budworm larvae, Choristoneura-fumiferana (Clem.) (Lepidoptera: Tortricidae). *Journal of Chemistry and Ecology*, 14, 753-764.

Sukumaran, R.K.; Singhania, R.R.; Pandey, A. (2005). Microbial cellulases – Production, application and challenges. *Journal of Scientific & Industrial Research,* 64, 832-844.

Summerell, B.A.; Salleh, B.; Leslie, J. F. (2003). A utilitarian approach to *Fusarium* identification. *Plant Disease*, 87, 117-128.

Tabata, N.; Tomoda, H.; Yamaguchi, Y.; Masuma, R.; Bamberger, M.J.; Omura, S. (1999). Inhibition of cholesteryl ester transfer protein by fungal metabolites, L681,512.*The Journal of Antibiotics*, 52, 1042-1045.

Takasu, K.; Mizutani, S.; Noguchi, M.; Makita,K.; Ihara, M. (1999). Stereocontrolled total synthesis of (±)-Culmorin via the intramolecular double Michael addition. *Organic Letters*, 3, 391-393.

Takemura, H., Shim, J.Y.; Sayama, K.; Tsubura, A.; Zhu, B.T.; Shimoi, K. (2007). Characterization of the estrogenic activities of zearalenone and zeranol in vivo and in vitro. *Journal of Steroid Biochemistry & Molecular Biology,* 103 170-177.

Tesmer, J.J.; Tesmer, V.M.; Lodowski, D.T.; Steinhagen, H.; Huber, J. (2010). Structure of human G protein-coupled receptor kinase 2 in complex with the kinase inhibitor balanol. *Journal of Medicinal Chemistry*, 53, 1867-1870.

Thigpen, J.T.; Vaughn, C.; Stuckey, W.J. (1981). Phase II trial of anguidine in patients with sarcomas unresponsive to prior chemotherapy: A Southwest Oncology Group Study. *Cancer Treatment Reports* 65, 881-882.

Thompson, W.L.; Wannemacher, R.W.Jr. (1986). Structure-function relationships of 12, 13-epoxytrichothecene mycotoxins in cell culture: comparison to whole animal lethality. *Toxicon* 24, 985-994.

Tomaszewski, J.; Miturski, R.; Semczuk, A.; Kotarski, J.; Jakowicki, J. (1998). Tissue zearalenone concentration in normal, hyperplastic and neoplastic human endometrium. *Ginekologia Polska*, 69, 363-366.

Tomoda, H.; Doi, T. (2008). Discovery and combinatorial synthesis of fungal metabolites Beauveriolides, Novel antiatherosclerotic agents. *Accounts of Chemical Research*, 41, 32-39.

Tomoda, H.; Nishida, H.; Huang, X.-H.; Masuma, R.; Kim, Y.K.; Omura, S. (1992). New cyclodepsipeptides, enniatins D, E and F produced by Fusarium sp. FO-1305 *Journal of Antibiotics*, 45, 1207-1215.

Trisuwan, K.; Khamthong, N.; Rukachaisirikul, V.; Phongpaichit, S.; Preedanon, S.; Sakayaroj, J. (2010). Anthraquinone, cyclopentanone, and naphthoquinone derivatives from the sea fan derived fungi Fusarium spp. PSU-F14 and PSU-F135. *Journal of Natural Products*, 73, 1507-1511.

Troncoso, C.; González, X.; Bömke, C.; Tudzynski, B.; Gong, F.; Hedden, P.; Rojas, MC. (2010). Gibberellin biosynthesis and gibberellin oxidase activities in Fusarium sacchari, Fusarium konzum and Fusarium subglutinans strains. *Phytochemistry,* 71, 1322-1331.

Tudzynski, B. (2005). Gibberellin biosynthesis in fungi: genes, enzymes, evolution, and impact on biotechnology. *Applied Microbiology and Biotechnology* 66, 597-611.

Tuomi, T.; Reijula, K.; Johnsson, T.; Hemminki, K.; Hintikka, E.L.; Lindroos, O.; Kalso, S.; Koukila-Kähkölä, P.; Mussalo-Rauhamaa, H.; Haahtela, T. (2000). Mycotoxins in crude building materials from water-damaged buildings. *Applied and Environmental Microbiology* 66, 1899-1904.

Tyson, J.L.; Henderson, R.C.; Fullerton, R.A.; Jamieson, L.E.; Froud, K.J. (2005). Distribution and new host records for *Cosmopora aurantiicola* and *Cosmopora flammea*: entomopathogens of Diaspididae in New Zealand. *New Zealand Plant Protection,* 58, 283-287.

Ueda, M,; Kubo, T,; Miyatake, K.; Nakamura, T. (2007). Purification and characterization of fibrinolytic alkaline protease from *Fusarium* sp. BLB. Biotechnologically Relevant Enzymes and Proteins, *Applied Microbiology and Biotechnology,* 74:331–338.

Usami, Y.; Aoki, S.; Hara, T.; Numata, A. (2002). New dioxopiperazine metabolites from a *Fusarium* species separated from a marine alga. *The Journal of Antibiotics,* 55, 655-659.

Vasko, R.C.; Rodriguez, R.A.; Cunningham, C.N.; Ardi, V.C.; Agard, D.A.; McAlpine, S.R. (2010). Mechanistic studies of sansalvamide a-amide: an allosteric modulator of Hsp90. *ACS Medicinal Chemistry Letters*, 1, 4-8.

Vershinin, A. (1999). Biological functions of carotenoids - diversity and evolution. *BioFactors*, 10, 99-104.

Vigier, B.; Reid, L.M.; Seifert, K.A. (1997): Distribution and prediction of *Fusarium* species associated with maize ear rot in Ontario. *Canadian Journal of Plant Pathology,* 19, 60–65.

Visconti, A.; Blais, L.A.; ApSimon, J.W.; Greenhalgh, R.; Miller, J.D. (1992). Production of enniatins by *Fusarium acuminatum* and *Fusarium compactum* in liquid culture: isolation and characterization of three new enniatins, B2, B3, and B4. *Journal of Agricultural and Food Chemistry*, 40, 1076-1082.

Visconti, A.; Solfrizzo, M. (1995). Visoltricin, a novel biologically active compound produced by Fusarium tricinctum. *Food Additives and Contaminants,* 12, 515-519.

Vlata, Z.; Porichis, F.; Tzanakakis, G.; Tsatsakis, A.; Krambovitis, E. (2006). A study of zearalenone cytotoxicity on human peripheral blood mononuclear cells. *Toxicology Letters*, 165, 274-281.

Waller, J.M.; Brayford, D. *(1990). Fusarium diseases in the tropics. tropical pest management,* 36, 181-194.

Wang, E.; Norred, W.P.; Bacon, R.T.; Riley, R.T.; Merril Jr.; A.H. (1991). Inhibition of sphingolipid biosynthesis by fumonisins: implications for diseases associated with Fusarium moniliforme. *The Journal of Biological Chemistry*, 266, 14486-144990.

Wang, G.-Y.-S.; Abrell, L. M.; Avelar, A.; Borgeson, B.M.; Crews, P. (1998). New hirsutane based sesquiterpenes from salt water cultures of a marine sponge-derived fungus and the terrestrial fungus Coriolus consors. *Tetrahedron*, 54, 7335-7342.

Wang L.H.; Fu Y.; Shi Y.X.; Wang, W.G. (2011). T-2 toxin induces degenerative articular changes in rodents: link to Kaschin-Beck disease. *Toxicologic Pathology,* 39, 502-507.

Wang, Q.X.; Li, S.F.; Zhao, F.; Dai, H.Q.; Bao, L.; Ding, R.; Gao, H.; Zhang, L.X.; Wen, H.A., Liu, H.W. (2011). Chemical constituents from endophytic fungus Fusarium oxysporum. *Fitoterapia,* 82, 777-781.

Wang, S.J.; Guo, X.; Zuo, H.; Zhang, Y.G.; Xu, P.; Ping, Z.G.; Zhang, Z.; Geng, D. (2006). Chondrocyte apoptosis and expression of Bcl-2, Bax, Fas, and iNOS in articular articular cartilage in patients with Kashin-Beck disease. *The Journal of Rheumatology*, 33, 615-619.

Wannemacher, R.W.; Wiener, S.L. (1997). Trichothecene Mycotoxins (From *Medical Aspects of Chemical and Biological Warfare*, p. 655-676, Sidell, F.R.; Takafuji, E.T. eds.

Wehner, F.C.; Marasas, W.F.O.; Thiel, P.G. (1978). Lack of mutagenicity to Salmonella typhimurium of some Fusarium mycotoxins. *Applied and Environmental Microbiology,* 35, 659- 662.

Wheeler, M.H.; Stipanovic, R.D.; Puckhaber, L.S. (1999). Phytotoxicity of equisetin and epi-equisetin isolated from Fusarium equiseti and F. pallidoroseum. *Mycological Research,* 103, 967-973

Wiebe, M.G. (2002). Mycoprotein from *Fusarium venenatum*: a well-stablished product for human consumption. *Applied Microbiology and Biotechnology*, 58, 421-427.

Wood, T.M.; Mc Crae, S.T. (1986). Studies of two low molecular weight endo-1,4-_-xylanase, constitutively synthesized by the cellulolytic fungus Trichodrma konongii. *Carbohydrates Research*, 148, 321-330.

Wright, J.L.C.; Vining, L.C.; McInnes, A.G.; Smith, D.G.; Walter, J.A. (1977). Use of ^{13}C in biosynthetic studies. The labelling pattern in tenellin enriched from isotope-labelled acetate, methionine, and phenylalanine. *Canadian Journal of Biochemistry,* 55, 678-685.

Yang, C.S. (1980). Research on esophageal cancer in China: a review. *Cancer Research*, 40, 2633-2644.

Yang, S.X.; Gao, J.M.; Zhang, Q.; Laatsch, H. (2011). Toxic polyketides produced by *Fusarium* sp., an endophytic fungus isolated from Melia azedarach. *Bioorganic & Medicinal Chemistry Letters*, 21, 1887-1889.

Yoder, W.T & Christianson, L.M. (1998). Species-specific primers resolve members of *Fusarium* section *Fusarium*. Taxonomic status of the edible "Quorn" fungus reevaluated. *Fungal Genetics and Biology*, 23, 68-80.

Yu, J.-H.; Keller, N. (2005). Regulation of secondary metabolism in filamentous fungi. *Annual Review of Phytopathology*, 43, 437-458.

Yu, S.-R.; Liu, X.-J.; Wang, Y.-H.; Liu, J. (1995). A survey of moniliformin contamination in rice and corn from Keshan Disease endemic and non-KSD areas in China. *Biomedical and Environmental Sciences* 8, 330-334.

Zhan, J.; Burns, A.M.; Liu, M.X.; Faeth, S.H.; Gunatilaka, A.A. (2007). Search for cell motility and angiogenesis inhibitors with potential anticancer activity: beauvericin and other constituents of two endophytic strains of *Fusarium oxysporum. Journal of Natural Products,* 70, 227-232.

Zhang, A.; Cao, J.L.; Yang, B.; Chen, J.H.; Zhang, Z.T.; Li, S.Y.; Fu, Q.; Hugnes, C.E.; Caterson, B. (2010). Effects of moniliformin and selenium on human articular cartilage metabolism and their potential relationships to the pathogenesis of Kashin-Beck disease. *Journal of Zhejiang University-Science B,* 11, 200-208.

Zhao, D.; Feng, Q.; Yan, X.. Li, C.; Pan, Y.; Cui, Q. (1993). Ultrastructural study of moniliformin induced lesions of myocardium in rats and mice. *Biomedical and Environmental Sciences* 6, 37-44.

Zinedine, A.; Soriano, J.M.; Moltò, J.C.; Magñes, J. (2007). Review on the toxicity, occurrence, metabolism, detoxification, regulations and intake of zearalenone: an estrogenic mycotoxin. Food and Chemical Toxicology, 45, 1-18.

In: Fusarium
Editors: Tony Ferarri Rios and Emilia Reyes Ortega

ISBN: 978-1-61942-539-2

Chapter 2

FUSARIUM: EPIDEMIOLOGY, ENVIRONMENTAL SOURCES AND PREVENTION

***Edlayne Gonçalez*[*,1], *Christiane Ceriani Aparecido*[2] *and Joana D'arc Felicio*[1]**
[1]Centro de Pesquisa e Desenvolvimento de Sanidade Animal and
[2]Centro de Pesquisa e Desenvolvimento de Sanidade Vegetal,
Instituto Biológico, Av. Cons. Rodrigues Alves, São Paulo, SP, Brazil

ABSTRACT

Fusarium is one of the most ubiquitous, abundant, and important genera of soil microfungi, and about 1000 species have been described. Identification of *Fusarium* species is difficult due to great variation in morphological and nonmorphological characteristics, including virulence. The genus contains many species of environmental, agricultural and human health importance. Fusaria have long been recognized as pathogens of many plant species. Wheat and other small grain cereals may be attacked by a wide range of *Fusarium* spp. and on different plant organs. However, infestation of the ears appears to be the most critical, leading to Fusarium head blight (FHB), also known as scab. *F. graminearum* and *F. culmorum* are the predominant *Fusarium* species infecting wheat.

Predominance of a specific species in a region is mostly influenced by climatic conditions especially temperature and moisture requirements. Plant diseases caused by *Fusarium* spp. include vascular wilts, seedling damping-off, and rots of stems, crowns, and roots.

Some species of *Fusarium* also produce mycotoxins resulting in food contamination. In general, biotic and abiotic stresses (heat, water and insect damage) cause plant stress and predispose plants in the field to mycotoxin contamination. The main mycotoxin classes of concern produced by *Fusarium* species include the trichothecenes, moniliformin, fumonisins and zearalenone.

Traditionally crop rotations have been proven to be an effective strategy to control many soil-borne diseases, but because most of these pathogens can survive for a long

* Corresponding author: Telephone: +55-11-50871753; fax: +55-11-50871754. E-mail address: goncalez@biologico.sp.gov.br (E. Gonçalez).

period of time, the effectiveness of crop rotation is limited once a disease outbreak occurs. Biological control and natural products (like essential oil) were an alternative strategy to control *Fusarium* disease.

INTRODUCTION

The *Fusarium* genus is one of the most important in world plant pathology, often causing wilt in different crops of economic importance [1]. In recent years it has acquired also important by the mycotoxins production, substances responsible for diseases in humans and animals [2]. Example of *Fusarium* mycotoxins are: trichothecenes, zearalenone, fumonisins and moniliformin.

It is noteworthy that the vast majority of species are soil saprophytes and plant pathogens, while a minority, pathogenic for human. The first major revision of the genus was carried out in 1935 by Wollenweber and Reinking, has been described in the publication "Die Fusarien" with 65 species, 55 varieties and 22 *formae speciales*, grouped into 16 sections. Since then, about 1000 species have been described [3].

Fungi of the genus *Fusarium* are widespread pathogens on maize and small-grain cereals (soft and durum wheat, barley, oats, rye and triticale) in all cereal-growing areas of the world. They are, together with the species *Microdochium nivale*, the cause of plant diseases such as root, stem and ear rot, the latter also known as *Fusarium* head blight (FHB) or scab. The infection with *Fusarium* commonly results in substantial reduction of crop yield and hence in economic losses. Despite the fact that several *Fusarium* species may be implicated in plant disease, three species are known to be of particular importance internationally: *F. graminearum* (*Gibberella zeae*), *F. culmorum* and *F. avenaceum* (*Gibberella avenacea*) [4]. While *F. graminearum* is often the prevalent species in many of the affected regions, the others can be highly pathogenic or are frequently found in association with the disease [5]. Thus, management of crop residues in field after harvesting is necessary to minimize these adverse effects, particularly if the pathogens involved can infect subsequent crops [6].

The factors contributing to development of Fusarium head blight and ear rot and subsequent mycotoxin contamination include fungal and host genotypes, temperature, relative humidity, amount and timing of rainfall, fungal sporulation, wind patterns, niche competition, insect damage to the plant, and cultural practices [7].

Prevention methods for *Fusarium* disease and mycotoxins production include: crop rotation, fungicides and seed resistant. Alternative treatments using bacteria and natural products are been studied.

Understanding the epidemiology of the disease system and the ecology of the pathogens are keys to developing the management components and strategies that will be most effective against FHB. Research into the environmental risk factors leading to mycotoxin development and contamination of grain will be necessary for the success of predictive modeling systems, as well as for integrative management of mycotoxins and disease [5].

This chapter will approach some aspects of epidemiology, environmental source, prevention and control methods for *Fusarium* genus.

EPIDEMIOLOGY AND ENVIRONMENTAL SOURCE

Fusaria have long been recognized as pathogens of many plant species. Wheat and other small grain cereals may be attacked by a wide range of *Fusarium* spp. and on different plant organs [8].

Fusarium head blight (FHB), also called ear blight or scab, is a major fungal disease affecting several gramineous hosts including wheat (*Triticum aestivum* L.) and barley (*Hordeum vulgare* L.). The disease occurs throughout much of the world and is associated with several *Fusarium* spp. including *Fusarium graminearum* Schwabe (teleomorph: *Gibberella zeae* (Schwein.) Petch) and *F. culmorum* (W.G. Smith) Sacc., as well as *F. poae* (Peck) Wollenw., *F. avenaceum* (Fr.) Sacc., *F. sporotrichoides* Scherb., and *Microdochium nivale* (Fr.) Samuels and I.C. Hallett. While *F. graminearum* is often the prevalent species in many of the affected regions, the others can be highly pathogenic or are frequently found in association with the disease [5]. FHB is a pre-harvest disease, but *Fusarium* species can grow post-harvest if wet grain is not dried efficiently and quickly. The trend could also reflect a changing trend of dominance of *Fusarium* spp. in different parts of the world. Several reservoirs of the parasite complex responsible for head blight are known. The primary reservoir of inoculum is debris from the previous crop [8]. Other sources of inoculum include numerous plant hosts. Seeds are the other major source of inoculum, making it possible for the disease to begin in autumn in the case of grains contaminated with mycelium [9].

Wheat, maize, and other grain crop residues are generally regarded as the primary source of inoculum for *Fusarium* species [7].

The genus is typically inhabitant of soil surviving in the perithecia form, when there is a sexual phase and, in the mycelium and / or conidia form, when it comes to the asexual phase in plant tissues, or even in the chlamydospores form (resistance structures) associated with the soil or organic debris. The spread in the case of spores can occur by wind, and fragments of hyphae through the soil. The fungus can start their infection penetrating the roots directly or by injuries. In general, stem rot caused by *Fusarium* is associated with the attack of nematode pests that weaken the ground and injure plant roots, facilitating the penetration of the fungus. The rot begins to manifest a few weeks after pollination, becoming more severe at this point, when pollination is preceded by a dry spell and then the rainy season. Drought stress and injury to plants as well as hot and humid conditions just seem to be highly conducive to fungal genus and the optimum temperature range for the development of diseases from 20° to 28°C, although some species are favored by warmer temperatures, from 15°C. The fungus initiates the infection by directly penetration of the roots or injuries. Soil humidity, particularly during winter, decreases the pathogen survival rate [10], however some species are favored by high humidity, while others need dry periods alternating with very wet periods. Soils with low pH (acid) also favor the microorganism. *Fusarium* spp. in soil is also affected by crop sequence, fertilizer use and method of straw disposal [11].

Many species can be associated with the seed, which is an important vehicle for the dissemination of the pathogen is through them that the pathogens can be transported great distances and could be introduced into new areas and, from them can be produced already infected plants that act as new sources of inoculum [12]. There are also other agents of dissemination such as rain, wind, humans, animals, insects and farm implements used during cultivation. Temperature and rainfall may have a significant effect on the distribution and abundance of *Fusarium* species [13].

Several *Fusarium* species, but primarily *F. solani* and some *formae speciales* of *F. oxysporum*, cause, instead of vascular wilts, rotting of seeds and seedlings (damping-off), rotting of roots, lower stems, crowns and necrosis of fruits (Figures 01-A-E).

They affect many different kinds of vegetables and flowers, herbaceous or perennial and plantation crops [14]. These diseases occur worldwide and cause severe losses by reducing the growth of infected plants in yield or greenhouse. Most of the wilt is caused by *Fusarium oxysporum*. Different host plants are attacked by f*ormae speciales* (f. sp.) and races of this specie. The fungus that attacks tomato is designated *F. oxysporum* f. sp. *lycopersici*; cucurbits - *F. oxysporum* f. sp. *conglutinans*; banana - *F. oxysporum* f.sp. *cubense*; cotton - *F. oxysporum* f.sp. *vasinfectum*; carnation - *F. oxysporum* f. sp. *dianthii*; and so on. Fusarial wilts are most severe under warm soil conditions and in greenhouses. The fungus *F. solani* causes root rot and can produces only asexual spores, although under certain conditions it produces its sexual stage or teleomorphic stage, *Netria haematococca* wich produces the infective structures (ascospores) within fruiting bodies called perithecia (Figure 02 – A-B).

In the case of parasite species and / or saprobe fungi, the majority are the habitat of plant roots, plant debris and debris that may be deposited up to 1m depth of soil. There are also species that parasitize the conductive vessels of plants [15]. With respect to infective units, these are dispersed by the action of rain water, wind, man, and insects. In infected plants can be observed symptoms such as wilt, root rot and collar rot [16]. The species that produce toxins such substances affect the permeability of cell membrane disrupting cell metabolism, resulting in wilt and then the death of them. The wilt also can be caused by mass production of conidia in the vascular tissue, which results in blocking the xylem, preventing water movement. *Fusarium oxysporum* produces diseases in tomatoes, sweet potatoes, pears and was responsible for the disease of bananas in Panama, producing the abandonment of thousands of acres in Central America [17, 18]. *Fusarium culmorum* (W.G. Sm.) Sacc. is a principal causal agent of brown foot rot of cereals, especially wheat, in the UK. The incidence and severity of the disease depend on suitable weather conditions, and are greatest in warm, dry conditions. The most likely source of inoculum for foot rot caused by *F. culmorum* is the soil, where it survives as chlamydospores or as mycelium on infested plant debris [11].

A B

Figure 1. Symptoms caused by Fusarium fungi on different hosts. A - D. Wilt in chrysanthemum with destruction, respectively, of the root system (A,B) and conductive vessels (C, D). E – Necrosis in pineapple fruit.

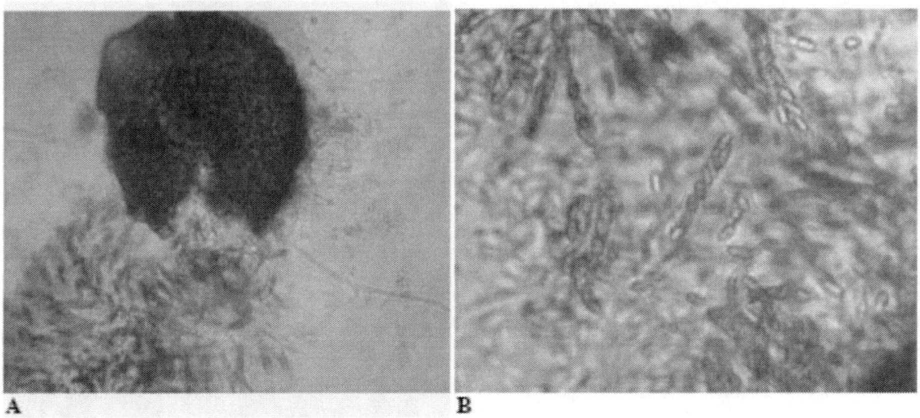

Figure 2. Sexual stage (*Nectria haematococca*) of *Fusarium solani*. Red perithecia with ascas (A) e ascospores (B).

Generally, cultures of *Fusarium* spp. are characterized by the rapid growth of the colony, cotton-like mycelium velvety to slightly opaque or slightly brighter. After 7-10 days of incubation at 25°C (optimum temperature), the fungus may have characteristics of mycelium pigmentation ranging from pink to purple, white, cream, salmon (Figure 03), which are important characteristics for species identification [19].

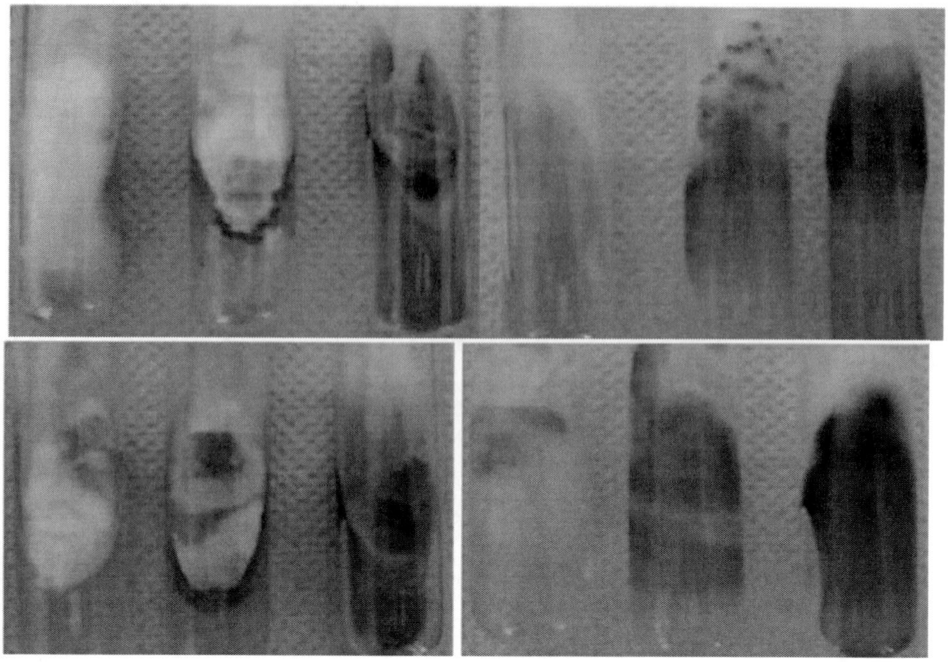

Figure 3. Variability in mycelium pigmentation of *Fusarium* spp.

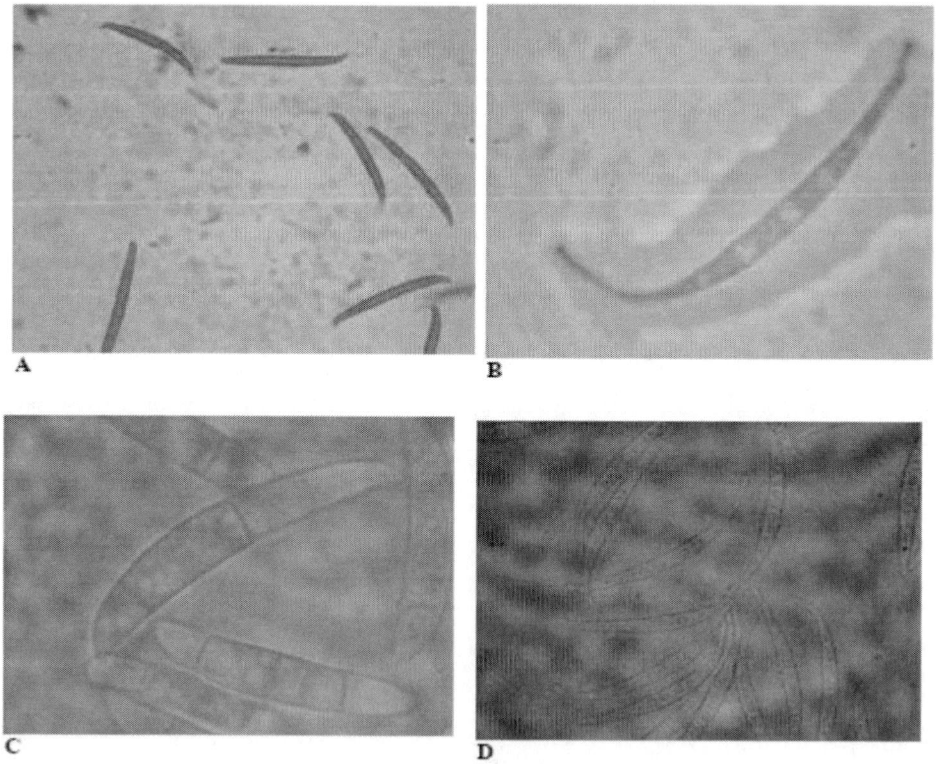

Figure 4. Variability in the microscopic characteristics of *Fusarium*. A –D. Macroconidia. E – F. Conidiogenic cells (monophialide – E; poliphialide – F). G-H. Microconidia. H – Mesoconidia (arrow). I – J. Clamidospores.

The main characteristic of this genus is the conidia production of different types: hyaline macroconidia (Figure 04A-D), which may have very characteristic sickle-shaped, with tapered ends slightly curved or inclined, with multiple transverse septa and pronounced structure called a foot cell. They emerge from conidiogenic cells (phialides), which may present singly or grouped in masses by means of mucilaginous material, growing directly from vegetative mycelia, in this case being referred sporodochia [19], the microconidia (Figure. 04G-H) which are small and may have ellipsoid shape, allantois, pyriform, ovoid or subglobose, single-celled, sometimes may present a septum. Are produced on single (Figure 04E) or polyphialidic (Figure 04F) conidiogenous cells heads and arranged in chains and mucoid, with a truncated base [20, 21]; the mesoconidia (Figure 04H - arrow), structures similar in shape, but the macroconidia size less, without basal cell in the form of foot and not causing the heads mucoid [22]. Under adverse conditions, some species are capable of producing resistance structures called chlamydospores be able survive in the soil for at least 20 years (Figure 04I-J). These structures can be single or grouped in chains or not, the walls are thick (round or wrinkled), and on the location, can be apical, interim or terminal [23]. The production or not all these structures, as well as color and shape of the mycelium growth are characteristics used for taxonomic identification of species. When grown in the laboratory after 7-10 days of incubation at 25 °C the fungus may have characteristics of pigmentation of the mycelium booing colored pink, purple, white or yellow, and these characteristics are also important for species identification [19, 24].

Invasion of the kernel by *Fusarium* destroys the starch granules and cell walls, and affects endosperm storage proteins resulting in a poor quality product. *F. graminearum* and *F. culmorum* produce the trichothecenes deoxynivalenol (DON), 3-acetyldeoxynivalenol (3A-DON) and nivalenol (NIV). Quality loss due to mycotoxin contamination has led to important economic losses [25]. In plants, DON retards the germination and growth of wheat, and inhibits the growth of the grain and the coleoptile tissues. The phytotoxicity of mycotoxins is generally estimated by means of coleoptile elongation tests. In adult plants, deoxynivalenol seems to circulate in the phloem, with the concentration of this molecule in the plant following a descending gradient from the rachis, through lemmas and grains to the peduncle [9].

The fungal *F. verticillioides* (= *F. moniliforme*), *F. roseus*, *F. tricinctum* and *F. nivale*, are ubiquitous soil organisms, which may infect cereals directly in the field thereby, increasing fumonisins, trichotecenes and zearalenone levels (depending on the species) during growth, ripening of grain and at harvesting [26]. The contamination of maize grain by *F. moniliforme* Sheldon and *F. proliferatum* (Matsushima) Nirenberg has received significant attention because of their pre- and post- harvest colonization of moist maize and concomitant production of the fumonisin group of mycotoxins [27].

F. verticillioides can invade maize grain via three pathways: systemic growth through seed transmission or in roots, stalks, or leaves; air- or splash-borne infection by macroconidia and microconidia produced on crop residues and tassels that infect ears through silks or insect-caused wounds; and insects as vectors of conidia. Among these pathways, systemic growth from contaminated seeds seems to be less harmful while the silk and insect routes are more relevant. Once the fungus enters the ear, fumonisins can be synthesized and contaminate kernels that eventually enter the food or feed chain [28].

Fusarium Toxins

Since the 1990s, interest in food health and safety has increased. For cereal products, one of the most important elements of alimentary risk is the possibility of mycotoxin accumulation on or in the grains [9].

Mycotoxin production of fungi is a complex process currently not fully understood. It is known that both the growth of fungi and their toxigenic potential are affected by environmental factors, such as temperature, water activity, pH and both nutrient composition and availability. Under field conditions there are many factors which influence the mycotoxin production: climate, agricultural practices, host plant and the presence of other microorganisms. Secondary metabolism of fungi is regulated by genetic mechanisms which respond to various stimuli from the environment. The same fungus can produce a very different range of mycotoxins under different culture conditions [29].

The main mycotoxin classes of concern produced by *Fusarium* species include the trichothecenes, fumonisins and zearalenone. Several trichothecenes are of concern, including diacetoxyscirpenol, T-2 toxin, deoxynivalenol, and nivalenol [7]. Table 1 showed the major *Fusarium* toxigenic species, their principal mycotoxins and the deleterious effects in animal and human [30].

Levels of the mycotoxins produced by *Fusarium* cannot be reduced after harvest by means of classical transformation procedures. These toxins are chemically and thermally stable. They may therefore be found in the raw material (grains) and in transformed products (flour, bread) or in products from animals fed with contaminated grain (meat, milk, eggs) [9].

Table 1. The major *Fusarium* toxigenic species and their principal mycotoxins and the deleterious effects [30]

Fungal species	Mycotoxins	Abr	Deleterius effects
Fusarium culmorum; F. gramineatum; F. sporomchioides	Deoxynivalenol	DON	Gastritis, vomition, and feed refusal known as "vomitoxin".
F. sporomchioides; F. graminearum; F. poae	T-2 toxin Diacetoxyscirpenol	T-2 DAS	Cause necrosis and ulceration of epithelial surfaces of the skin, mouth, and upper gastrointestinal tract, vomition, hemorrhagic enteritis, hematopoietic suppression
F. culmorum; F. graminearum; F. sporotrichioides	Zearalenone	ZEA	Infertility, reduced milk production and hyperoestrogenism in cows and swines
F. moniliforme (= *F. verticillioides*)	Fumonisins	FBs	Hepatic lesions in pigs and cattle; leukoencephalomalacia in Equine. Porcine pulmonary oedema. Implicated in oesophageal cancer in humans

Fumonisins are a group of mycotoxins produced by species within the Gibberella fujikuroi complex of species. The major producers are *F. verticillioides*, *F. proliferatum*, *F. subglutinans* and *F. nygamai* [31]. Corn is the commodity almost exclusively associated with clinically relevant fumonisin contamination, and both white and yellow corn is affected. Conditions favoring fumonisins production appear to include a period of drought during the growing season with subsequent cool, moist conditions during pollination and kernel formation. Low concentrations are commonly found in corn in many parts of the world [32]. They affect animals by interfering with their sphingolipid metabolism. Besides various animal diseases they are suspected to be responsible for certain forms of human esophageal cancer observed in Transkai (South Africa), China, and Northeast Italy [31] Fumonisins are recognized as the cause of equine leukoencephalomalacia (ELEM) [32] and porcine pulmonary edema (PPE) [33].

Currently, more than 40 naturally occurring trichothecenes are known from *Fusarium* species, with a combined total of nearly 150 variants produced by species of *Fusarium, Cephalosporium, Cylindrocarpon, Myriothecium, Stachybotrys,* and *Trichoderma* [7]. Trichothecenes, including T-2 toxin, diacetoxyscirpenol, deoxynivalenol and other minor or less common trichothecenes are produced by *Fusarium sporotrichioides* and *F. graminearum* in many temperate climates of the world (North America, Northern Europe, Russia and China). Corn, wheat and other cereal grains are common commodities susceptible to trichothecene production [32]. Deoxynivalenol, also commonly known as vomitoxin, is often the most prevalent trichothecene contaminating grain worldwide and is produced by *F. graminearum, Fusarium pseudograminearum,* and *Fusarium culmorum* [7].

Zearalenone is a fungal estrogenic metabolite non-steroid, chemically described as a lactone of the resorcilic phenic acid. It is produced by several species of fungi of the genus *Fusarium* including *F. culmorum, F. graminearum* and *F. crookwellense.* These species colonize cereals and tend to become particularly important during seasons of high humidity, accompanied by mild temperatures (Dilkin livro). The basis for the estrogenic effect is well established and is due to a close structural similarity between zearalenone and estradiol [34].

Moniliformin, a hydroxycyclobutenedione,was initially characterized as a culture extract from *F. proliferatum* strain NRRL 6322, which was identified at the time as *F. moniliforme,* hence the name moniliformin. The metabolite is only known to be produced by species of *Fusarium,* and while moniliformin is known to be toxic to poultry based on laboratory toxicity studies, it has not been associated with any acute or chronic disease outbreaks among animals. In addition to *F. proliferatum,* the *Fusarium* species knownto produce significant amounts of moniliformin include *F. avenaceum, F. fujikuroi, F. nygamai, F. pseudonygamai, F. subglutinans,* and *F. thapsinum* [7].

Prevention and Control by Fusarium Contamination

The steps of *Fusarium* prevention that are the most effective are those carried out before the fungal infestation and before mycotoxin production occur on plant material, when it comes to toxigenic species. The options include controlling phytopatogenic and toxigenic *Fusarium* are fungicides, biocontrol and cultural practices, on the other hand the use of resistant genotype is promising. The main goal is to prevent the introduction and establishment of *Fusarium* species in areas where it does not occur and where it occurs, to reduce inoculum available for dispersal, prevention of dispersal of inoculum and prevention of infection of spikelets if the inoculum is present. Recent research showed the development of plant genotypes resistant to infections by *Fusarium* species [35].

In the culture practices is an important measure of control as the crop management, preventing the survival of *Fusarium* in remains of successive crops of grasses. Among the alternatives viable, sustainable and efficient control highlights the practice of crop rotation and tillage with crop stubble incorporation [9, 24, 25, 36]. Available options of managing FHB include use of fungicides, cultural practices, resistant cultivars and biological agents. However, no wheat cultivar is completely resistant to FHB while fungicides are at most 70% effective against natural infection [8].

Chemical Control

The most common method for protecting plants against their pathogens is the use of genetic resistance [37]. However, not all plants are resistant to pathogens, and not all resistant material is adapted to different regions. Thus, the use of agrotoxic in disease control has been increasingly common in agriculture. The continued adoption of chemical control can lead to the emergence of pathogens resistant to the products used [38] and the contamination of food and the environment.

The efficacy of fungicide treatments is variable, and difficult to predict. It depends not only on the active ingredient but also on the method and date of application of the fungicide. If a fungicide treatment is to be effective, it must be applied several days before the attack and entirely cover all the ears. Efficacy also depends on the interaction between fungal development and climate, the virulence of the pathogenic strains and above all, the parasite complex present on the plot of land concerned. There are several relates of fungicides tested against *Fusarium* sp. in vitro, in the green-house or under field conditions. Fungicides that have been used in control of FHB include prochloraz, propiconazole, epoxyconazole, chlorothalonil, mancozeb, thiram metconazole, tebuconazole cyproconazole and azoxystrobin [25, 39,40].

Each of the fungal species implicated has a specific fungicide susceptibility profile for example *F. graminearum* is particularly susceptible to triazoles whereas *F. avenaceum* is more susceptible to strobilurins. *F. roseum*, a complex consisting of *F. graminearum, F. culmorum* and *F. avenaceum,* is much more susceptible to triazoles than *M. nivale*, which is essentially susceptible to strobilurins [9].

Jouany et al. [25] relate that the triazole, metconazole and tebuconazole fungicides are efficient against *Fusarium* spp. They reduce significantly both FHB and DON contamination. Conversely, the strobilurin fungicide azoxystrobin, which is active against *M. nivale*, resulted in a reduction of FHB but the level of DON contamination sometimes increased [41, 42, 43]. Azoxystrobin appear to have no direct effect on DON production [44], but could increase the level of *Fusarium* infection. Haidukowski et al. [39] observed that cyproconazole plus prochloraz and amixture of tebuconazole plus azoxystrobin significantly reduced FHB disease severity (25% and 77%, respectively) and DON content (32% and 89%, respectively) in grains of five different cultivars of soft and durum wheat. Tetraconazole showed a lower effectiveness when compared with the other treatments.

The soil mycoflora are treated with fungicides reduce the soil-borne pathogens [45]. However, the possible negative impact of the fungicides on the soil saprophyte microbial biomass [46] and contamination of groundwater with chemical residues are serious limitations of such applications [25]. As efficient control of fungi through fungicide diminishes this damage caused by it, but the high cost of ecological damage [47]. Many researches is developed new strategies the application using natural compounds, these natural fungicides have the advantage that left no residues but enhanced flavor. Various natural products, among which the herbal extracts, have shown the ability to control plant diseases, both for its antimicrobial activity directly and indirectly through the induction of resistance [48, 49].

Some essential oils have been showed inhibitory effects against phytopatogenic and toxigenic fungi (Table 2) [49, 50, 51] and some specifically for their effects against *Fusarium* spp. and their associated mycotoxins [52, 53, 54]. The use of natural substances volatiles with fungicides properties to grain, fruits and vegetables can be an ecologic alternative [1]. The natural protective effect of essential oils against plant pathogens can be exploited to control plant fungal infection and the associated mycotoxin production, but more studies should be performed to identify the components of essential oils with modulatory activity on the growth and toxin production of *Fusarium* spp. [25].

Cinnamomum zeilanicum essential oil has been inhibited the fungal growth of *F. graminearum* and *F. verticillioides.* The essential oil from *Ocimum gratissimum*, *Thymus vulgaris* and *Cymbopogon citratus* were the most effective and prevented conidial germination and the growth of *F. moniliforme* on corn meal agar at 800 ppm [55]. Same authors relate the inhibition of mycotoxins production *Fusarium* species by essential oils. *Sygygium aromaticum*, *Cymbopogon citratus*, *Cymbopogon martini*, *Cinnamomum zeylanicum* and *Origanum vulgare* essential oils were inhibitor on ZEA production by *F. graminearum*. All factors (essential oils, temperature, water activity, treatment timing) as well as their interactions had a significant impact on ZEA production. The *S. aromaticum*, *C. citratus* and *C. martinii* oils prevented ZEA production, while cinnamon essential oil had no significant effect, and *O. vulgare* essential oil stimulated the production [53]. The fumonisin B_1 (FB_1) biosynthesis was inhibited by thymol [56].

The cinnamon, clove and oregano essential oils were effective in inhibiting growth of *F. proliferatum* at 20 °C. The growth of *F. solani* was completely inhibited by oregano, thyme, dictamnus and marjoram essential oils at relatively low concentrations (85–300 μg/ml). Thymol was the main component of oregano oil, while thyme, dictamnus, and marjoram oils were rich in carvacrol. Lavender, rosemary, sage, and pennyroyal essential oils presented less inhibitory activity. The growth of the tested microorganisms was affected at concentrations up to 1000 μg/mL. Lavender oil was characterized by the high content of linalool and linalyl acetate, while eucalyptol was the main component of sage and rosemary oils. Pennyroyal oil was found rich of *cis*-menthone and pulegone [57].

The sensitivity of *Fusarium* strains to *Chelidonium majus* L. extracts were tested the authors verified that aqueous extracts had considerable inhibitory action but methanolic extracts showed the best results. Root extracts were inhibitorier than shoot extracts. On the basis of growth inhibition the *Fusarium* strains were aggregated into five classes, the extremes being *F. culmorum* plus *F. graminearum* (quite resistant) and *F. oxysporum* (very sensitive), with the other seven strains occupying the three intermediate classes. The high resistance of most *Fusarium* strains to conventional fungicides led us to propose *C. majus* as a good source of substances useful for the treatment of fungal infections [58].

Table 2. Extract and essential oil from plant species with antifungal activity in toxigenic and phytopatogenic fungal and inhibition of mycotoxins production

PLANTS		MOLDS	MYCOTOXIN	REFERENCE NUMBER
Allium sativum	Extract and ajoene	*F. oxysporum*		59,60
Baldea boldus		*F. oxysporum*		61
Carum carvi	Essential oil	*F. verticillioides*	fumonisin	62,63
Chenopodium ambrosioides	Essential oil	*F. oxysporum*		64
Chenopodium majus	Aqueous extract	*F. culmorum* *F. oxysporum* *F. graminearum*		58
Cinnamomum zeylanicum	Essential oil	*F. verticillioides* *F. graminearum*	zearalenone DON	53, 65, 66
Citrus reticulata	Essential oil	*F. oxysporum*		67
Cuminun cyminum	Essential oil	*F. solani* *F. oxysporum F. verticillioides* *F. poae* *F. equiseti* *F. dimerium* *F. lateritium*		54, 68
Cymbopogon citratus	Essential oil Extract	*F. verticillioides* *F. proliferatum*		52, 55, 62, 69, 80, 81
Equisetum giganteum		*F. oxysporum* *F. verticillioides*		61
Flourensia microphylla, Flourensia cernua, and Flourensia retinophylla.	Essential oil	*F. oxysporum*		82
Foenicculum vulgare	Essential oil	*F. verticillioides* *F. solani* *F. oxysporum* *F. poae* *F. equiseti*	fumonisin	63
Heracleum persicum	Essential oil	*F. solani* *F. oxysporum F. verticillioides* *F. poae* *F. equiseti*		54
Hyptis suaveolens	Essential oil	*F. oxysporum*		83
Leptospermum scoparium	Essential oil	*F. circinatum F. verticillioides*		84
Lippia turbinata	Essential oil	*F. verticillioides*	fumonisin B_1	85, 86
Melissa officinalis		*F. verticillioides*		61
Menta viridicum	Essential oil	*F. verticillioides*	fumonisin	63

PLANTS		MOLDS	MYCOTOXIN	REFERENCE NUMBER
Ocimum basilicum	Essential oil	*F. verticillioides*	fumonisin	63
Ocimum gratissimum	Essential oil	*F. verticillioides*	fuminisin	55, 62, 87
Origanum vulgare	Essential oil	*F. verticillioides* *F. proliferatum* *F. graminearum* *F. solani*	fuminisin B_1	52, 57, 88, 89, 90
Pagostemon patchouli	Essential oil	*F. circinatum*		84
Passiflora coerulea	Essential oil	*F. verticillioides*		61
Pimenta dioica		*F. oxysporum* *F. verticillioides*		91
Pimpinella anisum	Essential oil	*F. verticillioides*	fumonisin	63
Thymus vulgaris	Essential oil	*F. oxysporum* *F. verticillioides* *F. solani*	fumunisin	55, 57, 63, 81, 92
Zataria multiflora	Essential oil	*F. solani* *F. oxysporum F. verticillioides* *F. poae* *F. equiseti*		54

Biocontrol

Because of the low effectiveness of fungicides to control members of the *Fusarium* spp., there have been efforts to identify biological antagonists, which could be used in integrated pest management (IPM) strategies [25].

Development of alternative control strategies as biocontrol can provide additional management tools as supplements to other control Fusarium wilt disease, for use with other control measures [93]. A variety of bacteria or fungal, yeast and moulds are able to degrade or remove mycotoxins from foods and also can restrict several fungal growths.

Fusarium wilt disease in melon (*Cucumis melo* L.) was performed utilizing organic fertilizer (BIOS) and different antagonistic microbes. BIOs decreased the incidence of Fusarium wilt disease and increased melon yield. The disease incidence of treatments with double application (BIOs applied both in the nursery and in the pot soil) was 20%, much lower than control (80%). Moreover, double application of amix termed BIOII resulted in a 17.4-fold increase in plant fresh weight compared to control. Application of BIOs strongly reduced the number of pathogen colony-forming units in stems and roots of melon. Many antagonistic strains have been proved to be effective biocontrol agents in controlled laboratory or greenhouse conditions. *Rhizoctonia* spp., *Bacillus* spp., *Penicillium* spp., *Aspergillus* spp., *Trichoderma* spp. are a few among the long lists [94].

There are many examples in the literature of using nonpathogenic *Fusarium* spp. to control Fusarium wilts [95]. In Weld and greenhouse studies, *F. oxysporum* strain CS-20 has demonstrated potential for reducing the incidence of Fusarium wilt on tomato, muskmelon and basil [93]. To successfully integrate biocontrol organisms into production systems, the compatibility of these organisms with pesticides used in the production system must be known.

Currently available and affordable methods of Fusarium head blight control are only partially effective. Therefore, there is an opportunity for biological control to become a component in the integrated management of this disease. There are several approaches: spraying of antagonists at flowering time treatment of infected seed with antagonists, and combating inoculum production on crop residues by application of organisms that are specialized in degradation or have a high colonization potential. Antagonist application at flowering time is most studied. However, as with fungicides against Fusarium head blight, timing of application will largely influence the effect [24].

Wagacha and Muthomi [8] relate that isolates of *Clostachys rosea* have been shown to consistently suppress sporulation of *F. culmorum* and *F. graminearum* on wheat straw and of *F. culmorum, F. graminearum, F. proliferatum* and *F. verticillioides* on maize stalks. A strain of *F. equiseti* has been shown to consistently decrease DON (470%) on wheat inoculated with *F. culmorum* with similar performance to a standard fungicide tebuconazole [96, 97]. Diamond and Cooke [98] reported a 60% reduction in FHB symptoms relative to control treatment after 25 days on ears pre-inoculated with *Phoma betae* and challenged with *F. culmorum*. They further reported a significant increase in number of grains per ear of wheat ears preinoculated with *Pythium ultimum* and *Phoma betae*. Two strains of *Pseudomonas fluorescens* have been reported to inhibit growth of *F. culmorum* both in vivo and in vitro [99].

The genus *Bacillus* is known for the production of over 24 antibiotics, several of which are fungicidal with potential control of plant diseases [100]. *Bacillus mojavensis* was patented and it produced Leu(7)-surfactin, which was toxic to *F. verticillioides* [100]. The bacterium, a strain from a subgroup of *Bacillus subtilis* operates under the principle of competitive exclusion, reducing the fumonisin concentration in plant [101]. *B. subtilis* CE1 inhibited *F. verticillioides* growth and fumonisin B_1 accumulation in vitro was assayed [102]. *Bacillus subtilis* can inhibit the fungal growth during their endophytic phase. *Bacillus amyloliquefaciens* and *Microbacteriumoleovorans* reduced the *F. verticillioides* count and significantly decreased fumonisin B_1 and B_2 levels in maize grains [103].

Propionibacteria cultures (*Propionibacterium freudenreichii* ssp.) were tested for their ability to inhibit the growth and mycotoxin production of *F. verticillioides*, *F. graminearum*, *F. culmorum* and *Alternaria alternata*. The strongest inhibition of growth and toxin production was observed in the presence of cultures containing viable cells and supernatants obtained from propionibacteria cultures [104].

Bacterial isolates from Mississippi maize field soil and maize rhizosphere samples were evaluated for their potential as biological control agents against *Aspergillus flavus* and *F. verticillioides*. It was identified 221 strains that inhibited growth of both fungi. The most prevalent genera isolated from rhizosphere samples were *Burkholderia* and *Pseudomonas*, and the most prevalent genera isolated from nonrhizosphere soil were *Pseudomonas* and *Bacillus* [105].

Cavaglieri et al. [106] investigated the potential for the biological control of *F. verticillioides* by *Enterobacter cloacae*, *Microbacterium oleovorans*, *Pseudomonas solanacearum* and *Bacillus subtilis*, and the combination *E. cloacae* and *M. oleovorans* has the best results.

Eleven competitive rhizobacterial isolates (Arthrobacter globiformis RC1, Azotobacter armeniacus RC2, A. armeniacus RC3, A. globiformis RC4, A. globiformis RC5, A. armeniacus RC6, Pseudomonas solanacearum RC7, Bacillus subtilis RC8, B. subtilis RC9, P.

solanacearum RC10, B. subtilis RC11) were tested in against F. verticillioides. All bacterial antagonists reduced fumonisin B_1 production at all levels tested [107].

In addition to B. subtilis, an isolate of a species of the fungus Trichoderma spp. is being tested for postharvest control of fumonisin production in kernels during storage [100]. Culture filtrates of Trichoderma viride and Trichoderma harzianum were inhibitory of F. verticillioides [104]. T. viride can suppress fumonisin B_1 production by F. verticillioides, thereby functioning to control mycotoxin production and the fungal growth [108].

The ability of Lactobacillus and Propionibacterium strain to remove common Fusarium toxins, trichothecenes (deoxynivalenol, 3-acetyldeoxynivalenol, nivalenol, fusarenon, diacetoxyscirpenol, T-2 toxin and HT-2 toxin), from liquid media was studied. Lactobacillus rhamnosus strain GG and Propionibacterium freudenreichii spp. shermanii JS were more efficient than Lactobacillus rhamnosus strain LC-705 in removing the toxins from the liquid media [109].

F. graminearum infects wheat heads at flowering, causing the disease Fusarium head blight (FHB). The pathogen often produces deoxynivalenol resulting in grain size and quality reduction [110]. Deoxynivalenol is a trichothecene secondary metabolite produced by Fusarium species infecting cereal crops. As a mycotoxin, DON causes losses in livestock production and poses a health risk to humans consuming contaminated cereal products [111]. Microbial isolates obtained from wheat anthers were screened for their ability to utilize tartaric acid, a compound that is poorly utilized by F. graminearum and could be utilized in formulations of biological control agents. Four strains that utilized tartaric acid and three that did not were effective in reducing FHB disease severity by up to 95% in greenhouse and 56% in field trials [110].

A mixed microbial culture capable of metabolizing deoxynivalenol was obtained from soil samples by an enrichment culture procedure. strain E3-39 was classified as a bacterium belonging to the Agrobacterium-Rhizobium group removed exogenously supplied deoxynivalenol from culture medium after incubation for 1 day and it 39 was also capable of transforming 3-acetyldeoxynivalenol but not nivalenol and fusarenon-X [112].

Bacterial communities isolated from 17 of 20 samples of soils and waters with widely diverse geographical origins utilized T-2 toxin as a sole source of carbon and energy for growth. These isolates readily detoxified T-2 toxin as assessed by a *Rhodotorula rubra* [113]. *Lactobacillus paracasei* subsp. *tolerans*, isolated from a traditional sourdough bread culture and previously shown to have antifungal activity against *F. proliferatum* M 5991 and M 5689 and *F. graminearum* R 4053 in a liquid medium [114]. Addition of *Bacillus coagulans* to *F. culmorum* culture inhibits the DON production. The strongest inhibition of fungal growth was noticed when *Bacillus coagulans* was co-inoculated at the beginning of culture. Estimation of ergosterol level as a determinant of fungal growth showed the greatest degree of *Fusarium* spp. Inhibition [115]. Very researches are necessary to know the real possibility of microbial antagonist on the *Fusarium* spp. control infection.

RESISTANT CULTIVARS

During pathogen-host interaction occurs the activation of defense system of plant in various ways, resulting in the production of toxic substances, preventing the establishment of these. Some compounds produced by plants have antimicrobial activity, while others restrict the development of pathogens by the formation of structural barriers [116]. According Durrant and Dong [117] the systemic acquired resistance (SAR), characterized by the expression of genes related to the production of PR proteins (pathogenesis related), considered as important factors in plant resistance to various classes of pathogens, including fungi, bacteria, viruses and oomycetes. Genes involved in the different routes of synthesis of defense chemicals, as well as the interactions between the pathways involving salicylic acid (SA) and jasmonate / ethylene.

Several substances can induce resistance of the plant fungal infection. Study of the efficiency of abiotic inducers to reduce the severity of Fusarium wilt were carried out by several authors using acibenzolar-S-methyl (ASM), β-aminobutyric acid (BABA) and chitosan [118, 119, 120]..

The morphological characteristics also help protect against infection. In wheat, tall plants without awns, and plants with lax ears tend to have lower rates of natural infection than do plants that are short, or have a high spikelet density [121].

Quantitative trait loci (QTLs) for resistance are often coincident for genes controlling morphological characteristics. As the majority of common wheat cultivars are susceptible to Fusarium head blight (FHB) [120], the few available resistance sources are poorly adapted [24].

Schroeder and Christensen [122] classified FHB resistance as type I (resistance against initial infection) and type II (resistance to the spread of pathogen within a spike) and described additional types of FHB resistance in wheat [121]. Due to the poor agronomic characters of many FHB-resistant varieties, genotype, environment interaction and the high cost and tediousness of field screening, progress in breeding for FHB resistance using traditional methods has been slow [123].

Although several species of *Fusarium* are capable of inciting FHB in wheat, *F. graminearum* and *F. culmorum* are the dominant species [124]. Breeding resistant cultivars is the method of choice for disease control. Currently a key strategy for successful FHB resistance breeding involves the identification of resistance genes and their introduction into regionally adapted varieties [125]. However, breeding for FHB resistance is difficult for several reasons: (1) the most resistant germplasm is of exotic origin and poorly adapted to local growing conditions, (2) the inheritance is oligogenic or polygenic and (3) screening for FHB resistance is difficult [123].

Kang and Buchenauer [126] reported that FHB resistant cultivars were able to develop active defense reactions during infection and spreading of *F. culmorum* in the host tissues. The researchers also reported lower accumulation of DON in the tissues of infected spikes of resistant wheat cultivars. Miller et al. [127] found that resistant cereal lines inoculated with *F. graminearum* contained lower concentrations of DON in the grain than susceptible cultivars. This suggests that it could be possible that a resistance mechanism that neutralizes DON production [128].

Effects of dominance of genes probably influence FHB resistance, but additive effects appear to be important, and resistance genes can be accumulated [129, 130]. Recent studies on chromosomal location of FHB resistance genes/loci in the wheat genome have used RFLP and AFLP techniques and recombinant inbred lines as a mapping population [24].

According to Snijders [24], there is an urgent need for head blight resistance in future wheat varieties, whether it is attained by either conventional breeding, with the use of molecular markers or a transgenic approach, or a combination of approaches. A higher level of Fusarium head blight resistance will guarantee lower risks for the farmer of crop loss due to reduced grain yield, low quality and mycotoxin contamination.

Cultural Control

Various studies have identified different elements concerning the effects of agricultural practices (crop rotation, crop management) in fungal infection. The effects of crop rotation have been studied in detail. They depend on the preceding crop, whether that crop is a potential host for the pathogens responsible, and the frequency of the crop concerned in the rotation. For example the shorter the rotation, the higher the frequency of head blight [9].

According to Jouany [25] the head blight is most frequent when the susceptible crop occurs frequently within the rotation. The density of crop residues left in the field, their nutritional value [88] and pathogen competition may also modify the effect of crop rotation [131, 132].

Cultural management options include crop rotation, appropriate use of fertilizers, irrigation, weed control, proper land preparation, and timely harvesting.

Fusarium species that infect cereals are capable of surviving saprophytic ally on crop debris and ploughing to bury crop debris removes the source of inoculum from the soil surface, which could be available for dispersal [25, 124, 133].

The pathogen (*Fusarium*) can exist in soil for several years through the production of chlamydospores [91], and is thus difficult to control. Traditionally crop rotations have been proven to be an effective strategy to control many soil borne diseases, but because most of these pathogens (like *F. oxysporum* f. sp. *melonis*) can survive for a long period of time, the effectiveness of crop rotation is limited once a disease outbreak occurs [135].

Soil solarization is an effective strategy too but is not readily applicable for intensive vegetable farming systems where time to solarize soil is very limited. Furthermore, soil solarization is often limited by local climate constrains [136].

Use of reduced tillage practices increases incidence and severity of FHB. The effect of soil preparation (tillage), nitrogen fertilization levels, crop rotation, previous crop, intensity of cultivation and weed control play an important role in FHB infection [8]. It is critical to avoid planting wheat adjacent to fields with large amounts of small grain or corn residue remaining on the soil surface. No-till planting wheat into corn residues substantially increases FHB infection. Rotation to a legume crop between corn and small grain crops provides time for the residues to break down and the pathogen population decline [9].

Conclusion

Fungi of the genus *Fusarium* are important pathogens of small grains and cereals, especially in the temperate regions. The pathogen is a soil-inhabiting fungus that is a competitive saprophyte and facultative parasite. Infection with *Fusarium* is influenced by initial natural inoculum supply, resistance level of the plant, variety and climatic conditions. Meteorological, environmental and agronomic factors could affect *Fusarium* disease and mycotoxins production. For this reason, it is difficult to predict the occurrence of fungal diseases and toxin contamination in grains and cereals. A decreased disease incidence in the field ensures a decreased mycotoxin content of the grain. As a result of the pathogen persistence, effective control of this disease has generally been achieved only by the use of disease-resistant cultivars. Integrated management is a good approach for *Fusarium* spp. and mycotoxins controls.

REFERENCES

[1] Booth, C. *The Genus Fusarium*. Commonw Mycol. Inst., Kew, England, 1971, 58 p.

[2] Hawksworth, DL; Kirk, PM; Sutton, BC; Pegler, DN. *Ainsworth and Bisby's dictionary of the fungi*. 8.ed. Oxon, UK, CAB International, 1995, 650p.

[3] Monteiro, SA. Análise genômica e sequenciamento automático de rDNA em populações de *Fusarium oxysporum*. 2004. 112p. Dissertação (Mestardo em Química e Biotecnologia). Universidade Federal de Alagoas, Maceió, 2004.

[4] Uhlig, S; Jestoi, M; Parikka, P. *Fusarium avenaceum* - The North European situation. *International Journal of Food Microbiology*, 2007, 119, 17-24.

[5] Osborne, LE; Stein, JM.. Epidemiology of Fusarium head blight on small-grain cereals. *International Journal of Food Microbiology*, 2007, 119, 103–108.

[6] Suárez-Estrella, F; Vargas-García, MC; López, MJ; Moreno, J. Survival of *Fusarium oxysporum* f.sp. melonis on plant waste. *Crop Protection*, 2004, 23, 127–133.

[7] Glenn, AE.. Mycotoxigenic *Fusarium* species in animal feed. *Animal Feed Science and Technology*, 2007, 137, 213–240.

[8] Wagacha, JM; Muthomi, JW. *Fusarium culmorum*: Infection process, mechanisms of mycotoxin production and their role in pathogenesis in wheat. *Crop Protection*, 2007, 26, 877–885.

[9] Champeil, A; Doré, T; Fourbet, JF. *Fusarium* head blight: epidemiological origin of the effects of cultural practices on head blight attacks and the production of mycotoxins by *Fusarium* in wheat grains. *Plant Science*, 2004, 166, 1389–1415.

[10] Sutton, JC. Epidemiology of wheat head blight and maize ear rot caused by *Fusarium graminearum*. *Canadian Journal Plant Pathology*, 1982, 4, 195–209.

[11] Bateman, GL; Murray, G. Seasonal variations in populations of *Fusarium* species in wheat-field soil. *Applied Soil Ecology*, 2001, 18, 117–128.

[12] Neergaard, P. *Seed pathology*. London: McMillan, 1979, v.1, 839p.

[13] Burgess, L. General ecology of the Fusaria. In Nelson, PE; Toussoun, A R; Cook, J. (editors). *Fusarium: Diseases,Biology and Taxonomy*, The Pennsylvania State University: University Park and London, 1981. pp. 225-235.

[14] Camargo, LEA; Pereira, OAP; De Carvalho, RV. Doenças do milho. In: Kimati, H; Amorim, L; Bergamin Filho, A; Camargo, LEA; Rezende, JAM (editors). *Manual de fitopatologia: doenças das plantas cultivadas*. 4. ed. São Paulo: Agronômica Ceres, 2005, v.2, pp. 477-488.

[15] Agrios, GN. *Plant Pathology*. 5th ed., San Diego: Academic Press, 2005, 922 p.

[16] Fernandes, FT; Oliveira, E; Casela, CR; Ferreira, AS; Pinto, NFJA. Doenças de Milho, Podridões do colmo e das raízes. Sistema de Produção 1. *Embrapa Milho e Sorgo*. 2006.
[17] Alexopoulos, CJ; Mims, CW; Blackwell, M. *Introductory Mycology*. 4th ed. United States: JW. and Sons Inc., 1996, 880 p.
[18] Gordon, TR; Martyn, RD. The evolutionary biology of *Fusarium oxysporum*. *Annual Review of Phytopathology*, 1997, 35, 111–128.
[19] Guarro J; Gené J. *Fusarium* infections. Criteria for the identification of the responsible species. *Mycoses*, 1992, 35, 109-114.
[20] Nelson, PE; Dignani, MC; Anaissie, EJ. Taxonomy, biology and clinical aspects of *Fusarium* species. *Clinical Microbiology Review*, 1994, 7, p. 479-504.
[21] Guarro, J; Gené, J. Opportunistic Fusarial infections in humans. *Eurupean Journal of Clinical Microbiology and Infectious Disease*, 1995, 14, 741-754.
[22] Pascoe, IG. *Fusarium* morphology I: Identification and characterization of a third conidial type, the mesoconidium. *Mycotaxon*, 1990, 37, 121-160.
[23] Samson, RA; Von Reenen-Hoekstra, ES. *Introduction to food-borne fungi*. 3th ed. Baarn, Delft. Centralbureau voor Schmmelculturs, 1995. 248 p.
[24] Nelson, PE; Toussoun, TA.; Cook, RJ. (eds.). *Fusarium*: Diseases, Biology, and Taxonomy. Pennsylvania State Press, University Park, PA.1981. 457 p.
[25] Snijders, CHA. Resistance in wheat to *Fusarium* infection and trichothecene formation. *Toxicology Letters*, 2004, 153, 37–46.
[26] Jouany, JP. Methods for preventing, decontaminating and minimizing the toxicity of mycotoxins in feeds. *Animal Feed Science and Technology*, 2007, 137, 342-362.
[27] Marín, S; Homedes, V; Sanchis, V; Ramos, AJ; Magan, N. Impact of *Fusarium moniliforme* and *F. proliferatum* colonisation of maize on calorific losses and fumonisin production under different environmental conditions. *Journal of Stored Products Research*, 1999, 35, 15-26.
[28] Maiorano, A; Reyneri, A; Sacco, D; Magni, A; Ramponi, C. A dynamic risk assessment model (FUMAgrain) of fumonisin synthesis by *Fusarium verticillioides* in maize grain in Italy. *Crop Protection*, 2009, 28, 243–256.
[29] Kokkonen, M; Ojala, L; Parikka, P; Jestoi, M. Mycotoxin production of selected Fusarium species at different culture conditions. *International Journal of Food Microbiology*, 2010, 143, 17–25.
[30] Felicio, JD; Freitas, TE; Rossi, MH; Gonçalez, E. Mycotoxins detection by chromatography. *Recent Patents on Food and Agriculture*, 2011 (in press).
[31] Niessen, L. PCR-based diagnosis and quantification of mycotoxin producing fungi. *International Journal of Food Microbiology*, 2007, 119, 38–46.
[32] Osweiler, GD. Equine Mycotoxicosis. In Gonçalez, E; Felicio, JD; Aquino, S (editors). *Mycotoxicoses in Animals Economically Importants*, Nova Science Publisher, NY, 2011, pp. 1-30.
[33] Mallmann, CA; Dilkin, P; Rauber, RH; Giacomini, LZ. Mycotoxicosis in Swine. In Gonçalez, E; Felicio, JD; Aquino, S (editors). *Mycotoxicoses in Animals Economically Importants*, Nova Science Publisher, NY, 2011, pp. 64-91.
[34] Osweiler, GD. Mycotoxins contemporary issues of food animal health and productivity. *Veterinary Clinics of North America: Food and Animal Practice*, 2000, 16, 511–530.

[35] Gonçalez, E; Aquino, S; Felicio, JD. Chemical and biological approaches for mycotoxin control: a review. *Recent Patents on Food Nutrition and Agriculture*, 1, 2009, 155-161.

[36] Riley, RT; Norred, WP. Mycotoxin prevention and decontamination: a case study on maize. *Food Nutrition and Agriculture,* 23, 1999, 31-36.

[37] Lyon, GD; Relinski, T; Newton, AC. Novel disease control compounds: the potential to immunize" plants against infection. *Plant Pathology*, 44, 1995, 407-427.

[38] Ghini, R; Kimati, H. Resistência de fungos a fungicidas. Jaguariúna: Embrapa Meio Ambiente, 2000. 78p.

[39] Haidukowski, M; Pascale, M.; Perrone, G; Pancaldi, D; Campagna, C; Visconti, A. Effect of fungicides on the development of *Fusarium* head light, yield and deoxynivalenol accumulation in wheat inoculated under field conditions with *Fusarium graminearum* and *Fusarium culmorum*. *Journal of the Science of Food and Agriculture*, 85, 2005, 191–198.

[40] Matthies, A; Buchenauer, H. Effect of tebuconazole (Folicur) and prochloraz (Sportak) treatments on *Fusarium* head scab development, yield, and deoxynivalenol (DON) content in grains of wheat following artificial inoculation with *Fusarium culmorum, Journal of Plant Diseases and Protection,*107, 2000, 33–52.

[41] Pirgozliev, SR; Edwards, SG; Hare, MC; Jenkinson, P. Strategies for the control of *Fusarium* head blight I cereals. *European Journal of Plant Pathology*, 109, 2003, 731–742.

[42] Jennings, P; Turner, JA; Nicholson, P. Overview of *Fusarium* ear blight in the UK; effect of fungicide treatment on disease control and mycotoxin production. In: Proceedings of the Brighton Crop Protection Conference; Pests and Diseases, British Crop Protection Council, Farnham, vol. 2, 2000, pp. 707–712.

[43] Simpson, DR; Weston, GE; Turner, JA; Jennings, P; Nicholson, P. Differential control of head blight pathogens of wheat by fungicides and consequences for mycotoxin contamination of grain. *European Journal of Plant* Pathology, 107, 2001, 421–431.

[44] Pirgozliev, SR; Edwards, SG; Hare, MC; Jenkinson, P. Effect of dose rate of azoxystrobin and metconazoleon the development of *Fusarium* head blight and the accumulation of deoxynivalenol (DON) in wheat grain. *European Journal of Plant Pathology*, 108, 2002, 469–478.

[45] Elmholt, S; Frisvad, JC; Thrane, U. The influence of fungicides on soil mycoflora with special attention to tests of fungicide effects on soil-borne pathogens. In: Altman, J. (Ed.), Pesticide Interactions in Crop Production: Beneficial and Deleterious Effects. CRC Press, New York, 1993, pp. 227–243.

[46] Knudsen, IMB; Elmholt, S; Hockenhull, J; Jensen, DF. Distribution of saprophytic fungi antagonistic to *Fusarium culmorum* in two differently cultivated field soils, with special emphasis on the genus Fusarium. *Biological Agriculture and Horticulture*, 12, 1995, 61–79.

[47] Mcmullen, M; Jones, K; Gallenberg, D. Scab of wheat and barley: a reemerging disease of devastating impact. *Plant Disease*, 81, 1997, 1340-1348.

[48] Stangarlin, JR, Schwan – Estrada, KRF, Cruz, M. E. S., Nozak, M. H. Plantas medicinais. *Biotecnologia Ciencia and Desenvolvimento*, 11, 1999, 16-21.

[49] Felicio, JD; Aquino, S; Gonçalez, E. Methods for preventing and control mycotoxins and mycotoxigenic fungi. In: Mycotoxicoses in Animal Ecomonomically Important, Nova Science Publishers, Inc., New York. 2011, pp 139-178.

[50] Guynot, ME; Marin, S; Seto, L.; Sanchis, V; Ramos, AJ. Screening for antifungal activity of some essential oils against common spoilage fungi of bakery products. Food *Science Technology International,* 11, 1992, 25–32.

[51] Nogueira, JHC; Gonçalez, E; Galetti, SSR ; Facalani, R; Marques, MOM; Felicio, JD. *Ageratum conyzoides* essential oil as aflatoxin suppressor of *Aspergillus flavus. International Journal of Food Microbiology*, 137, 2010, 55-60.

[52] Velluti, A; Marín, S; Gonzalez, P; Ramos, AJ; Sanchis, V. Initial screening for inhibitory activity of essential oils on growth of *Fusarium verticillioides, F. proliferatum* and *F. graminearum* on maize-based agar media. *Food Microbiology, 21, 2004, 649-656.*

[53] Marın, S; Velluti, A; Ramos, AJ; Sanchis, V. Effect of essential oils on zearalenone and deoxynivalenol production by *Fusarium graminearum* in non-sterilized maize grain. *Food Microbiology*, 2004, 21, 313–318.

[54] Naeini, A; Ziglari, T; Shokri, H; Khosravi, AR. Assessment of growth-inhibiting effect of some plant essential oils on different *Fusarium* isolates *Journal of Medical Mycology, 20, 2010, 174-178.*

[55] *Nguefack, J; Leth, V; Zollo, PHA; Mathur, SB. Evaluation of five essential oils from aromatic plants of Cameroon for controlling food spoilage and mycotoxin producing fungi.* International journal of food microbiology, *2004, 94, 329-334.*

[56] Dambolena, JS; López, AG; Cánepa, MC; Theumer, MG; Zygadlo, JA; Rubinstein, HR. Inhibitory effect of cyclic terpenes (limonene, menthol, menthone and thymol) on *Fusarium verticillioides* MRC 826 growth and fumonisin B1 biosynthesis. *Toxicon*, 2008, 51, 37-44.

[57] Daferera, DJ; Ziogas, BN; Polissiou, MG. The effectiveness of plant essential oils on the growth of *Botrytis cinerea*, *Fusarium* sp. and *Clavibacter michiganensis* subsp. *michiganensis*. *Crop Protection*, 22, 2003, 39-44.

[58] Matos, J; Baeta, OC; Silva, MJ; Pinto Ricardo CP. Sensitivity of *Fusarium* strains to *Chelidonium majus* L. extracts. *Journal of Ethnopharmacology*, 46, 1999, 151-158.

[59] Singh, UP; Pandey; VN; Wagner, KG; Singh, KP. Antifungal activity of ajoene, a constituent of garlic (Allium sativum). *Canadian Journal of Botany*, 68, 1990, 1354-1356.

[60] *Singh, UP*; Pathak, KK; Khare, MN; *Singh,* RB. Effect of leaf extract of garlic on Fusarium oxysporum f.sp. ciceri, Sclerotinia sclerotiorum and on gram seeds. *Mycologia,* 71, 1979, 556-563.

[61] Rizzo, I; Vedoya, G; Maurutto, S; Haidukowski, M; Varsavsky, E. Assessment of toxigenic fungi on Argentinean medicinal herbs. *Microbiological Research*, 2004, 159, 113–120.

[62] Nguefack, J; Dongmo, JBL; Dakole, CD; Leth, V; Vismer, HF; Torp, J; Guemdjom, EFN; Mbeffo, M; Tamgue, O; Fotio, D; Zollo, PHA; Nkengfack, AE. Food preservative potential of essential oils and fractions from *Cymbopogon citratus, Ocimum gratissimum* and *Thymus vulgaris* against mycotoxigenic fungi. *International Journal of Food Microbiology*, 2009, 131, 151-156.

[63] Soliman, KM; Badeaa, RI. Effect of oil extracted from some medicinal plants on different mycotoxigenic fungi. *Food and Chemical Toxicology*, 2002, 40, 1669-1675.

[64] Kumar, R; Mishra, AK; Dubey, NK; Tripathi, YB. Evaluation of *Chenopodium ambrosioides* oil as a potential source of antifungal, antiaflatoxigenic and antioxidant activity. *International Journal of Food Microbiology*, 2007, 115, 159-164.

[65] Singh, HNP; Prasad, MM; Sinha, KK. Efficacy of leaf extracts of some medicinal plantas against disease deve lop ment in banana. *Letters in Applied Microbiology,* 1993, 17, 269-271.

[66] Marın, S; Velluti, A; Ramos, AJ; Sanchis, V. Effect of essential oils on zearalenone and deoxynivalenol production by *Fusarium graminearum* in non-sterilized maize grain. *Food Microbiology, 21, 2004, 313-318.*

[67] Chutia, M; Bhuyan, PD; Pathak, MG; Sarma, TC; Boruah, P. Antifungal activity and chemical composition of *Citrus reticulata* Blanco essential oil against phytopathogens from North East India Original. *Food Science and Technology*, 42 *2009, 77-780.*

[68] Hashem, M; Moharam, AM; Zaied, AA; Saleh, FEM. Efficacy of essential oils in the control of cumin root rot disease caused by *Fusarium* spp. *Crop Protection, 29, 2010, 1111-1117.*

[69] Mishra, AK; Dubey, NK. Evaluation of some essential oils for their toxicity against fungi causing deterioration of stored food commodities. *Applied and Environmental Microbiology*, 1994, 60, 1101-1105.

[70] Chinaphuti, A; Aukkasarakul, S. Inhibition of *Aspergillus flavus* growth and aflatoxin production using herbal extracts. *Journal of Biotechnology*, 2008, *136, 152.*

[71] Razzaghi-Abyaneh, M; Shams-Ghahfarokhi, M; Yoshinari, T; Rezaee, M; Jaimand, K; Nagasawa, H; Sakuda, S; Tzortzakis, NG; Economakis, CD. Antifungal activity of lemongrass (*Cympopogon citratus* L.) essential oil against key postharvest pathogens. *Innovative Food Science and Emerging Technologies*, 2007, 8, 253-258.

[72] Rodríguez, D. J.; Hernández-Castillo, D.; Angulo-Sánchez, J.L.; Rodríguez-García, R.; Quintanilla, J.A. V.; Lira-Saldivar R.H. Antifungal activity in vitro of *Flourensia* spp. extracts on *Alternaria* sp., *Rhizoctonia solani*, and *Fusarium oxysporum. Industrial Crops and Products, 25, 2007, 111-116.*

[73] Sharma, N; Tripathi, A. Integrated management of postharvest *Fusarium* rot of gladiolus corms using hot water, UV-C and *Hyptis suaveolens* (L.) Poit. Essential oil *Postharvest Biology and Technology*, 47, 2008, 246-254.

[74] Lee, Y. S; Kim, J; Lee, SG; Oh, E; Shin, SC; Park, IK. Effects of plant essential oils and components from Oriental sweetgum (Liquidambar orientalis) on growth and morphogenesis of three phytopathogenic fungi. *Pesticide Biochemistry and Physiology*, 93, 2009, 138-143.

[75] Juglal, S; Govinden, R; Odhav, B. Spice oils for the control of co-occurring mycotoxin producing fungi. *Journal of Food Protection*, 2002, 65, 683–687.

[76] Bluma, R; Amaiden, MR; Etcheverry, M. Screening of Argentine plant extracts: Impact on growth parameters and aflatoxin B_1 accumulation by *Aspergillus* section *Flavi. International Journal of Food Microbiology*, 2008, 122, 114-125.

[77] Dambolena, JS; Zunino, MP; López, AG; Rubinstein, HR; Zygadlo, JA; Mwangi, JW; Thoithi, GN; Kibwage, IO; Mwalukumbi, JM; Kariuki, ST. Essential oils composition of *Ocimum basilicum* L. and *Ocimum gratissimum* L. from Kenya and their inhibitory

effects on growth and fumonisin production by *Fusarium verticillioides. Innovative Food Science and Emerging Technologies*, 11, 2010, 410-414.

[78] Paster, N; Menasherov, M; Ravid, U; Juven, B. Antifungal activity of oregano and thyme essential oils applied as fumigants against fungi attacking stored grain. *Journal Food Protection*, 1995, 58, 81–85.

[79] Novak, J; Christina, B; Langbehn, B; Pank, F; Skoula, M; Gotsiou Y. Ratios of *cis-* and *trans-*sabinene hydrate in *Origanum marjoram* L. and *Orignaum midrophyllum* (Bentham) Vogel. *Biochemical Systematics and Ecology*, 2000, 28, 697–704.

[80] Aligianis, N; Kalpoutzakis, E; Mitaku, S; Chinou, IB. Composition and antimicrobial activity of the essential oil from *Origanum* species. *Journal of Agricultural and Food Chemistry*, 2001, 49, 4168–4170.

[81] Zabka, M; Pavela, R; Slezakova, L. Antifungal effect of *Pimentadioica* essential oil against dangerous pathogenic and toxinogenic fungi. *Industrial Crops and Products*, 30, 2009, 250-253

[82] Kumar, A; Shukla, R; Singh, P; Prasad, CS; Dubey, NK. Assessment of *Thymus vulgaris* L. essential oil as a safe botanical preservative against post harvest fungal infestation of food commodities. *Innovative Food Science and Emerging Technologies*, 2008, 9, 575-580.

[83] Fravel, DR; Deahl, KL; Stomme J.R. Compatibility of the biocontrol fungus Fusarium oxysporum strain CS-20 with selected fungicides, *Biological Control,* 34, 2005, 165–169.

[84] Zhao, Q; Dong, C; Yang, X; Mei, X; Ran, W; Shen, Q; Xu, Y. Biocontrol of Fusarium wilt disease for Cucumis melo melon using bio-organic fertilizer. *Applied Soil Ecology* 47, 2011, 67–75.

[85] Luongo, L; Galli, M; Corazza, L; Meekes, E; De Haas, L;V an der Plas, CL; Kohl, J. Potential of fungal antagonists for biocontrol of Fusarium spp. in wheat and maize through competition in crop debris. *Biocontrol Science and Technology,* 15, 2005, 229–242.

[86] Dawson, W.A.J.M., Jestoi, M., Rizzo, A., Nicholson, P., Bateman, G.L. Field evaluation of fungal competitors of Fusarium culmorum and F. graminearum, causal agents of ear blight of winter wheat, for control of mycotoxin production in grain. *Biocontrol Science and Technology* 14, 2004, 783–799.

[87] Diamond, H; Cooke, BM. Preliminary studies on biological control of Fusarium ear blight complex of wheat. *Crop Protection,* 22, 2003, 99–107.

[88] Kurek, E; Jaroszuk, CS; Shtssls, JZ. Rye (Secale cereale) growth promotion by Pseudomonas fluorescens strains and their interactions with Fusarium culmorum under various soil conditions. *Biological Control,* 26, 2003, 48–56.

[89] Snook ME; Mitchell T; Hinton DM; Bacon CW. isolation and characterization of leu(7)-surfactin from the endophytic bacterium *bacillus mojavensis* rrc 101, a biocontrol agent for *fusarium verticillioides. Journal Agriculture and Food Chemistry*, 2009, 57, 4287-4292.

[90] Bacon, CW; Yates, IE; Hinton, DM; Meredith, F. Biological control of *Fusarium moniliforme* in maize. *Environmental Health Perspectives*, 2001, 109, Suppl 2, 325-332.

[91] Cavaglieri, L; Orlando, J; Rodríguez, MI; Chulze, S; Etcheverry, M. Biocontrol of *Bacillus subtilis* against *Fusarium verticillioides* in vitro and at the maize root level. *Research in Microbiology*, 2005, 156, 748-754.
[92] Sartori, M; Nesci, A; Etcheverry, M. Impact of osmotic/matric stress and heat shock on environmental tolerance induction of bacterial biocontrol agents against *Fusarium verticillioides*. *Research in Microbiology, 161, 2010, 681-686.*
[93] Gwiazdowska, D; Czaczyk, K; Filipiak, M; Gwiazdowski, R. Effects of Propionibacterium on the growth and mycotoxin production by some species of *Fusarium* and *Alternaria*. *Polish Journal of Microbiology*, 2008, 57, 205-212.
[94] Palumbo, JD; O'Keeffe, TL; Abbas, HK. Isolation of maize soil and rhizosphere bacteria with antagonistic activity against *Aspergillus flavus* and *Fusarium verticillioides*. *Journal of Food Protection*, 2007, 70, 1615-1621.
[95] Cavaglieri, L; Orlando, J; Etcheverry, M. In vitro influence of bacterial mixtures on *Fusarium verticillioides* growth and fumonisin B1 production: effect of seeds treatment on maize root colonization. *Letters Applied of Microbiology*, 2005, 41, 390-396.
[96] Cavaglieri, L; Passone, A; Etcheverry, M. Screening procedures for selecting rhizobacteria with biocontrol effects upon *Fusarium verticillioides* growth and fumonisin B1 production. *Research in Microbiology,* 2004, 155, 747-754.
[97] Yates, IE; Meredith, F; Smart, W; Bacon, CW; Jaworski, AJ. *Trichoderma viride* suppresses fumonisin B1 production by *Fusarium moniliforme*. *Journal of Food Protection*, 1999, 62, 1326-1332.
[98] El-Nezami, HS; Chrevatidis, A; Auriola, S; Salminen, S; Mykkänen, H. Removal of common *Fusarium* toxins in vitro by strains of *Lactobacillus* and *Propionibacterium*. *Food Additives and Contaminants*, 2002, 19, 680-686.
[99] Schisler, DA; Khan, NI; Boehm, MJ. Biological control of *Fusarium* head blight of wheat and deoxynivalenol levels in grain via use of microbial antagonists. *Advances in Experimental Medicine and Biology*, 2002, 504, 53-69.
[100] Völkl, A; Vogler, B; Schollenberger, M; Karlovsky, P. Microbial detoxification of mycotoxin deoxynivalenol. *Journal of Basic Microbiology*, 2004, 44, 147-156.
[101] Shima, J; Takase, S; Takahashi, Y; Iwai, Y; Fujimoto, H; Yamazaki, M; Ochi, K. Novel detoxification of the trichothecene mycotoxin deoxynivalenol by a soil bacterium isolated by enrichment culture. *Applied and Environmental Microbiology*, 1997, 63, 3825-3830.69.
[102] Beeton, S; Bull, AT. Biotransformation and detoxification of T-2 toxin by soil and freshwater bacteria. *Applied and Environmental Microbiology*, 1989, 55, 190-197.
[103] Hassan, YI; Bullerman, LB. Antifungal activity of *Lactobacillus paracasei* subsp. tolerans against *Fusarium proliferatum* and *Fusarium graminearum* in a liquid culture setting. *Journal of Food Protection*, 2008, 71, 2213-2216.
[104] Czaczyk, K; Trojanowska, K; Mueller, A. Antifungal activity of *Bacillus coagulans* against *Fusarium* spp. *Acta Microbiologuca Polonica*, 2002, 51, 275-283.
[105] Rodrigues, AAC; Neto, EB; Coelho, RSB. Induced resistance to *Fusariumoxysporum* f. sp. *tracheiphilum* in cowpea plants: effectiveness of abiotic inducers and elicited enzymatic activity. *Fitopatologia Brasasileira,* 31, 493-499, 2006.
[106] Durrant, WE; Dong, X. Systemic Acquired Resistance. *Annual Review of Phytopathology*, 42, 2004, 185-209.

[107] Benhamou, N.; Kloepper, J.W.; Tuzun, S. Induction of resistance against *Fusarium* wilt of tomato by combination of chitosan with an endophytic bacterial strain: ultrastructure and cytochemistry of the host response. *Planta*, 204, 1998, 153-168..

[108] Cole, DL. The efficacy of acibenzolar-S-methyl, an inducer of systemic acquired resistance, against bacterial and fungal disease of tobacco. *CropProtection,* 18, 1999, 267-273.

[109] Jakab, G; Cottier, V; Toquin, V; Rigoli, G; Zimmerli, L; Métraux, JP; Mauch-Mani, B. β-aminobutyric acid-induced in plants. *European Journal of Plant Pathology*, 107, 2001, 29-37.

[110] Mesterhazy, A, Selection of head blight resistant wheats through improved seedling resistance. *Plant Breeding,* 98, 1987, 25–36.

[111] Mesterhazy, A; Logrieco, A; Bailey, JA; Corazza, L. Cooke, BM, 2002. Role of deoxynivalenol in aggressiveness of *Fusarium graminearum* and *F. culmorum* and in resistance to Fusarium head blight. *European Journal of Plant Pathology* 108, 675–684.

[112] Schroeder, HW, Christensen, JJ, 1963. Factors affecting resistance of wheat to scab caused by *Gibberella zeae. Phytopathology,* 53, 831–838.

[113] Mardi, M; Buerstmayr, H; Ghareyazie, B; Lemmens, M; Mohammadi, SA; Nolz, R; Ruckenbaue, P. QTL analysis of resistance to Fusarium head blight in wheat using a 'Wangshuibai'-derived population. *Plant Breeding,* 124, 2005, 329—333.

[114] Parry, DW; Jenkinson, P; McLeod, L. Fusarium ear blight (scab) in small grain cereals -a review. *Plant Pathology*, 44, 1995, 207–238.

[115] Bai, GH; Shaner, G. Scab of wheat: prospects for control. *Plant Disease.* 78, 1994, 760–766.

[116] Kang, Z; Buchenauer, H. Immunocytochemical localization of Fusarium toxins in infected wheat spikes by *Fusarium culmorum. Physiological and Molecular Plant Pathology,* 55, 1999, 275–288.

[117] Miller, JD, Young, JC, Sampson, DR. Deoxynivaleno and Fusarium head blight resistance in spring cereals. *Journal of Phytopathology.* 111, 1985, 359–367.

[118] Snijders, CHA; Krechting, CF. Inhibition of deoxynivalenol translocation and fungal colonization in Fusarium head blight resistant wheat. *Canadian Journal of Botanic*, 70, 1992, 1570–1576.

[119] Snijders, CHA. Genetic variation for resistance to Fusarium head blight in bread wheat. *Euphytica*, 50, 1990, 171–179.

[120] Bai, GH; Kolb, FL; Shaner, G; Domier, LL. Amplified fragment length polymorphism markers linked to a major quantitative trait locus controlling scab resistance in wheat. *Phytopathology,* 89, 1999, 343–348.

[121] Dill-Macky, R; Jones, RK. The effect of previous crop residues and tillage on *Fusarium* head blight of wheat. *Plant Disease,* 84, 2000, 71–76.

[122] Jenkinson, P; Parry, DW. Isolation of *Fusarium* species from common broad-leaved weeds and their pathogenicity to winter wheat. *Mycology Research*, 98, 1994, 776–780.

[123] Jones, RK. Assessments of Fusarium head blight of wheat and barley in response to fungicide treatment. *Plant Disease,* 84, 2000, 1021–1030.

[124] Garret, SD. Pathogenic Root-Infecting Fungi. Cambridge University Press, London, UK. 1970.

[125] King, SR; Davis, AR; Liu, W; Levi, A., Grafting for disease resistance. *HortScience,* 43, 2008, 1673–1676.

[126] Tamietti, G; Valentino, D. Soil solarization as an ecological method for the control of fusarium wilt of melon in Italy. *Crop Protection*. 25, 2006, 389–397.

In: Fusarium
Editors: Tony Ferarri Rios and Emilia Reyes Ortega
ISBN: 978-1-61942-539-2

Chapter 3

FUSARIUM WILT IN CHICKPEA: GENERAL ASPECT AND MOLECULAR BREEDING

***Patricia Castro*[1], *Josefa Rubio*[2], *Teresa Millán*[1], *Juan Gil*[1], *and María José Cobos*[3]**

[1]Departamento de Genética, Universidad de Córdoba, Campus Rabanales Edificio C5 2ª Planta, Córdoba, Spain
[2]Área de Mejora y Biotecnología, IFAPA, Apdo,Córdoba, Spain
[3]Departamento de Mejora Genética, IAS-CSIC, Córdoba, Spain

ABSTRACT

Fusarium wilt is a widespread and serious chickpea disease caused by the soil-borne fungus *Fusarium oxysporum* f.sp. *ciceris*. Eight physiological races of the pathogen (0, 1A, 1B/C, 2, 3, 4, 5 and 6) have been reported. Persistence of the pathogen in soil and its capacity to survive there for years even in the absence of host renders its control difficult. The use of resistant cultivars is the most effective, economical and environmentally friendly method of managing the disease. Therefore breeding for resistance to Fusarium wilt has been an important objective of many chickpea research programmes. Systematic screening of germplasm collections have identified useful resistance sources that have been used successfully to breed resistant cultivars. Besides, genetic studies have revealed that resistance genes to Fusarium wilt is governed by one two or three genes depending on the race. Genetic linkage analyses and quantitative trait loci (QTL) mapping have identified molecular markers that could be useful for marker-assisted selection (MAS) and gene pyramiding. On the other hand, comparative genomics and synteny studies with model species and closely related legumes promises to further advance the knowledge of the chickpea genome and provide breeders with additional genes and selectable markers for use in MAS. In this chapter we will review the current state of molecular breeding for Fusarium wilt in chickpea. Special emphasis is given on the application of marker technology, and QTL analysis for MAS in this species.

1. INTRODUCTION

Chickpea (*Cicer arietinum* L.) is the second most important grain legume after beans in terms of total harvested area cultivated worldwide. It is grown across the world, in both semi-arid and sub-tropical areas, India being the largest producer with the 67.5% of total world production (FAOSTAT, 2010). Chickpea is self-pollinated, diploid (2n = 2x =16) and annual, it has a relatively small genome (740 Mbp; Arumuganathan and Earle 1991). It belongs to the legume family (Fabaceae) and is the only cultivated species within the genus *Cicer*. Indeed, chickpea is mainly used for food as it is an excellent source of proteins, particularly for vegetarian diets (Taylor and Ford 2007).

Breeders have classified chickpea into two main types: 'desi' and 'kabuli'. Desi types have pink flowers and seeds are generally small, angular and relatively dark with a thick, rough coat. They are mostly cultivated in India, Pakistan, East Africa and, more recently, Australia. Kabuli types have white flowers, and large, light-coloured seeds with a thin, smooth coat. These latter types are mainly cultivated in the Mediterranean basin, Near East, Central Asia and America. The two types represent different genetic backgrounds, differing in important agronomic traits such as disease resistance, cold tolerance and growth habit (Singh 1987).

Though the average chickpea yield has increased in recent years, reaching 952.5 kg/ha, it is still relatively low. Instability in production and low yields are due in great part to biotic and abiotic stress. Two major sources of biotic stress worldwide in chickpea are the fungal diseases Ascochyta blight, caused by *Ascochyta rabiei* (Pass.) Lab., and Fusarium wilt, which is caused by the vascular deuteromycetes pathogen *Fusarium oxysporum* Schlechtend.: Fr. f. sp. *ciceris* (Padwick) Matuo & K. Sato and will be the primary *focus* of this chapter.

The pathogen *F. oxysporum* f. sp. *ciceris (Foc)* penetrates the vascular bundles of chickpea roots, stopping or reducing water uptake to the foliage. Infected plants eventually wilt and die. The disease causes considerable annual yield losses, estimated to range from 10-90% (Singh and Reddy 1991), and is present in all chickpea growing areas of the world from the Indian subcontinent, Iran, Syria, Ethiopia, Tunisia, Turkey and Spain, to Mexico, Peru, and USA (Nene et al. 1989; Halila and Strange 1996).

Fusarium wilt of chickpea produces microconidias, macroconidias and chlamydospores. The microconidias (2.5-4.5 x 5-11 μm) are oval or cylindrical, straight or curved, and macroconidias (3.5-4.5 x 25-65 μm) are 3 to 5 μm septate or fusoid, while the chlamydospores are formed in 15-day-old cultures singly, in pairs or in chains, and are smooth or rough-walled. The fungus grows best at a temperature of 25°C and pH of 6.0 (Singh et al. 2007).

The pathogen is a facultative saprophyte, and infects systemically. Both seed and soil borne, it can survive in the soil for long periods (up to 6 years), even in the absence of a host, making it difficult to control (Haware et al. 1996). On the other hand, infected seeds play a major role in the spread of the disease to uninfected areas (Kraft et al. 1994). Consequently, the most practical and cost-effective method for management of Fusarium wilt of chickpea is the use of resistant cultivars.

2. Pathogenic Races: Geographical Distribution and Symptomatology

For efficient disease management through the use of resistant cultivars, it is important to understand the nature of the variation and distribution of pathogenic races. Variation in the virulence among isolates of a given *forma specialis* has led to the designation of pathogenic races. Eight distinct physiological races have been classified for *F. oxysporum* f. sp. *ciceris* (0, 1A, 1B/C, 2, 3, 4, 5 and 6). Races 1A, 2, 3 and 4 were found in India (Haware and Nene 1982), while races 0 and 5 were identified in Spain (Cabrera de la Colina et al. 1985) and Tunisia (Halila and Strange 1996) and race 6 in California, USA (Phillips 1988). Race 1 was later divided into two sub-races referred to as 1A (from India) and 1B/C (from Spain) based on their pathogenicity on differential host lines (Trapero-Casas and Jiménez-Díaz 1985; Jimenez Diaz et al. 1993). Race 1B/C has also been reported in California, Turkey, Syria and Tunisia. Briefly, races 0, 1B/C, 5 and 6 are mainly found in the Mediterranean region and the USA (Phillips 1988; Jiménez Díaz et al., 1989, 1993; Halila and Strange 1996; Jimenez-Gasco et al. 2001), while races 1A, 2, 3 and 4 are found in India (Haware and Nene 1982).

Based on the disease symptoms they induce in pathogenicity tests, these races have been divided into two groups: those causing wilting and those causing yellowing (Trapero-Casas and Jimenez-Diaz 1985). Races 0 and 1B/C cause the yellowing syndrome, while races 1A and 2 through 6 induce the wilting syndrome and are economically the most damaging races (Haware and Nene 1982; Jiménez-Díaz et al., 1993). The yellowing pathotypes induce progressive foliar yellowing with vascular discoloration, while the wilting ones cause leaves to become flaccid and vascular discoloration, leading to complete wilt within three to four weeks of inoculation. Although, traditionally, yellowing and wilting syndromes have been considered race-specific, evidence is emerging to indicate that both types of symptoms can be caused by a single race depending on the host genotype (Sharma and Muehlbauer 2007). In addition, some susceptible chickpea lines exhibit a phenomenon known as late wilting, which is characterized by an extended latent period (the time between inoculation and appearance of the first visible wilt symptoms) followed by a normal rate of disease development (Upadhyaya et al. 1983a). This late wilting has been described in response to both races 1 (Upadhyaya et al. 1983a, b; Singh et al. 1987a, b) and 2 (Gumber et al., 1995). Sharma et al. (2005) reported another phenomenon called slow wilting, where the first signs of disease appear at about the same time as in other susceptible lines showing early wilting, but disease incidence increases slowly over time. The disease usually progresses to reach a final intermediate rate of incidence. The occurrence of slow wilting has been reported in response to races 2 and 3 (Sharma et al. 2005), race 5 (Cobos et al. 2009; Castro et al., 2010) and, more recently, race 0 (Halila et al., 2010).

This diversity and geographic distribution of races of *F. oxysporum* f. sp. *ciceris* makes the identification of races of the pathogen in a given area of chickpea production important for resistance breeding as well as for efficient use of available resistant cultivars. Similarly, detailed characterization of the resistance of chickpea lines to specific races of *F. oxysporum* f. sp. *ciceris* is essential for the deployment of resistant varieties.

3. Inheritance of Wilt Resistance

For the identification and integration of resistance genes on linkage maps and their use in marker-assisted selection (MAS), detailed knowledge of the inheritance of race-specific Fusarium wilt resistance is essential.

Resistance to Fusarium wilt in chickpea has been reported to be race-specific and governed by major resistance genes (Upadhyaya et al. 1983a, b, Tullu et al. 1999, Sharma et al. 2004). Some years ago, two genes were reported to control resistance to race 0, one of them present in the germplasm line JG62 (ICC 4951) and another in CA2139 (Rubio et al. 2003). Early studies of the genetics of resistance to *Foc* showed that resistance to race 1 was governed by three independent genes (*h1*, *h2*, *H3*). Late wilting is caused by any one of these three genes, and complete resistance occurs when a combination of any of the two late wilting genes are present (*h1h2* or *h1H3* or *h2H3)* (Upadhyaya et al. 1983b; Singh et al. 1987a, b; Sharma and Muehlbauer 2007). By contrast, other authors have observed monogenic inheritance of resistance to race 1. Sharma et al. (2005) reported a single gene in WR315 governing resistance to race 1 when crossed with C-104. Indeed, WR315 has been reported to be a source of resistance to all Fusarium wilt races (Sharma and Muehlbauer 2007). Recently, Gowda et al. (2009) also suggested monogenic inheritance for resistance to race 1. This finding could be due to the presence of a single gene of the three aforementioned segregating in the populations studied by these authors.

Initial research on the genetics of resistance to race 2 found that it was governed by a single recessive gene (Pathak et al. 1975). Nevertheless, more recent studies have reported two (Gumber et al. 1995) or three genes (Kumar 1998) conferring resistance. In such case, a combination of any two of these genes in the recessive state confers resistance, whereas when only one of these genes is recessive late wilting occurs (Kumar 1998). The reaction of F_2 and F_3 populations derived from a cross between WR315 (resistant) and C-104 (susceptible) showed the digenic (*a* and *b*) nature of resistance to this race in WR315 (Kumar 1998). Having a recessive genotype for both these genes (*aabb*) conferred resistance in WR315, while *AABB* led to susceptibility in C-104. Crosses between two tolerant parental lines, K850 and GW5/7, revealed involvement of a third gene, *C*, for resistance to this race. However, Sharma et al. (2005) noted that the segregation of the F_2 and the recombinant inbred line (RIL) derived from the same cross (C-104 x WR-315) for resistance to race 2 suggested that a single recessive gene in WR315 governs resistance to this race. The reason for the discrepancies between these studies is unknown. It could be due to different strains of race 2 being used, or due to environmental conditions (Sharma et al. 2005). More recently, Gowda and co-workers (2009) have pointed out that resistance to race 2 was monogenic in the RIL population employed in their study (JG62 x Vijay).

Resistance to race 3 has been found to be monogenic (Sharma et al. 2005; Gowda et al. 2009). Nevertheless, its dominant or recessive nature remains unknown as two RIL populations derived from the crosses C-104 (susceptible) x WR315 (resistant) and JG62 (susceptible) x Vijay (resistant) were used for the genetic studies. With this type of segregating population it is not possible to infer whether a trait is dominant or recessive because alleles segregate in a 1:1 ratio. Resistance to race 4 was observed to be recessive and monogenic in studies of a F_2 population derived from the same cross mentioned above, namely C-104 x WR315 (Tullu et al. 1998), while it was recessive and digenic in JG62 (susceptible) x Surutato-77 (resistant) (Tullu et al. 1999). Therefore, to date, two genes seem to be involved in resistance to race 4. The pattern of inheritance of resistance to race 5 has been found to be monogenic in several studies using ICC4958 (Tekeoglu et al. 2000), WR315 (Sharma et al. 2005; Iruela et al. 2007) and ICCL81001 (Cobos et al. 2009) as the sources of resistance. Evaluation of F_1 plants in our laboratory has shown resistance to race 5 to be recessive (Dr. J. Gil, personal communication).

So far, the genetics of resistance to races 1B/C and 6 remains unknown. Similarly, the genetics of slow wilting resistance in chickpea has not yet been determined though it has been proposed that the process is related to minor genes (Sharma and Muehlbauer 2007). Sharma et al. (2005) observed a slow wilting reaction for race 2 in the germplasm line FLIP84-92C(3) as well as some RILs from the cross C104 x WR315 for races 2 and 3. Cobos et al. (2009) reported that some RILs in the population derived from the inter-specific cross ICCL81001 x *C. reticulatum* (Cr5-9) also showed slow wilting for race 5 and suggest that, apart from a major gene, an additional genomic area influences the disease reaction. Similarly, Castro et al. (2010) reported that near-isogenic lines (NILs) to Fusarium wilt carrying both ICCL81001 and WR315 as sources of resistance showed a slow wilting response. They suggested that a combination of the major gene *foc-5* with other gene(s) could be required to complete resistance, while the absence of these unknown genes leads to slow wilting reactions. Hence, it seems that for wilt management we need to consider two types of host resistance, vertical resistance and slow wilting resistance. The NILs developed by Castro et al. (2010) could be useful test subjects with which to study these two types of resistance. In any case, further research is required to elucidate the genetics and mechanisms of wilt resistance.

Table 1. Disease score scales employed in different studies to phenotype resistance and susceptibility to *Fusarium oxysporum* f. sp. *ciceris* in chickpea lines

Disease scoring scale (% of wilted plants)			Reference
Resistant	Intermediate reaction	Susceptible	
0-20	21-50	> 50	Haware and Nene 1982; Phillips 1988
0-33	34-66	67-100	Jiménez-Díaz et al. 1989; Jiménez-Gasco et al. 2001
0-10	11-90	91-100	Tullu 1996 ; Sharma et al. 2004, 2005 ; Castro et al. 2010
0-30	-	70-100	Iruela et al. 2007
< 20	-	> 80	Gowda et al. 2009

Differences in results between different studies can be attributed to the evaluation techniques and disease scoring scales used to phenotype chickpea lines. Some authors (e.g., Gumber et al. 1995; Kumar 1998 and Gowda et al. 2009) have studied the disease using sick pots; with this approach, the time of inoculation may not be uniform across the plants and, further, it cannot be ensured that all plants are inoculated. As the appearance of wilt symptoms depends upon the time from the infection of root surface cells to the accumulation of sufficient numbers of propagules in the vascular bundles, it is very likely that plants infected early will wilt sooner than those infected later (Sharma and Muehlbauer 2007). By contrast, other authors (e.g., Sharma et al. 2005; Castro et al. 2010) have inoculated individual plants by dipping their roots, previously wounded by cutting, in pathogen inoculum containing a fixed concentration of spores (1 x 10^6 spores/ml). This method ensures that inoculation and penetration of the pathogen is uniform, as well as that plants are at the same stage of vegetative growth at the time of inoculation (Figure 1).

Another factor that has hindered progress with the genetic analysis is the use of different disease scoring scales, to phenotype resistance and susceptibility (Table 1). The range of terminology used, including "medium", "tolerant", "moderately susceptible" and "intermediate" has further complicated the process of race identification (Sharma et al. 2005). Evidently, it would be desirable to develop a standardized rating scale to be used by all the chickpea community.

Figure 1. Chickpea lines inoculated with *Fusarium oxysporum* f. sp. *ciceris* race 5.

Table 2. Reaction of differential lines against *Fusarium oxysporun* f. sp. *ciceris* races 1A, 2, 3, 4 and 5

Differential line	Race 1A[a]	Race 2	Race 3	Race 4	Race 5	Reference
JG-62[b]	S	S	S	S	S	Sharma et al. 2005
P-2245[b]	S	S	S	S	S	"
'Sanford'	R	S	S	S	S	"
CRIL1-53	S	R	R	R	R	"
CRIL1-94	R	S	R	I	I	"
CRIL1-17	R	R	R	S	R	"
CRIL1-36	I	S	S	S	R	"
WR315	R	R	R	R	R	"
NIL RIP8-94-11	S	S	R	S	S	Castro et al. 2010

[a] R = Resistant (0-10% wilt), I = intermediate (11-89% wilt), S = Susceptible (90-100% wilt).

[b] JG-62 and P-2245 are resistant and susceptible, respectively, to race 0 (Tullu 1996; Rubio et al. 2003).

4. Differential Sets for Race Identification

It is difficult to identify *F. oxysporum* f. sp. *ciceris* races by morphology alone. The classical method of race identification involves inoculating differential chickpea cultivars with a particular *F. oxysporum* f. sp. *ciceris* isolate and determining its pathogenicity in each.

Since resistance to Fusarium wilt in chickpea is race-specific and inheritance is oligogenic, differential sets have been developed for chickpea wilt in order to identify race-specific resistance. The interpretation of results can be complex, as on occasions the differentiation is based on intermediate reactions. Nevertheless, some lines have shown reactions which make it possible to distinguish between races. Initially, even though several authors established differential lines (Haware and Nene 1982; Jimenez-Diaz et al. 1989; Tullu 1996), no clear set was found for races 2 and 3 (Sharma et al. 2004). Later, Sharma et al. (2005) established eight selected lines that could differentiate between races 1A, 2, 3, 4 and 5. In addition, Castro et al. (2010) have recently characterized four pairs of near-isogenic lines for disease reaction to these *Foc* races (1A, 2, 3, 4 and 5). They proposed the NIL RIP8-94-11 as a clear differential line for the identification of race 3 as it shows a clear-cut resistance reaction to race 3 and susceptibility to races 1A, 2, 4 and 5 (Table 2).

To date, differentiation between races 1B/C and 6 is still difficult. Landa et al. (2006) characterized the disease reaction of 6 chickpea cultivars (PV-1; JG-62; C-104; JG-74; CPS-1 and ICCV-4) to these races but they also show moderate reactions. It would be desirable to find clear differential lines with resistance to races 1B/C and 6 to facilitate identification of all races of the pathogen.

A new approach to the identification of the pathogen and its pathogenic races is a DNA-based diagnostic method (Kelly et al. 1994; Jiménez-Gasco et al. 2001; Jiménez-Gasco and Jiménez-Díaz 2003). Random amplified polymorphic DNA (RAPD) markers related to races 0, 1A, 5 and 6 have been reported (Jiménez-Gasco et al. 2001). To facilitate precise identification of these races the RAPD markers have been converted into sequence characterized amplified regions (SCARs) which are robust markers and should avoid errors that can occur in the comparison of amplification patterns between laboratories using RAPD markers (Jiménez-Gasco and Jiménez-Díaz 2003). Nevertheless, whether these assays will be able replace the traditional approach based on host reaction to identify the pathogen and its races is still to be confirmed. Furthermore, some authors have suggested these assays may need to be refined before being used routinely by pathologists or breeders (Sharma and Muehlbauer 2007).

5. Sources of Resistance (Germplasm Colections) and Breeding

Resistance breeding programs were initiated more than 90 years ago and the first wilt resistant cultivar 'Karachi' was introduced in 1923 (Infantino et al. 1996). Since then, wilt resistant cultivars have been released in many countries.

Examples include the wilt resistant cultivars Surutato-77 and Sonora-80, developed by plant breeders in Mexico by transferring resistance genes from desi accessions L-41 and L-1186 to kabuli 'Macarena' and 'Breve Blanco' (Singh, 1987). In addition, plant breeders at

ICRISAT (International Crops Research Institute for the Semi-Arid Tropics) developed four short duration kabuli lines, namely ICCV2-5, carrying resistance to race 1 of the pathogen (Kumar et al. 1985). Such cultivars have, however, only been partially successful due to location-specific races of the pathogen (Singh and Reddy 1991).

The screening of international germplasm has led to the identification of additional sources of resistance to *F. oxysporum* f. sp. *ciceris*. In particular, a number of researchers have reported examples of high levels of resistance of Fusarium wilt and more than 200 wilt-resistant lines have been identified (Nene and Haware 1980; Zote et al. 1983; Singh et al. 1986; Ahmad and Sharma 1990; Reddy et al. 1990; Jiménez-Díaz et al. 1991; Iqbal et al. 1993; Iftikhar et al. 1997; Iqbal et al. 2005; Sharma et al. 2005, Chaudhry et al. 2006, 2007, among others).

The pyramiding of resistance genes in elite cultivars is expected to provide durable resistance against this disease. So far, most sources of wilt resistance available are in desi types, while resistance sources in kabuli types are limited. Haware et al. (1992) evaluated a world collection with 13,500 germplasm accessions for race 1, identifying 160 resistant accessions but only 10 of them were kabuli type. Gaur et al. (2006) evaluated 50 large-seeded kabuli chickpea germplasm accessions from ICRISAT's genebank and found two accessions, ICC14194 and ICC17109, originating from Mexico, with complete resistance. These new Fusarium wilt resistance sources will be very useful in breeding extra-large kabuli varieties with Fusarium resistance and typical kabuli type seeds. Indeed, desi x kabuli crosses have been widely used at ICRISAT for enhancing Fusarium wilt resistance of kabuli chickpea. As mentioned before, the desi type accession WR315 is considered to be resistant to all known races of *Foc*-carrying favourable alleles for races 0, 1, 2, 3, 4 and 5. It has been widely used in inheritance and mapping studies and is considered to be a very interesting source of resistance to Fusarium wilt for chickpea breeding programs.

Recently, a collection of 321 genotypes from various sources was evaluated under greenhouse and field conditions for a mixture of various isolates at Pakistan's National Agricultural Research Centre (Islamabad, Pakistan) at seedling and reproductive stages; 82 genotypes were found to have steady resistance at both stages (Ahmad et al. 2010). These authors propose to exploit these genotypes for the development of resistant cultivars against wilt.

Wilt resistance has also been found in some wild *Cicer* spp (Nene and Haware 1980, Kaiser et al. 1994; Infantino et al. 1996; Singh et al. 1998). Introgression of resistance genes from wild species could be accelerated in the cultigen using molecular markers to facilitate the selection of resistance genes and to eliminate undesirable traits.

6. Integrated Management of Fusarium Wilt

Management of Fusarium wilt of chickpea is difficult and no single control measure is fully effective. Considering the nature of the damage caused and the survivability of the fungus, the use of resistant varieties seems to be the only economical and practical solution. However, integrated management strategies including the efficient use of chemicals, encouragement of beneficial biological agents to reduce pathogen inoculum, and modification of traditional practices, as well as the introduction of resistant varieties, may all contribute to maintaining plant health (Bendre and Barhate 1998).

With respect to chemical treatments, Gupta et al. (1997) screened six different fungicides against *F. oxysporum* f. sp. *ciceris* and reported carbendazim at 100 mg/ml to be most effective in inhibiting the growth of the fungus in vitro. Nikam et al. (2007) found that treatment of seed with thiram (0.15%) + carbendazim (0.1%) were relatively effective against wilt. Recently, Subhani et al. (2011) studied the effects of six fungicides (Benomyl, Derosal, Ridomil, Cabrio Top, Vitavax), finding Derosal and Benomyl to the most effective (100% reduction in disease incidence), followed by Vitavax, while the Cabrio Top was the least effective.

Biological agents such as *Trichoderma* can also be a useful component of integrated chickpea Fusarium wilt management. Prasad et al. (2002) found that soil application of *T. viride* and *T. harzianum* one week before sowing was effective in reducing wilt and wet root rot of chickpea. The consortium *T. viride* + *T. harzianum* + *T. hamatum* seemed to have a synergistic effect and was found very effective to control chickpea wilt (Nikam et al. 2007). Meki et al. (2011) studied 32 *Trichoderma* isolates grown with chickpea in the central highlands of Ethiopia and concluded that *T. harzianum* (T23) reduced wilt severity for race 3 and delayed disease onset. The biocontrol of Fusarium wilt of chickpea using arbuscular mycorrhizal fungi (AMF) *Glomus hoi* (Gh), *Glomus fasciculatum* (Gf) and *Rhizobium leguminosorum* Biovar. (Rl) was studied by Singh et al. (2010), and they concluded that suitable combinations of AMF and *Rhizobium* bacteria may increase plant growth and resistance to pathogens.

The efficacy of wilt management could be also improved by combining host resistance and biological control with modifying traditional farming practices (Landa et al. 2004). In particular, it was confirmed that sowing date was the factor with the greatest effect on Fusarium wilt and yield of chickpea. Traditionally, chickpeas in the Mediterranean region are sown in the spring, and these authors showed that in the Mediterranean environment bringing the date of chickpea sowing back from early spring to winter significantly delays the onset, reduces the final extent of disease caused by highly virulent *F. oxysporum* f. sp. *ciceris* race 5, and increases chickpea yields.

7. Location of Resistance Genes on the Chickpea Genetic Map

Molecular markers closely linked to a particular agronomic trait facilitate the detection of favourable alleles in breeding programs. Marker-assisted selection is particularly useful in the case of breeding for disease resistance in order to avoid complex and time consuming evaluations as well as for pyramiding different resistance genes in the same genotype. However, the efficacy of MAS relies on the saturation of genomic areas of interest with robust, highly polymorphic, easy to interpret and cost-effective markers (Collard and Mackill 2008).

In chickpea, great progress has been made in the development of genetic maps and identification of markers close to genes or genomic regions of interest in recent years. Even though maps are still incomplete, the chances of finding new polymorphic markers have been considerably increased, essentially due to the development of sequence tagged microsatellite site (STMS) markers (Hüttel et al. 1999; Winter et al. 1999; Sethy et al. 2003, 2006a, b; Lichtenzveig et al. 2005; Choudhary et al. 2006; Nayak et al. 2009). Given the possibility of exchanging information between maps based on STMS markers, it has been possible to build a consensus chickpea map (Millán et al. 2010) following the linkage group nomenclature previously proposed by Winter et al. (2000), which has been considered the map of reference. Nayak et al. (2009) completed the map published by Winter et al. (2000) adding 175 new markers and provided anchor points for comparing chickpea linkage groups with linkage groups in the model species *Medicago truncatula*. In addition, large-scale genic molecular markers (GMMs) have recently been used to develop a transcript map of chickpea (Gujaria et al. 2011). Those markers together with the new transcript dataset now available increase the likelihood of successfully using single nucleotide polymorphims (SNPs) in high-throughput genotyping in this crop (Hiremath et al. 2011).

To date, phenotyping RIL populations has made it possible to localize some of the genes conferring resistance to different races for Fusarium wilt and to identify useful markers for chickpea breeding programs (Table 3). In this chapter, in order to use a consistent nomenclature, the resistance genes will be designated according to nomenclature proposed for Fusarium wilt resistance genes by Tekeoglu et al. (2000). Both intra- and inter-specific populations have been used to localize different resistance genes and sources of resistance utilized in mapping population have been limited to the parentals: JG62 and CA2139 (race 0), WR315 (races 1, 2, 3, 4 and 5), Vijay (races 1, 2 and 3) and ICC4958 (races 4 and 5). On most occasions, populations segregate for single resistance genes and have been assessed considering resistance as a qualitative trait. However, in some studies the disease reaction has been evaluated as a quantitative trait, either because more than one major gene could be controlling the resistance reaction or to try to identify an accurate location of the resistance genes.

Table 3 shows that the two genes conferring resistance to race 0 (Rubio et al. 2003) have been mapped in an intraspecific integrated map composed of two RIL populations. The population derived from CA2156 x JG62 segregated for a single resistance gene present in JG62, whereas in the case of CA2139 x JG62 both parental lines were resistant. One gene (*foc0*$_1$) was mapped on linkage group (LG) 3 (Cobos et al. 2005), which corresponds to LG5 of the *Cicer* reference map (Winter et al. 2000). Furthermore, the association of LG5 with *foc-0*$_1$ was confirmed by quantitative trait loci (QTL) analysis in a RIL derived from CA2139 x JG62 where either of the two genes could be involved in the resistance reaction.

Halila et al. (2009) attempted to locate the second gene that offers resistance to race 0 in the CA2139 genotype. The genotypic information provided by markers tightly linked to *foc-0*$_1$ was useful to select ten resistant lines in which only the second gene *foc-0*$_2$ was present. Analysis with those resistant lines vs. susceptible ones revealed the association between *foc-0*$_2$ and STMS markers sited on LG2 of the *Cicer* map. With QTL analysis of the same populations used by Cobos et al. (2005), Halila et al. (2009) confirmed the location of *foc-0*$_1$ on LG5 and located *foc-0*$_2$ on LG2 for the first time. Using molecular markers, *foc-0*$_2$ has also been found in other resistant accessions (WR315, ICCL81001) (Cobos et al. 2009; Halila et al. 2010).

As described in Section 3 of this chapter (Inheritance of wilt resistance), several authors have employed the same intra-specific RIL population (C-104 x WR315) to identify Fusarium wilt resistance genes, the WR315 genotype being the source of resistance. In this population, the first gene for which molecular markers were identified was one of the three loci controlling resistance to race 1 (*foc-1*). This gene was tagged by two RAPD markers (UBC-170_{550} and CS-27_{700}) 7.0 cM apart but sited on the same side of the gene. An allele-specific associated primer (ASAP) was designed from the CS27 marker that amplified a fragment linked to the allele for susceptibility to race 1 (Mayer et al. 1997). This marker was the starting point for subsequent detection of several different *Foc* genes clustered on LG2 of the chickpea genetic map. Tullu et al. (1998) revealed that the loci controlling resistance to race 4 and race 1 were closely linked to each other (5.0 cM) and linked to markers UBC-170_{550} and CS-27_{700} sited 9.0 cM apart from the race 4 resistance locus. These authors were the first to suggest a possible cluster of several Fusarium wilt resistance genes. Later, Sharma and Muehlbauer (2005) and Sharma et al. (2004) localized genes controlling race 2 (*foc-2*) and race 3 (*foc-3)* resistance respectively. Both were located on LG2 linked to STMS markers (Table 3). Linkage between *foc-3*, *foc-1* and *foc-4* was established by Sharma et al. (2004), *foc-1* and *foc-4* being found to be closely linked and sited at a distance of 8.9 cM from *foc-3*.

A gene controlling resistance to race 5 was also located on LG2, by Iruela et al. (2007), using a different intraspecific population (ILC3279 x WR315). Recently, Gowda et al. (2009) confirmed the location of resistance genes *foc-1, foc-2* and *foc-3* and were able to identify flanking STMS markers tightly linked to the three loci. They used a RIL population derived from the cross between JG62 x Vijay, JG62 being the susceptible parental line (Table 3).

Populations derived from inter-specific crosses have been used for mapping purposes in order to increase the chance of finding polymorphic markers. Table 3 includes several race-specific genes resistant to Fusarium wilt that were located using wide crosses between the cultivated chickpea and *C. reticulatum.*

A quantitative trait locus (QTL_{Foc02}) coincident with *foc-0_2* was located on LG2 by Cobos et al. (2009). QTL_{Foc02} was found to be responsible for 73% of the phenotypic variation of the disease reaction. On this occasion, Fusarium wilt was considered as a quantitative trait in order to locate the gene more precisely on the linkage group. Co-location of QTL_{Foc02} and *foc-0_2* suggests that the resistant genotype ICCL81001 could share the same resistance gene with the resistant parental line CA2139 employed by Cobos et al. (2005). Additionally, these authors identified a QTL related to race 5 (QTL_{Foc5}) explaining 46.5% of total phenotypic variation in the same region. Even though QTL_{Foc02} and QTL_{Foc5} are closely interlinked, they should be considered different loci because not all the resistant lines to race 0 were also resistant to race 5. As previously mentioned (Section 3), the observation of a slow wilting response for this race and the association of a second region on LG5 with *Foc* race 5 resistance (12.7% of the phenotypic variation) suggested that more than one gene might control the response to race 5. This view was supported by Castro et al. (2010) who observed a slow wilting reaction in two groups of near-isogenic lines. They suggested that the combination of *foc-5* with other gene(s) appears to be required for full resistance.

Another wide cross that has often been used for Fusarium wilt mapping is the RIL derived from the cross ICC4958 (resistant) x PI489777 (susceptible) (Table 3). Ratnaparkhe et al. (1998a) were the first to find an inter simple sequence repeats (ISSR) marker (UBC-855_{500}) closely linked to the gene for resistance to race 4 at a distance of 5.2 cM; the marker co-segregated with CS-27_{700} and mapped on LG6 of the chickpea genetic map reported by Simon and Muehlbauer (1997). Furthermore, Ratnaparkhe et al. (1998b) revealed that the gene for resistance to race 5 was close to the ISSR marker UBC-825_{1200} and linked to the race 4 gene (at a distance of 5.0 cM). The development of an extended and comprehensive interspecific map including STMS developed by Winter et al. (2000) made it possible to tag the genes for resistance to races 4 and 5 with STMS TA96 and TA27 at distances of 3.4 and 3.5 cM, respectively, on LG2. The ASAP marker CS27, previously developed by Mayer et al. (1997), was also sited in this cluster. This linkage group has been saturated by Benko-Iseppon et al. (2003) using DNA amplification fingerprinting (DAF) markers and more recently by Nayak et al. (2009) using simple sequence repeats (SSR) and SNP markers.

Summarizing, results from both intra- and inter-specific crosses consistently find clustering of genes controlling resistance to races 0, 1, 2, 3, 4 and 5 on LG2 of the chickpea map. Another interesting area of the genome is LG5, where a second resistance gene for race 0 and a putative QTL related to race 5 are located. Within the cluster on LG2, sub-clusters of Fusarium wilt resistance genes have been differentiated. Several authors identify one containing *foc-1* and *foc-4* and a second comprising *foc-3* and *foc-5* (Winter et al. 2000; Benko-Iseppon et al. 2003; Castro et al. 2010). However, Sharma and Muehlbauer (2005) grouped *foc-4*, *foc-2* and *foc-3* in the same sub-cluster and *foc-5* and *foc-1* in a different one. On the other hand, Gowda et al. (2009) reported that the closest genes were *foc-1* and *foc-2* (6.8 cM), while *foc-1* was flanked on the opposite side by *foc-3* at a distance of 22 cM; this is consistent with the proposal of Castro et al. (2010), from studying NILs, that *foc-2* and *foc-3* are delimiting points at opposite ends of a genomic region that includes the remaining *Foc* genes and TA59 marker.

Even though the order of STMS markers on LG2 was conserved across different chickpea maps reported by various authors (Winter et al. 2000; Tekeoglu et al. 2000; Udupa and Baum 2003; Cho et al. 2004; Sharma et al. 2004; Abbo et al. 2005; Iruela et al. 2007; Halila et al. 2009), there seem to be a few small discrepancies. Segregation distortion among interspecific crosses might explain the differences in gene order and genetic distances between wilt resistance genes. Additionally, the number of markers studied, use of software packages based on different procedures and variation in population size could also contribute to the discrepancies. Fine mapping using large populations together with the development of the chickpea physical map (Zhang et al. 2010) could help determine the actual distances between genes.

Furthermore, it is interesting to note that apart from resistance *Foc* genes, QTLs and genes for resistance to Ascochyta blight of chickpea and several potential resistance and pathogenesis-related genes have also been detected on LG2 (Hüttel et al. 2002; Benko-Iseppon et al. 2003; Pfaff and Kahl 2003; Udupa and Baum 2003; Cho et al. 2004; Cobos et al. 2006; Iruela et al. 2007; Palomino et al. 2009). Therefore, this LG may be considered a hot spot for pathogen defence in chickpea (Millán et al. 2006) in line with the clusters of disease resistance genes reported for other cultivated species (Gebhardt and Valkonen 2001; Hulbert et al. 2001; Scott et al. 2004).

Markers closely linked to genes for Fusarium wilt resistance to race 0 (*foc0*$_1$ /*foc0*$_2$) would allow pyramiding of both genes for resistance to race 0 in a single individual in order to provide durable resistance against *Foc* race 0. In some cases, markers closely linked to a gene or QTL are not polymorphic in other segregating populations. For resistance to *Foc* race 0, however, the large number of markers available would be likely to enable us to solve this problem.

Table 3. Markers identified linked to race-specific *Fusarium oxysporum* f. sp. *ciceris* resistance genes using both intra- and inter-specific populations. The wilt resistance genes were designated according to nomenclature proposed for Fusarium wilt resistance genes by Tekeoglu et al. (2000)

Population	Cross	LG	Resistance gene	Indicative markers	Resistance gene distance (cM)	References
Intraspecific	CA2156xJG62 + CA2139xJG62	5	*foc-0₁*	$OPJ20_{600}$	3.0	Cobos et al. 2005
				TR59	2.0	
		2	*foc-0₂*	TA59	QTL (6.14)	Halila et al. 2009
				TA27		
	C-104 x WR315	2	*foc-1*	UBC-170_{550}	7/ 7.2*	Mayer et al. 1997, Sharma and Muehlbauer 2005
				CS-27_{700}	7.0	
		2	*foc-2*	TA96	1.5	Sharma and Muehlbauer 2005
				TA27	1.5	
				TR19	4.9	
				CS27	1.5	
		2	*foc-3*	TA 96	0.6	Sharma et al. 2004
				TA 194	14.3	
		2	*foc-4*	UBC-170_{550}	9.0	Tullu et al. 1998, Sharma and Muehlbauer 2005
				CS-27_{700}	9/15.2*	
	JG62 x Vijay	2	*foc-1*	TA110	2.1	Gowda et al. 2009
				H3A12	3.9	
		2	*foc-2*	H3A12	2.7	"
				TA96	0.2	
		2	*foc-3*	H1B064	0.2	"
				TA194	0.7	
	ILC3279xWR315	2	*foc-5*	TA110	6.5	Iruela et al. 2007
Interspecific	ICCL81001 x Cr5-9	2	*foc0₂*	TA59	QTL (18.8)	Cobos et al. 2009
		2	*foc-5*	TA59	QTL (10.4)	"
	ICC4958 x PI489777	2	*foc-4*	UBC-855_{500}	5.2	Ratnaparkhe et al. 1998a
				UBC-825_{1200}	5.0	Ratnaparkhe et al. 1998b
				CS27	7.2	Tekeoglu et al. 2000
				TA96	3.4	Winter et al. 2000
				R-2609-1	2.0	Benko-Iseppon et al. 2003
		2	*foc-5*	UBC-857_{800}	23.2	Ratnaparkhe et al. 1998b
				CS27	4.0	Tekeoglu et al. 2000
				TA27	3.5	Winter et al. 2000

Values in parenthesis mean LOD value in case of QTL detection; * distance values for the same markers to the resistance gene for different authors.

MAS could become an efficient strategy to accelerate the process of pyramiding of resistance genes. However, a higher density of markers in the area of these genes is still necessary to detect polymorphisms for MAS in different genetic backgrounds and to perform fine mapping that could contribute to the understanding of the molecular aspects of resistance. Furthermore, new sources of resistance should be included in mapping populations to make it possible to identify and then target other still unmapped resistance genes.

8. Future Prospect

Improving resistance to Fusarium wilt by traditional breeding is difficult as the assessment of large numbers of germplasm lines for resistance to specific races of the pathogen is tedious, time consuming and, hence, expensive, as well as being affected by inoculum load and environmental conditions. The recent emergence of new biotechnological techniques makes possible today both the use of molecular markers to replace screening tests in breeding populations and the introgression of genes from related species to confer resistance in cultivated species. Further, the screening of plants carrying wilt resistance genes can be facilitated with MAS. However, the efficacy of MAS depends upon the closeness of the marker to the gene. Although some markers associated with wilt resistance have been identified, marker density on LG2 and LG5 is still low. In addition, the wilt resistance genes in chickpea seem to form a cluster with resistance genes lying a few hundred to a few thousand kb apart. Hence, markers flanking the individual genes are needed for effective MAS (Sharma and Muehlbauer 2007). An ideal situation is that the marker is part of the gene. However, such markers, called diagnostic markers, are not yet available for wilt. Nevertheless, on the basis of the current availability of markers, one approach to selecting resistant lines for wilt in chickpea breeding programs would be to use these markers in early generations (F_2). Later, the selected lines ($F_{2:3}$) would have to be screened for wilt in controlled conditions or in infected fields in order to confirm their resistance. This approach increases the frequency of favourable alleles while reducing the population size that has to be screened to identify resistant lines.

A possible strategy for developing molecular marker to map genomic regions where marker density is low involves identifying SNPs. Up to now, the use of this kind of markers has been hampered by the limited number of chickpea sequences deposited in public databases. However, current efforts are changing this situation. Chickpea genomic research has progressed rapidly during recent years with the availability of genomic tools such as the bacterial artificial chromosome (BAC) (Rajesh et al. 2004; Lichtenzveig et al. 2005) and cDNA libraries (Coram and Pang 2005; Gupta et al. 2011), yielding a large collection of expressed sequence tags (ESTs) and increasing the number of sequences in public databases. Moreover, new sequencing (e.g., 454, Illumina, Solexa) and genotyping technologies can be expected to play an important role in the understanding of chickpea genetics. These technologies have the ability to sequence in great detail, and this will make it easier to discover SNPs in the narrow germplasm base of this species (Varshney et al. 2009). This progress will enable biotechnologists to more rapidly and precisely target genes that underlie Fusarium wilt resistance and advance the development of molecular markers for breeding applications. Further, the availability of high-throughput, cost-effective genotyping methods will increase the use of genomic tools in chickpea breeding programs.

Challenges for the future include obtaining higher levels of durable resistance for wilt through the pyramiding of resistance genes. In addition, implementation of marker-assisted selection schemes will offer plant breeders a means to improve selection efficiency, reducing the time and effort required to develop new cultivars resistant to Fusarium wilt.

ACKNOWLEDGMENTS

This work was funded by the Spanish project RTA2010-00059, financed in part by the EU funds FEDER. The authors thank support form tha University of Cordoba and Agrarian and Fishing Technologies Research and Education Institute (IFAPA), both member of the Campus of Internationa Excellence (ceiA3).

REFERENCES

Abbo S, Molina C, Jungmann R, Grusak MA, Berkovitch Z, Reifen R, Kahl G, Winter P, Reifen R, 2005. Quantitative trait loci governing carotenoid concentration and weight in seeds of chickpea (*Cicer arietinum* L.). *Theor Appl Genet* 111: 185-195.

Ahmad Q, Sharma RN, 1990. Source of resistance to Fusarium wilt of chickpea in Bihar.Internat. *Chickpea Newslet* 23:20.

Ahmad MA, Iqbal SM, Ayub N, Ahmad Y, Akram A, 2010. Identification of resistant sources in chickpea against Fusarium wilt. *Pak J Bot* 42(1): 417-426.

Arumuganathan K, Earle ED, 1991. Nuclear DNA content of some important plant species. *Plant Mol Biol Rep* 9: 208-218.

Bendre NJ, Barhate BG, 1998. *A souvenir on Disease Management in Chickpea.* MPKV, Rahuri during 10th Dec.

Benko-Iseppon AM, Winter P, Hüttel B, Staginnus C, Muehlbauer FJ, Kahl G, 2003. Molecular markers closely linked to fusarium resistance genes in chickpea show significant alignments to pathogenesis-related genes located on Arabidopsis chromosomes 1 and 5. *Theor Appl Genet*107: 379-386.

Cabrera de la Colina J, Trapero-Casas A, Jiménez-Díaz RM, 1985. Races of *Fusarium oxysporum* f.sp. *ciceri* in Andalucía, Southern Spain. *Int. Chickpea News*l 13: 24-26.

Castro P, Pistón F, Madrid E, Millán T, Gil J, Rubio J, 2010. Development of chickpea near isogenic lines for fusarium wilt. *Theor Appl Genet* 121: 1519-1526.

Chaudhry MA, Ilyas MB, Muhammad F, Ghazanfar MU, 2007. Sources of resistance in chickpea germplasm against Fusarium wilt. *Mycopath* 5(1): 17-21.

Chaudhry MA, Muhammad F, Afzal M, 2006. Screening of chickpea germplasm against Fusarium wilt. *J Agric Res* 44 (4).

Cho S, Chen W, Muehlbauer FJ, 2004. Pathotype-specific genetic factors in chickpea (*Cicer arietinum* L.) for quantitative resistance to ascochyta blight. *Theor Appl Genet* 109:733–739.

Choudhary S, Sethy NK, Shokeen BA, Bhatia S, 2006. Development of sequence tagget microsatellite site markers for chickpea (*Cicer arietinum* L.). *Molecular Ecology Notes* 6: 93–95.

Cobos MJ, Fernández MJ, Rubio J, Kharrat M, Moreno MT, Gil J, Millán T, 2005. A linkage map of chickpea (*Cicer arietinum* L.) based on populations from Kabuli × Desi crosses: location of genes for resistance to fusarium wilt race 0. *Theor Appl Genet* 110: 1347-1353.

Cobos MJ, Rubio J, Strange RN, Moreno MT, Gil J, Millán T, 2006. A new QTL for Ascochyta blight resistance in a RIL population derived from an interspecific cross in chickpea. *Euphytica* 149: 105-111.

Cobos MJ, Winter P, Kharrat M, Cubero JI, Gil J, Millán T, Rubio J, 2009. Genetic analysis of agronomic traits in a wide cross of chickpea. *Field Crops Research* 111: 130-136.

Collard BCY, Mackill DJ, 2008. Marker-assisted selection: an approach for precision plant breeding in the twenty-first century. *Phil Trans R Soc* B 363: 557-572.

Coram T, Pang ECK, 2005. Isolation and analysis of candidate ascochyta blight defence genes in chickpea. Part I. Generation and analysis of an expressed sequence tag (EST) library. *Physiol Mol Plant Pathol* 66: 192-200.

FAOSTAT (2010): http://faostat.fao.org/ (Updated 17 May 2011).

Gaur PM, Pande S, Upadhyaya HD, Rao BV, 2006. Extra-Large Kabuli Chickpea with High Resistance to Fusarium Wilt. An Open Access *Journal published by ICRISAT.* Vol 2 (1).

Gebhardt C, Valkonen J PT, 2001. Organization of genes controlling disease resistance in the potato genome. *Annu. Rev. Phytopathol.* 39, 79-102.

Gowda SJM, Radica P, Kadoo NY, Mhase LB, Gupta VS, 2009. Molecular mapping of wilt resistance genes in chickpea. *Mol Breeding* 24: 177-183.

Gujaria N, Kumar A, Dauthal P, Dubey A, Hiremath P, Prakash AB, Farmer A, Bhide M, Shah T, Gaur PM, Upadhyaya HD, Bhatia S, Cook DR, May GD, Varshney RK, 2011. Development and use of genic molecular markers (GMMs) for construction of a transcript map of chickpea (*Cicer arietinum* L.). *Theor Appl Genet* 122: 1577-1589.

Gumber RK, Kumar J, Haware MP, 1995. Inheritance of resistance to Fusarium wilt in chickpea. *Plant Breeding* 114: 277–279.

Gupta S, Chakraborti D, Sengupta A, Basu D, Das S, 2011. Primary Metabolism of Chickpea Is the Initial Target of Wound Inducing Early Sensed *Fusarium oxysporum* f. sp. *ciceris* Race I. PLoS ONE, February Volume 5 Issue 2.

Gupta SK, Upadhyaya JP, Ojha KH, 1997. Effect of fungicidal seed treatment on the incidence of chickpea wilt complex. *Ann Pl Prot Sci* 5: 184-187.

Halila I, Cobos MJ, Rubio J, Millán T, Kharrat M, Marrakchi M, Gil J, 2009. Tagging and mapping a second resistance gene for Fusarium wilt race 0 in chickpea. Eur J Plant Pathol 124: 87-92.

Halila I, Rubio J, Millán T, Gil J, Kharrat M, Marrakchi M, 2010. Resistance in chickpea (*Cicer arietinum* L.) to Fusarium wilt race ‘0’. *Plant Breeding* 129: 563-566.

Halila MH, Strange RN, 1996. Identification of the casual agent of wilt of chickpea in Tunisia as *Fusarium oxysporum* f. sp. *ciceris* race 0. *Phytopathol Mediterr* 35: 67-74.

Haware MP, Nene YL, 1982. Races of *Fusarium oxysporum* f. sp. *ciceris*. *Plant Disease* 66: 809-810.

Haware MP, Nene YL, Pundir RPS, Narayana Rao J, 1992. Screening of world chickpea germplasm for resistance to fusarium wilt. *Field Crops Research* 30: 147-154.

Haware MP, Nene YL, Natarajan M, 1996. Survival *of Fusarium oxysporum* f. sp. *ciceris* in the soil in the absence of chickpea. *Pytopathol Mediterr* 35: 9-12.

Hiremath PJ, Farmer A, Cannon SB, Woodward J, Kudapa H, Tuteja R, Kumar A, BhanuPrakash A, Mulaosmanovic B, Gujaria N, Krishnamurthy L, Gaur PM, KaviKishor PB, Shah T, Srinivasan R, Lohse M, Xiao Y, Town CD, Cook DR, May GD, Varshney RK, 2011. Large-scale transcriptome analysis in chickpea (*Cicer arietinum* L.), an orphan legume crop of the semi-arid tropics of Asia and Africa *Plant Biotechnology Journal* pp. 1-10.

Hüttel B, Santra D, Muehlbauer FJ, Kahl G, 2002. Resistance gene analogues of chickpea (*Cicer arietinum* L.): isolation, genetic mapping and association with a Fusarium resistance gene cluster. *Theor Appl Genet* 105: 479-490.

Hüttel B, Winter P, Weising K, Choumane W, Weigand F, Kahl G, 1999. Sequence-tagged microsatellite-site markers for chickpea (*Cicer arietinum* L.). *Genome* 42: 210-217.

Hulbert SH, Webb CA, Smith SM, Sun Q, 2001. Resistance gene complexes: evolution and utilization. *Annu. Rev. Phytopathol.* 39: 285-312

Iftikhar K, Kahlid T, Ilyas MB, 1997. Field screening of chickpea germplasm for the sources of resistance against Fusarium wilt. Pak J Phytopathol 9(1): 31-33.

Infantino A, Porta-Puglia A, Singh KB, 1996. Screening wild *Cicer* species for resistance to Fusarium wilt. *Plant Disease* 80: 42-44.

Iqbal SM, Haq IU, Bukhari A, Ghafoor A, Haqqani AM, 2005. Screening of chickpea genotypes for resistance against Fusarium wilt. *Mycopath* 3(1-2): 1-5.

Iqbal MJ, Iftikhar K, Ilyas MB, 1993. Evaluation of chickpea germplasm for resistance against wilt disease (*Fusarium oxysporum*). *J Agric Res* 31(4): 449- 453.

Iruela M, Castro P, Rubio J, Cubero JI, Jacinto C, Millan T, Gil J, 2007. Validation of a QTL for resistance to ascochyta blight linked to resistance to fusarium wilt race 5 in chickpea (*Cicer arietinum* L.). *Eur J Plant Pathol* 119: 29-37.

Jiménez-Díaz RM, Alcala-Jim´´enez AR, Hervas A, Trapero-Casas JL, 1993. Pathogenic variability and hosts resistance in the *Fusarium oxysporum* f. sp. *ciceris/Cicer arietinum* pathosystem. In: Proc Eur Semin Fusarium Mycotoxins, Taxonomy, Pathogenicity and Host Resistance, 3rd Hodowsla Roslin Aklimatyazacja i Nasiennictwo. *Plant Breeding and Acclimatization Institute,* Radzikov, Poland, pp. 87–94.

Jiménez-Díaz RM, Singh KB, Trapero-Casas A, Trapero-Casas JL, 1991. Resistance in kabuli chickpeas to Fusarium wilt. *Plant Dis* 75: 914-918.

Jiménez-Díaz RM, Trapero-Casas A, Cabrera de la Colina J, 1989. Races of *Fusarium oxysporum* f. sp. *ciceris* infecting chickpea in southern Spain. In: Tjamos EC, Beckman CH (eds) *Vascular wilt diseases of plants.* NATO ASI Series, vol. H28. Springer Verlag, Berlin, pp. 515-520.

Jiménez-Gasco MM, Jiménez-Díaz RM, 2003. Development of a specific polymerase chain reaction-based assay for the identification of *Fusarium oxysporum* f. sp. *ciceris* and its pathogenic races 0, 1A, 5 and 6. *Phytopathology* 93: 200-209.

Jiménez-Gasco MM, Pérez-Artes E, Jiménez-Díaz RM, 2001. Identification of pathogenic races 0, 1B/C, 5, and 6 of *Fusarium oxysporum* f. sp. *ciceris* with random amplified polymorphic DNA (RAPD). *Eur J Plant Pathol* 107: 237-248.

Kaiser WJ, Alcalá-Jiménez AR, Hervas-Vargas A, Trapero-Casas JL, Jiménez-Díaz RM, 1994. Screening of wild *Cicer* species for resistance to races 0 and 5 of *Fusarium oxysporum* f. sp. *ciceris*. *Plant Disease* 78: 962-967.

Kelly AG, Alcalá-Jiménez AR, Bainbridge BW, Heale JB, Pérez-Artes E, Jiménez-Díaz RM, 1994. Use of genetic fingerprinting and random amplified polymorphic DNA to characterize pathotypes of *Fusarium oxysporum* f. sp. *ciceris* infecting chickpea. *Phytopathology* 84: 1293-1298.

Kraft JM, Haware MP, Jiménez-Díaz RM, Bayaa B, Harrab M, 1994. Screening techniques and sources of resistance to root rots and wilts in cool season food legumes. In: Muehlbauer FJ, Kaiser WJ (eds) *Expanding the production and use of cool season food legumes.* Dordrecht, Kluwer Academic Publ, Netherlands, pp. 268-289.

Kumar J, Haware MP, Smithson JB, 1985. Registration of four short duration fusarium-wilt resistant kabuli chickpea germplasms. *Crop Sci.* 25: 576-577.

Kumar S, 1998. Inheritance of resistance to Fusarium wilt (race 2) in chickpea. *Plant Breeding* 117: 139-142.

Landa BB, Navas-Cortes JA, Jiménez-Díaz RM, 2004. Integrated Management of *Fusarium* Wilt of Chickpea with Sowing Date, Host Resistance, and Biological Control. *Phytopathologia* 94(9): 946-960.

Landa BB, Navas-Cortes JA, Jiménez-Gasco MM, Katan J, Retig B and Jimenez-Diaz RM, 2006. Temperature response of chickpea cultivars to races of *Fusarium oxysporum* f. sp. *ciceris*, causal agent of Fusarium wilt. *Plant Disease* 90: 365-374.

Lichtenzveig J, Scheuring C, Dodge J, Abbo S, Zhang HB, 2005. Construction of BAC and BIBAC libraries and their applications for generation of SSR markers for genome analysis of chickpea, *Cicer arietinum* L. *Theor Appl Genet* 110: 492-510.

Mayer MS, Tullu A, Simon CJ, Kumar J, Kaiser WJ, Kraft JM, Muehlbauer FJ, 1997. Development of a DNA marker for fusarium wilt resistance in chickpea. *Crop Sci* 37: 1625-1629.

Meki S, Ahmed S, Sakhuja P K, 2011. Control of chickpea wilt (*Fusarium oxysporum* f. sp. *ciceris*) using *Trichoderma* spp. in Ethiopia. *Archives of Phytopathology and Plant Protection* 44(5): 432-440.

Millán T, Clarke HJ, Sidiqque KHM, Buhariwalla HK, Gaur PM, Kumar J, Gil J, Kahl G, Winter P, 2006. Chickpea molecular breeding: new tools and concepts. *Euphytica* 147: 81-103.

Millán T, Winter P, Jüngling R, Gil J, Rubio J, Cho S, Cobos MJ, Iruela M, Rajesh PN, Tekeoglu M, Khal G, Muehlbauer FJ, 2010. A consensus genetic map of chickpea (*Cicer arietinum* L.) based on 10 mapping populations. *Euphytica* 175: 175-189.

Nayak SN, Zhu H, Varghese N, Datta S, Choi HK, Horres R, Jüngling R, Singh J, Kavi Kishor PB, Sivaramakrishnan S, Hoisington DA, Kahl G, Winter P, Cook DR, Varshney RK, 2009. Integration of novel SSR and gene-based SNP marker loci in the chickpea genetic map and establishment of new anchor points with *Medicago truncatula* genome. *Theor Appl Genet* 120: 1415-1441.

Nene YL, Haware MP, 1980. Screening chickpea for resistance to wilt. *Plant Disease* 64: 379-380.

Nene YL, Haware MP, Reddy NMV, Philps JP, Castro EL, Kotasthane SR, Gupta O, Singh G, Shukia P, Sah RP, 1989. Identification of broad based and stable resistance to wilt and root-rots in chickpea. *Indian Phytopathol* 42: 499-505.

Nikam PS, Jagtap GP, Sontakke PL, 2007. Management of chickpea wilt caused by *Fusarium oxysporum* f. sp. *ciceris*. *African Journal of Agricultural Research* 2(12): 692-697.

Palomino C, Fernández-Romero MD, Rubio J, Torres A, Moreno MT, Millán T, 2009. Integration of new CAPS and dCAPS-RGA markers into a composite chickpea genetic map and their association with disease resistance. *Theor Appl Genet* 18: 671–682.

Pathak MM, Singh KP, Lal SB, 1975. Inheritance of resistance to wilt (*F. oxysporum* f. sp. *ciceris*) in gram. *Indian J Farm Sci* 3: 10-11.

Pfaff T, Kahl G, 2003. Mapping of gene-specific markers on the genetic map of chickpea (*Cicer arietinum* L.). *Mol Genet Genom* 269: 243-251.

Phillips JC, 1988. A distinct race of chickpea wilt in California. *Int Chickpea Newsl* 18: 19-20.

Prasad RD, Rangeshwaran R, Anuroop CP, Rashni HJ, 2002. Biological control of wilt and root rot of chickpea under field conditions. *Ann Pl Prot Sci* 10(1): 72-75.

Rajesh PN, Coyne C, Meksem K, Sharma KD, Gupta V, Muehlbauer FJ, 2004. Construction of a *HindIII* bacterial artificial chromosome library and its use in identification of clones associated with disease resistance in chickpea. *Theor Appl Genet* 108: 663-669.

Ratnaparkhe MB, Santra DK, Tullu A, Muehlbauer FJ, 1998a. Inheritance of inter-simple-sequence-repeat polymorphisms and linkage with a fusarium wilt resistance gene in chickpea. *Theor Appl Genet* 96: 348-353.

Ratnaparkhe MB, Tekeoglu M, Muehlbauer FJ, 1998b. Inter-simple-sequence-repeat (ISSR) polymorphisms are useful for finding markers associated with disease resistance gene clusters. *Theor Appl Genet* 97: 515519.

Reddy MV, Raju TN, Pundir RPO, 1990. Additional source of resistance to wilt and root rots in chickpea. *Internat Chickpea Newslett* 22: 36-38.

Rubio J, Moussa E, Kharrat M, Moreno MT, Millán T, Gil J, 2003. Two genes and linked RAPD markers involved in resistance to *Fusarium oxysporum* f. sp. *ciceris* race 0 in chickpea. *Plant Breeding* 122: 188-191.

Sethy NK, Choudhary S, Shokeen B, Bhatia S, 2006a. Identification of microsatellite markers from *Cicer reticulatum*: molecular variation and phylogenetic analysis. *Theor Appl Genet*112: 347-357.

Sethy NK, Shokeen B, Bhatia S, 2003. Isolation and characterization of sequence tagged microsatellite sites markers in chickpea (*Cicer arietinum* L.). *Molecular Ecology Notes 3*: 428-430.

Sethy NK, Shokeen B, Edwards KJ, Bhatia S, 2006b. Development of microsatellite markers and analysis of intraspecific genetic variability in chickpea (*Cicer arietinum* L.). *Theor Appl Genet* 112: 1416-1428.

Sharma KD, Muehlbauer FJ, 2005. Genetics mapping of *Fusarium oxysporum* f. sp. *ciceris* race-specific resistance genes in chickpea (*Cicer arietinum* L.). In: *Abstract of the International food legume research conference-IV*. Indian Agricultural Research Institute, New Delhi, pp. 18-22.

Sharma KD, Muehlbauer FJ, 2007. Fusarium wilt of chickpea: physiological specialization, genetics of resistance and resistance gene tagging. *Euphytica* 157: 1-14.

Sharma KD, Chen W, Muehlbauer FJ, 2005. Genetics of chickpea resistance to five races of Fusarium wilt and a concise set of race differentials for *Fusarium oxysporum* f. sp. *ciceris. Plant Disease* 89: 385-390.

Sharma KD, Winter P, Kahl G, Muehlbauer FJ, 2004. Molecular mapping of *Fusarium oxysporum* f. sp. *ciceris* race 3 resistance gene in chickpea. *Theor Appl Genet* 108: 1243-1248.

Scott JW, Agrama HA, Jones JP, 2004. RFLP-based analysis of recombination among resistance genes to Fusarium wilt races 1, 2 and 3 in tomato. *J Am Soc Hortic Sci* 129(3): 394-400.

Simon CJ, Muehlbauer FJ, 1997. Construction of a chickpea linkage map and its comparison with maps of pea and lentil. *J Hered* 88: 115-119.

Singh G, Chen W, Rubiales D, Moore K, Sharma YR, Gan Y, 2007. Diseases and their management. In: SS Yadav, R Redden, W Chen, B Sharma (eds) *Chickpea Breeding and Management.* CABI Publ, Wallingford, UK, pp. 497-519.

Singh G, Kapoor S, Gill AS, Singh K, 1986. Chickpea varieties resistant to Fusarium wilt and foot rot. *Indian Journal of Agricultural Science* 56: 344-346.

Singh H, Kumar J, Smithson JB, Haware MP, 1987a. Complementation between genes for resistance to race 1 of *Fusarium oxysporum* f. sp. *ciceris* in chickpea. *Plant Pathol* 36: 539-543.

Singh H, Kumar J, Haware MP, Smithson JB, 1987b. Genetics of resistance to Fusarium wilt in chickpeas. In: Day PR, Jellis GJ (eds) *Genetics and plant pathogenesis.* Blackwell Scientific Publications, Oxford, UK, pp. 339–342.

Singh KB, 1987. Chickpea Breeding. In: Saxena MC and Singh KB (eds) *The chickpea.* CAB International, Wallingford, U.K. pp. 127-162.

Singh KB, Reddy MV, 1991. Advances in disease resistance breeding in chickpea. *Adv Agron* 45: 191-222.

Singh KB, Ocampo B, Robertson LD, 1998. Diversity for abiotic and biotic stress resistance in the wild annual *Cicer* species. *Genet Resour Crop Evol* 45: 9-17.

Singh PK, Singh M, Vyas D, 2010. Biocontrol of Fusarium wilt of chickpea using Arbuscular Mycorryzal Fungi and *Rhizobium leguminosorum* Biovar. *Caryologia* 63(4): 349-353.

Subhani MN, Sahi ST, Hussain S, Ali A, Iqbal J, Hameed K, 2011. Evaluation of various fungicides for the control of gram wilt caused by *Fusarium oxysporum* f. sp. *ciceris. African Journal of Agricultural Research* 6(19): 4555-4559Taylor PWJ, Ford R, 2007. Chickpea. *Genome Mapping and Molecular Breeding in Plants* 3: 109-121.

Tekeoglu M, Tullu A, Kaiser WJ, Muehlbauer FJ, 2000. Inheritance and linkage of two genes that confer resistance to fusarium wilt in chickpea. *Crop Science* 40: 1247-1251.

Trapero-Casas A, Jiménez-Díaz RM, 1985. Fungal wilt and root rot diseases of chickpea in Southern Spain. *Phytopathology* 37: 197-246.

Tullu A, 1996. Genetics of Fusarium wilt resistance in chickpea, *Dissertation.* Washington State University, Pullman, WA

Tullu A, Kaiser WJ, Kraft JM, Muehlbauer FJ, 1999. A second gene for resistance to race 4 of Fusarium wilt in chickpea and linkage with a RAPD marker. *Euphytica* 109: 43-50.

Tullu A, Muehlbauer FJ, Simon CJ, Mayer MS, Kumar J, Kaiser WJ, Kraft JM, 1998. Inheritance and linkage of a gene for resistance to race 4 of fusarium wilt and RAPD markers in chickpea. *Euphytica* 102: 227-232.

Udupa SM, Baum M, 2003. Genetic dissection of pathotype-specific resistance to ascochyta blight disease in chickpea (*Cicer arietinum* L.) using microsatellite markers. *Theor Appl Genet* 106: 1196-1202.

Upadhyaya HD, Haware MP, Kumar J, Smithson JB, 1983a. Resistance to wilt in chickpea. I. Inheritance of late wilting in response to race 1. *Euphytica* 32: 447-452.

Upadhyaya HD, Smithson JB, Haware MP, Kumar J, 1983b. Resistance to wilt in chickpea. II. Further evidence for two genes for resistance to race 1. *Euphytica* 32: 749-755.

Varshney RK, Close TJ, Singh NK, David A Hoisington DA, Cook DR, 2009. Orphan legume crops enter the genomics era!. *Current Opinion in Plant Biology* 12: 202-210.

Winter P, Benko-Iseppon AM, Hüttel B, Ratnaparkhe M, Tullu A, Sonnante G, Pfaff T, Tekeoglu M, Santra D, Sant VJ, Rajesh PN, Kahl G, Muehlbauer FJ, 2000. A linkage map of the chickpea (*Cicer arietinum* L) genome based on recombinant inbred lines from a *C. arietinum* x *C. reticulatum* cross: localization of resistance gene for fusarium wilt races 4 and 5. *Theor Appl Genet* 101: 1155-1163.

Winter P, Pfaff T, Udupa SM, Hüttel B, Sharma PC, Sahim S, Arreguin-Espinoza R, Weigand F, Muehlbauer FJ, Kahl G, 1999. Characterization and mapping of sequence-tagged microsatellite sites in the chickpea (*Cicer arietinum* L.) genome. *Mol Gen Genet* 262: 90-101.

Zhang X, Scheuring CF, Zhang M, Dong JJ, Zhang Y, Huang JJ, Lee MK, Abbo S, Sherman A, Shtienberg D, Chen W, Muehlbauer F, Zhang HB, 2010. A BAC/BIBAC-based physical map of chickpea, *Cicer arietinum* L. BMC *Genomics* 11: 501.

Zote KK, Khalikar PV, Dandnaik BP, 1983. *Source of resistance to chickpea wilt.* Internat Chickpea Newslett 8: 23.

In: Fusarium
Editors: Tony Ferarri Rios and Emilia Reyes Ortega
ISBN: 978-1-61942-539-2

Chapter 4

VASCULAR WILT OF TOMATO CAUSED BY *FUSARIUM OXYSPORUM* F. SP. *LYCOPERSICI*

F. I. Okungbowa*[1,*] *and H. O. Shittu [1,2]

[1]Department of Plant Biology and Biotechnology, University of Benin, Benin City, Nigeria

[2]Department of Molecular and Cellular Biology, University of Guelph, Guelph, Canada

ABSTRACT

The commercial tomato, *Lycopersicon esculentum* is one of the most popular vegetables in the world. It is cultivated for its culinary, industrial and ornamental values. It has also served as a model plant for various scientific studies. Tomato is affected by several economically important plant diseases, notably vascular wilt caused by *Fusarium oxysporum* f. sp. *lycopersici* (Fol). *Fusarium* is a soil borne fungus which infects the plant root and colonises the xylem vessels, resulting in wilting, stunting and death. *Fusarium* wilt of tomato continues to be a major cause of yield losses in endemic regions. As a result of its growth within the confinement of the vascular tissues of the host plant, chemical and cultural attempts at protecting tomato against *Fusarium* wilt have been rendered ineffective or impractical. The most effective method of control is breeding for resistant cultivars. Resistance is mediated by the I, I-2 and I-3 resistance genes in tomato against Fol. There is remarkably little knowledge available about the molecular mechanism or/and pathogenicity genes required by Fol to cause disease and how hosts combat or tolerate the pathogen. Understanding the process by which Fol infects, colonises and interacts with tomato host plant could lead to more efficient ways of controlling it. More so, the interaction between Fol and tomato, which has become a model system for the study of the molecular mechanism of disease susceptibility and resistance, could give a better insight into the control of wilt in many agriculturally important crops caused by other species of *Fusarium*. This present review highlights the nature and molecular basis of *Fusarium* wilt in tomato, as well as disease management.

---

[*] Corresponding author: fiokun2002@yahoo.com.

INTRODUCTION

Diseases have been a major concern in agriculture for a long time. In the eighteenth century, Duhumel de Monceau described a fungal disease and demonstrated that it could be passed from plant to plant. Presently, about 80% of plant diseases can be traced to fungal pathogens. Tomato, *Lycopersicon esculentum*, is one of the most popular and widely grown vegetables in the world (Heuvelink, 2005), ranking second in importance to potato in many countries (El Hariry *et al.*, 2004). It is cultivated for its fresh fruits, processed food for industrial uses and ornamental values (Kinet and Peet, 1997). It has also been used as a well established model plant for various scientific studies (Shittu and Robb, 2011). Tomato is affected by several economically important plant diseases, notably vascular pathogens causing wilts. In general, fungal vascular disease is caused by species of *Fusarium, Verticillium* and *Ophiostoma* (Agrios, 2005), with the greatest number of these diseases in higher plants caused by *Fusarium* and *Verticillium* species (Green, 1981). *Fusarium* wilt of tomato is caused by three races of *Fusarium oxysporum* f. sp. *lycopersici* (Fol).

Fusarium Schlechtendahl Emend. Snyder & Hansen is a filamentous fungus belonging to the Class Ascomycetes and Family Hypocreaceae (Pitt *et al.*, 1994). It is a soil-inhabiting fungus and is distributed worldwide. Some species of *Fusarium* are restricted to the tropics while others are predominant in the temperate zones. They are also found in desert, alpine, and arctic areas, where harsh climatic conditions prevail (Francis and Burgess, 1975). Many *Fusarium* species are abundant in fertile cultivated and rangeland soils but are relatively uncommon in forest soils (Jeschke *et al.*, 1990). Most species are harmless saprobes, and are relatively abundant members of the soil microbial community and also thrive on subterranean and aerial plant parts, plant debris, and other organic substrates. *Fusarium* species are often regarded as soil-borne fungi because of their abundance in soil and their frequent association with plant roots, as either parasites or saprophytes (Nelson *et al.*, 1994). The genus includes a number of economically important plant pathogenic species (Arif *et al.*, 2011). Its different species are considered to be some of the most important plant disease pathogens, with some species producing mycotoxins (such as fumonisins, zearalenones and trichothecenes) on plants which contaminate the seeds and enter the food chain, affecting human and animal health, and are thus hazardous to agricultural products, wildlife, livestock and humans (Balali and Iranpoor, 2006; Wang *et al.*, 2011). As a result of production of these mycotoxins which affect animal and human health, *Fusarium* species have been studied extensively in the last 20 years (Marasas *et al.*, 1984; Guarro and Gene, 1992; Lacmanova *et al.*, 2009). More recently, they have become important as pathogens of human patients with compromised immune systems (Marasas *et al.*, 1984).

Some *Fusarium* species have a teleomorphic state (Booth, 1981). However, many have active or passive means of dispersal of spores in the atmosphere and are common colonisers of aerial plant parts, where they may result in diseases of considerable economic importance (Burgess *et al.*, 1988). The wide spread distribution of *Fusarium* species has been attributed to the ability of these fungi to grow on a wide range of substrates and their efficient mechanisms for spore dispersal. (Nelson *et al.*, 1994). One of the economic importance of *Fusarium* species over the years is their role as plant pathogens, causing diseases such as crown rot, head blight, and scab on cereal grains; vascular wilts on a wide range of horticultural crops (for example, tomato, cucurbits and banana), root rots; cankers; and other diseases such as pokkah-boeng on sugarcane and bakanae disease of rice (Booth, 1971).

General Characteristics of *Fusarium* Species

The genus *Fusarium* currently contains over 20 species (Mueller and Beckman, 1988; Wang *et al.*, 2011). The commonest species include *Fusarium solani, F. oxysporum*, and *F. chlamydosporum* (Burgess *et al.*,1994). Macroscopic and microscopic features, such as colour of the colony, length and shape of the macroconidia, the number, shape and arrangement of microconidia, and presence or absence of chlamydospores are key features for the differentiation of *Fusarium* species (Larone, 1995; De Hoog *et al.*, 2000).

Fusarium species can grow on Sabouraud Dextrose Agar (SDA) at 25°C. The cultures produce woolly, cottony, flat or spreading colonies (Larone, 1995; Mui-Yun, 2003). Colony colour from the top is white, cream, tan, salmon, cinnamon, yellow, red, violet, pink, or purple. On the underside, the colour may be tan, red, dark purple, brown, or may also be colourless (Nelson *et al.*, 1983). During unfavourable conditions, the fungus produces sclerotia (singular = sclerotium). A sclerotium is the organised mass of hyphae that remains dormant during unfavourable conditions (Mui-Yun, 2003) and germinates later when favourable conditions return and becomes a source of infection. *Fusarium* species typically produce both macroconidia and microconidia from slender phialides (Nelson *et al.*, 1994). Macroconidia are hyaline, two- to several-celled, fusiform- to sickle-shaped, mostly with an elongated apical cell and pedicellate basal cell. Microconidia are 1- to 2-celled, hyaline, pyriform, fusiform to ovoid, straight or curved. Chlamydospores may be present or absent (Nelson *et al.*, 1994). Macroscopic and microscopic features, such as colour of the colony, length and shape of the macroconidia, the number, shape and arrangement of microconidia, and presence or absence of chlamydospores are key features for the differentiation of *Fusarium* species (De Hoog *et al.*, 2000). Molecular methods, such as 28S rRNA gene sequencing, may be used for rapid identification of *Fusarium* strains to species level (Hennequin *et al.*, 1999). Others are polymerase chain reaction (PCR) based rDNA detection (Lacmanova *et al.*,2009) and detection of protein banding patterns by SDS-PAGE and esterase isozyme electrophoresis (El-Kazzaz *et al.,* 2008).

Host Plants

Fusarium oxysporum is a complex species. It causes destructive vascular wilts in a wide variety of crops (Namiki *et al.,* 1994). Tomato and other solanaceous crops, sweet potato, legumes, cucurbits and banana are the most susceptible plants (Miller *et al.*, 2011) though it will also infect other herbaceous plants as well as cotton, ornamentals and palms (especially date palm). However, individual pathogenic strains within the species have a limited host range, and strains with similar or identical host ranges are assigned to intraspecific groups, called formae speciales (f. sp.) (Armstrong and Armstrong, 1981). The high level of host specificity of pathogenic strains in *F. oxysporum* led to the development of the formae speciales concept to enable better differentiation of these morphologically similar strains

(PADIL, 2011). Some of the formae speciales are further divided into subgroups, named races, on the basis of virulence to a set of differential cultivars within the same plant species (Armstrong and Armstrong, 1981). The formae speciales are distinguished by the ability of their members to cause a wilt disease on a limited taxonomic range of host plants (Lievens *et al.*, 2009). For example, Banana wilt; Panama wilt (*F. oxysporum* f.sp. *cubense*), *Fusarium* wilt of cotton (*F. oxysporum* f.sp. *vasinfectum*), *Fusarium* wilt of sweet potato; Stem rot (*F. oxysporum* f.sp. *batatas*), *Fusarium* wilt of tomato (*F. oxysporum* f.sp. *lycopersici*), *Fusarium* wilt of date palm (*F. oxysporum* f.sp. *albidinis*) and *Fusarium* yellows of common beans (*F. oxysporum* f.sp. *phaseoli*). *Fusarium oxysporum* also includes non-pathogenic strains capable of persisting through asymptomatic colonisation of plant roots and saprophytic growth on dead organic matter.

Fusarium Oxysporum f. sp. *Lycopersici*

On PDA, the mycelia of *F. oxysporum* f. sp. *lycopersici* (Fol) are delicate white (Plate 1A) to pink, often with purple tinge, and are sparse to abundant (Mui-Yun, 2003) with sporodochia often produced in the centre of the colony (Fig.1). Under the microscope, *F. oxysporum* can be differentiated from other *Fusarium* species by the presence of certain characteristic spores (Mui-Yun, 2003; Wokoma, 2008) and by molecular methods (Lievens *et al.*, 2009). On carnation leaf agar, a medium for the conidiation of *F. oxysporum* (Togawa, 1992), the fungus produces three types of spores (Plate1B - D). These are microconidia, macroconidia and chlamydospores (Mui-Yun, 2003). Microconidia are borne on simple, false heads on short monophialides (conidia in heads) arising laterally and are abundant, oval - ellipsoid, straight to curved, 5-12 x 2.2-3.5 mm, and non-septate. Macroconidia, sparse to abundant, are borne on branched conidiophores or on the surface of pale orange sporodochia and are thin walled, three- to five-septate, fusoid-subulate and pointed at both ends, with pedicellate base (having a foot cell). Three-septate spores are more common. Chlamydospores, both smooth and rough walled, are abundant and form terminally or on an intercalary basis, in pairs or chains (De Hoog *et al.*, 2000). *F. oxysporum* f. sp. *lycopersici* has been further sub divided into three physiological races (1, 2 and 3) based on their ability to overcome resistance genes in causing wilt in tomato (Keen, 1990; Elias and Schneider, 1991; Lievens *et al.*, 2009). Race 1 is the most widely distributed and has been reported from most geographical areas (Mui-Yun, 2003). Race 2 was first reported in Ohio (United States) in 1940, and Race 3 in Brazil, in 1966.

Fusarium Wilt Disease of Tomato

As previously mentioned, *Fusarium* wilt disease of tomato is caused by the vascular wilt fungus, *F. oxysporum* f. sp. *lycopersici*. The fungus is a soil-borne facultative parasite that causes economically important losses in a wide variety of crops (Burgess *et al.*, 1994). The fungus directly penetrates roots and colonises the vascular tissue (Inoue *et al.*, 2002). Other wilt pathogens of tomato are *Verticillium albo-atrum* (Pegg and Brady, 2002; Shittu *et al.*, 2009) and *Pseudomonas solanacearum* (Wokoma, 2008). The leaves of *Fusarium* infected plants turn yellow, while leaves of *Verticillium* infected plants will often have brown V-shaped lesions. Vascular system discoloration occurs for both *Fusarium* wilt (brown or red-brown) and *Verticillium* wilt (tan). The vascular wilt disease of tomato was first described by G.E. Massee (1895) in England. It is of worldwide importance where at least 32 countries have reported the disease, which is particularly severe in countries with warm climate (Mui-Yun, 2003). Some of these countries include United States, Australia, United Kingdom, Netherlands, Brazil, Mexico, Nigeria, Morocco, Israel, Iraq and Vietnam.

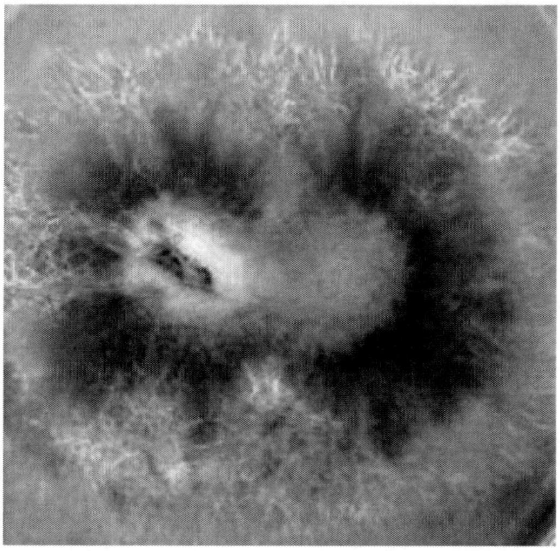

Figure 1. Culture of *F. oxysporum* showing purple pigmentation. (Mycology online). http://www.mycology.adelaide.edu.au/Fungal_Descriptions/Hyphomycetes_(hyaline)/*Fusarium*/.

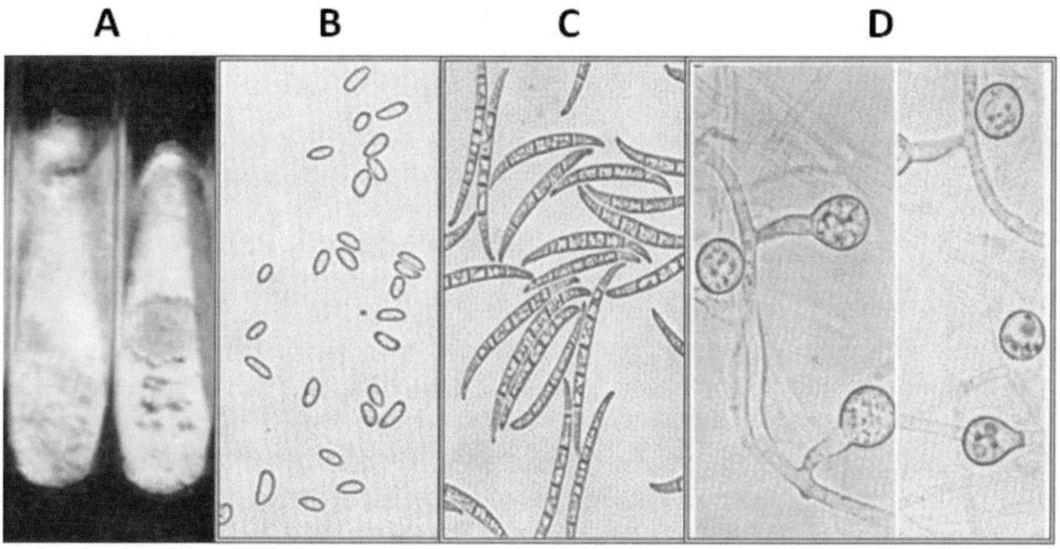

Plate 1. Some characteristic features of *Fusarium oxysporum* f. sp. *lycopersici*. Mycelium (A), Microconidia (B), Macroconidia (C), Chlamydospores (D). (Adapted from Toussoun and Nelson, 1976).

THE *FUSARIUM* COLONISATION CYCLE

Although *F. oxysporum* causes severe wilts in about 80 plant species, the mechanisms of pathogenicity and symptom induction by this fungus are poorly understood (Beckman, 1987). In soil, fungal colonisation of plant roots has been traditionally studied by indirect methods such as microbial isolation that do not enable direct observation of infection sites or of interactions between fungal pathogens and their antagonists (Olivain *et al.*, 2006). Scanning electron microscopy study to visualise the colonisation of tomato roots showed that there is competition for nutrients rather than for infection sites (Olivain *et al.*, 2006). Infection takes place through the roots and the disease develops quickly when soil temperatures are high (El Hariry *et al.*, 2004).

The fungus generally enters through wounds in the roots, although it can also enter roots with an intact epidermis (AAFC, Canada, 2006; Rodríguez-Molina, 2003). Dispersal is limited but *F. oxysporum* spores (macroconidia, microconidia and chlamydospores) are dispersed by contact and transportation through contaminated seeds, soil, plants, equipment and irrigation water, and as such it often enters a new plant through the roots when transplanting. In tomato, the stakes are a common dispersal mechanism of infected soil. Contaminated soil can also be carried on farm animals and people, as well as ploughs and in irrigation water. Certain weeds are symptomless carriers of Fol (Fassihiani, 2000).

There is often an association with *Fusarium* wilt and nematode colonisation, where the nematodes provide a potential entry point (wound) for the fungus and also the bacterial wilt pathogen (Mai and Abawi, 1987; Morrell and Bloom, 1981; AVRDC, 2005). It is common to find root knot nematode galls on roots of tomato infected by *Fusarium* wilt fungus (Morrell and Bloom, 1981; AVRDC, 2005). Nematodes shorten time for expression of wilt symptoms and increase incidence of wilt (Caperton *et al.*, 1986). Non-parasitic nematodes carry fungal spores internally which will not only increase their mobility, but also protects them from fungicides. They also cause some physiological changes which increase the rate of infection. One of such is that some nematodes modify plant tissue and makes it a better substrate for the fungus, thus increasing their growth and reproduction at the expense of the host. In addition, quantitative and qualitative changes in root exudates are induced by some nematodes, stimulating the germination, growth and reproduction of fungal propagules in the rhizosphere microflora which are antagonistic to plant pathogens. Depending on the species of nematode and fungus, concomitant infection may stimulate nematode reproduction (*Pratylenchus* and *Verticillium*) or inhibit reproduction (*Heterodera* and *Fusarium*) (Bergeson, 1972). Growth rate of roots and root branching are frequently suppressed by infection with root-knot nematodes. The altered root growth results in reduced root volume and surface area. Thus the root has a reduced capacity for water and mineral uptake as well as the synthesis of cytokinins, gibberellins, and other growth-determining metabolites (Mai and Abawi, 1987). There may also be abnormal vessel elements and disruption of the arrangement and continuity of the vascular tissues. Such changes markedly affect water flow in the xylem and disrupt nutrient absorption and translocation. It is believed that giant cells associated with nematode feeding serve as a nutrient sink. Furthermore, infection of roots by root-knot nematodes decreases the rate of photosynthesis because of the interference of the nematodes with the production of a photosynthesis regulating factor in root tissues. The concentration of auxins, ethylene, and cytokinins also differs considerably in galled, as compared to non-galled, root tissues.

The pathogen also persists in soil for long periods as chlamydospores in plant residues until germination is stimulated by host roots, root excretions from non-host roots, or contact with pieces of fresh non-colonised plant remains (Steinkellner *et al.*, 2008). During the disease process of *F. oxysporum* the following steps take place. Spores germinate in response to root exudates and produce penetration hyphae that attach to the root surface. Root exudates contain sugars, amino acids and organic acids which are generally known to be germination stimuli of spores (Nelson, 1991). However, what triggers the production of exudates and the actual role of exudates in the germination of spore are not yet understood. Germination in response to production of exudates does not seem to be host-specific as host plants and non-host plants are sometimes affected similarly. The spores germinate on the root surface under conditions (such as temperature of about 28°C and other environmental conditions) that favour optimal growth of tomato plants. Spore germination represents a pivotal step in the colonisation of a new host by filamentous fungi (Harris, 2005). Once dormancy is broken, spores undergo a defined set of morphogenetic events that lead to the formation of a polarized growth axis and the emergence of one or more germ-tubes (Hardham, 2001). Plate 2 below shows the different stages of germination initiation in *F. gramminearum* which is applicable to Fol. Expression of several germination related genes are also involved in the process (Xu, 2000).

Germination is followed by entry of pathogen into the host. A relatively high population level, compared to other soil-borne pathogens is necessary for infection as multiple infection sites are required for above-ground symptoms to occur. The optimum temperature for infection is 28°C (an infected plant may remain symptomless at lower temperatures), with the same moisture, nutrition and pH requirements that favour tomato plant growth. Low light intensity and short day length also favour the development of *Fusarium* wilt (AAFC, Canada, 2006). The short hyphae enter the host, sometimes through wounds (from equipment, handling during transplanting, and wounds from nematodes), but mostly by direct penetration of the host epidermis and cell wall (Inoue *et al.*, 2002; Wanjiru *et al.*, 2002; Mandeel, 2007). The penetration process is likely enhanced by certain hydrolysing enzymes secreted by *Fusarium* (Walter *et al.*, 2009). Next is invasive growth (colonisation) in host plant tissue (Rodríguez-Gálvez and Mendgen, 1995). Colonisation refers to the rapid upward and horizontal growth of the fungus in the host. Following penetration, *Fusarium* hyphae spread within the cell apoplast, which leads to significant cytological alterations (Walter *et al.*, 2009). The hyphae spread up into the stems and leaves through the water conducting vessels in the tomato plant. Effective colonisation is genetically controlled (Inoue *et al.*, 2002). Recently, certain genes which play a role in infection have been isolated (Di Pietro *et al.* 2001; Xu, 2000). During infection of tomato, the fungus, Fol secretes several unique proteins called 'secreted in xylem' (Six) proteins, into the xylem sap (see details under *Fusarium* virulence factors section). At least some of these proteins promote virulence towards tomato (Lievens *et al.*, 2009). Also, a gene that encodes a mitochondrial carrier protein that is required specifically for colonisation in the plant tissue by Fol has been isolated (Inoue *et al.*, 2002). In the advanced stages of the disease, the fungus grows out of the xylem tissue into the adjacent parenchyma, producing extensive hyphae and chlamydospores. As infection spreads, the plant vessels become plugged and damaged so that the water transport system to the tomato plants becomes blocked. This is the process which causes the wilt to occur.

The chlamydospores are returned to the soil when the diseased plant residues decay. They can remain viable in the soil in their dormant state for several years, and grow upon germination by parasitic or saprophytic colonisation of a new host. The saprophytic activity of *F. oxysporum* is not a means of survival, but rather a way to increase inoculum levels to a point where the host will succumb to disease.

SYMPTOMOLOGY AND EPIDEMIOLOGY

The molecular mechanisms of pathogenicity and symptom induction by *F. oxysporum* remain undefined (Beckman, 1987). The first indication of this disease is a yellowing and drooping of the lower leaves. Symptoms are often characterized by a yellowing on only one side of a leaf or branch; the affected leaves wilt and die though they remain attached to the stem (El Hariry *et al.,* 2004; Mueller and Beckman, 1988).The symptoms of *Fusarium* wilt of tomato include (in the order of appearance) leaf chlorosis (Fig. 2A), stunted growth, vascular browning (discolouration) of stem and root tissue (Fig. 2B, starting about 1/8 inch below the bark), and death. Vascular browning stops 5 to 25 cm above the soil level. Tomato plants wilt temporarily during the hotter parts of the day for a few days before dying. Prior to this, yellowing (due to toxins produced by pathogen) begins on the lower tomato leaves, which then turn brownish. Typically the wilting begins on one side of the tomato plant's leaves or shoot and spreads to the other side as infection progresses.

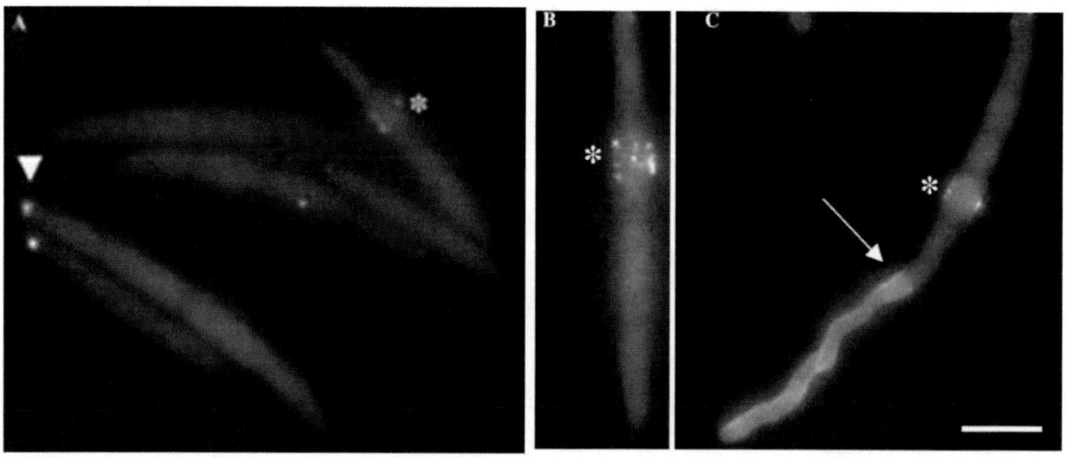

Plate 2. Distinct patterns of chitin deposition associated with germination in *Fusarium graminearum* (teleomorph *Gibberella zeae*) strain PH-1. Macroconidia germinating on cover slips were incubated at 28°C for 5 min in YMA media containing 5–10 mg/mL FITC-conjugated wheat germ agglutinin (WGA), then fixed and analyzed with fluorescence microscopy. Arrowhead= polar chitin patch; Asterisks = chitin patches on internal cells; Arrow = demarcation between chitin-rich and chitin-poor region of new germ tube. Bar = 5 10 mm. SOURCE: Harris (2005).

Full grown tomato plants have the biggest problem with severe wilting; it is not as severe in younger shoots (AVRDC, 2005). In seedling infection, the older leaves droop and curve downward, vascular tissue darkens, and the plant wilts and dies. In older diseased plants, the leaves yellow after blossoming. The plant then wilts during the hotter parts of the day and in a few days, collapses and dies. The vascular tissue in the stem is dark brown but the pith remains healthy. Fruit infection can occur, displaying the same brown vascular discolouration. *Fusarium* wilt of tomato can be associated with infection by root knot nematode (Morrell and Bloom, 1981). The effect of Fol wilt on the photosynthetic ability of tomato was studied by Nogues *et al.* (2002). During the first stage of disease development, Fol race 1 (Fol-1), wilt decreases the light-saturated rate of Carbon (IV) oxide (CO_2) assimilation followed by a decrease in the maximum carboxylation velocity and the capacity for ribulose-1, 5-bisphophate regeneration, and increases stomatal limitation of the CO_2 supply. Fol-1 wilt also decreases leaf area, thereby causing a reduction in the ability of the tomato plants to capture photosynthetically active radiation; hence there is a depression in the photosynthetic productivity of these infected plants.

Fusarium wilt disease of tomato is more prevalent in warmer climates (AVRDC, 2005). This is because high temperatures favour infection and colonisation (Morrell and Bloom, 1981). Infection commonly takes place at flowering. When the disease occurs in a tomato plant, the latter becomes stunted and may not produce adequate fruit. Typically, the plant either does not produce fruit at all, or if it does, the fruit is deformed. This leads to great economic losses. *Fusarium* wilt in tomato plants is favoured by temperature of 28°C, especially during weather that is wet which allows it to spread more quickly throughout the nearby tomato plants.

Figure 2. (A) Wilt of tomato caused by *Fusarium oxysporum* f.sp *lycopercisci* (El Hariry *et al.*, 2004). (B) Dark brown vascular discoloration (AVRDC, 2005).

Plant Defence against *Fusarium Species*

Resistance to a pathogen can be measured by components such as pathogen growth, disease symptoms and host fitness (Kover and Schaal, 2002). The best known defence mechanism against infection by vascular wilt pathogens is through quantitative resistance, involving numerous genes that are expressed constitutively or induced in response to attack by the pathogen. These genes provide the plant with defensive structures, degradative enzymes and toxic substances that slow or restrict the colonisation of the pathogen in the host tissue and reduce the damage caused by the pathogen (Agrios, 2005). In some crops, including tomato, monogenic resistance against *F. oxysporum* has been found (Michielse and Rep, 2009). Resistance genes encode specialized proteins that recognize pathogen effectors, resulting in incompatible reaction (Stukenbrock and McDonald, 2009). In tomato, I (for immunity) genes I, I-1, I-2 and I-3 have been implicated in resistance against Fol (Huang and Lindhout, 1997). Of these genes, only I-2 has been cloned and studied (Simons *et al.*, 1998).

In a plant resisting *Fusarium* species attack, both structural and biochemical defence responses are employed to restrict vascular colonisation of the pathogen to the basal part of plant. Structural defences can occur in both lateral and longitudinal directions (Beckman, 1987). The lateral defence response inhibits the pathogen from getting into the vascular system in the prevascular stage of attack (i.e. determinative phase 1). This is brought about by the contact cells, parenchyma cells which lie adjacent to vessel elements (Mueller and Beckman, 1984). If the pathogen escapes the prevascular phase and succeeds in penetrating the plant xylem vessels (i.e. vascular stage or determinative phase II), the plant resists fungal penetration and horizontal spread by forming other barriers through the deposition of callose (Beckman *et al.*, 1982), suberin and other coating materials on vascular cell walls surrounding the trapping sites (Bell and Mace, 1981). At this stage, infection commonly results in vessel occlusion by gummosis, gelation, vessel wall coatings and other deposits secreted by the neighboring parenchyma cells (VanderMolen *et al.*, 1977; Ludwig, 1952; Beckman, 2000). Tyloses, outgrowths of vascular parenchyma cells, are also formed in tomato to wall off the pathogens (Beckman and Talboys, 1981; Bishop and Cooper, 1984). All these obstructions block movement of the fungus and other transport, through the vessel element and may contribute to wilting of the plant. It should be noted that the classical hypersensitive response (HR) associated with defence against foliar pathogens (Pedley and Martin, 2003) does not occur in incompatible reaction between tomato and Fol.

Pathogenesis-related (PR) proteins such as glucanases and chitinases as well as phenylalanine ammonia lyase (PAL) and phenolic compounds are also important because they contribute to the production of phytoalexins (stress metabolites) that limit the growth of the fungus or inhibit the hydrolytic enzymes secreted by them. Some compounds that have been identified in tomato include chlorogenic acid and other cinnamic acid derivatives; scopolin; a number of flavonoids; tomatine, an antifungal steroid alkaloid; and rishitin, a terpenoid compound (Stoessl, *et al.*, 1976; Beckman, 1987; Keukens *et al.*, 1992). Fol has also been found to induce an increased production and deposition of cell wall strengthening materials such as lignin (Mandal and Mitra, 2007). The lignification or incorporation of lignin - like materials into cell walls and tissues is a common occurrence in plants following injury or infection, and such tissue is rarely penetrated by plant pathogenic microorganism.

FUSARIUM VIRULENCE FACTORS

Virulence factors are molecules such as toxins, secreted by pathogens to disrupt the proper functioning of host plant, thereby increasing its disease causing ability (Hogan *et al.*, 1996). Several fungal pathogens have been demonstrated to secrete various extracellular materials hypothesized to serve as virulence factors (Wanjiru *et al.*, 2002). For example, it has been shown that a secreted lipase encoded by FGL1 gene in *F. graminearum,* the causal agent of the *Fusarium* head blight disease is required for infection of cereals (Voigt, 2005). The mechanism by which vascular pathogens such as *Fusarium and Verticillium* species induce wilt symptoms in a susceptible host plant has been an issue addressed by many research studies and has provoked much scientific controversy for decades. The lifestyle of these pathogens is unusual in that, in the plant, the fungi are mainly confined to the fluid environment of the xylem vessels and exert their effect on host physiology indirectly (Pegg and Brady, 2002). The classic debate has been whether vascular occlusion or toxin activity causes wilting.

Vascular occlusion can result from the physical blockage of the plant xylem by either the pathogen biomass or vessel plugging by the host (Beckman, 1987; Pegg and Brady, 2002). Colonisation of the host xylem is effected by mycelia growth, but more rapidly by conidia carried in the xylem fluid (Di Pietro *et al.*, 2001; reviewed in Di Pietro *et al.*, 2003). Vascular occlusion as a result of vessel plugging occurs in a susceptible host plant in an attempt to defend itself against infection by *Fusarium* species. This is attributed to blocking of the vessels by gums, gels, tyloses and xylem vessel wall coatings (Moreau *et al.*, 1978; VanderMolen *et al.*, 1977).

If vascular pathogens such as *Fusarium* and *Verticillium* are to colonise the host successfully, they must have the capacity to prevent, delay, or disrupt one or more of the host defence responses. There is considerable evidence that this is the case (Beckman, 1987; Lairini *et al.*, 1996; Desjardins and Proctor, 2007). These can be achieved by avoiding or overcoming one or more of the host defence strategies such as recognition, callose deposition, vascular gelation, vascular blockage by tylose occlusion (vessel crushing) and stress metabolites (Beckman, 1987), which can be achieved through the secretion of toxins. Stress metabolites refer to secondary metabolites which are produced when a plant is subjected to environmental or pathogenic attack (Edreva, 2008). Toxins can be sub-divided broadly into small molecular weight (<1000 Da) and macromolecular weight (>1000 Da) compounds. In this model, toxic molecules as extracellular metabolites of the pathogen are envisaged to pass through xylem pit membranes and other parenchyma cells and there directly affect permeability, causing cellular dysfunctioning and death (Pegg and Brady, 2002).

Recent trend in *Fusarium* research using fluorescent proteins in visualizing colonisation of the pathogen during host infection has given more insight into pathogenesis of the fungus (Di Pietro *et al.*, 2003). Also, with the use of forward and reverse genetic approaches, the role of several genes which function as virulence factors of Fol has been identified (Michielse and Rep, 2009). Martin-Urdiroz and coworkers (2008) identified three genes (chs1, chs2 and chs3) in the genome of Fol, which encode structural chitin synthases of class I, class II and class III, respectively. Using targeted gene disruption by homologous recombination, the results suggest that chitinases may play a role in pathogenesis. Other studies also implicated chitinases as pathogenicity factors of Fol (Madrid *et al.*, 2003; Martin-Urdiroz *et al.*, 2008). The use of *Agrobacterium*-mediated insertional mutagenesis to generate and screen several transformants for loss or reduction of pathogenicity, coupled with complementation and gene knockout experiments, suggest that the role of some transcription factor, chitin synthase V, developmental regulator FlbA, phosphomannose isomerase and certain cellular processes are important in pathogenesis of Fol (Michielse *et al.*, 2009). During infection, *Fusarium* species secrete several cell wall degrading enzymes such as pectate lyases, proteases and xylanases

(Beckman, 1987), but their role as virulence factors is not yet understood (Michielse and Rep, 2009).

A defence strategy in tomato is the production and secretion of α-tomatine, an antifungal material which causes pores and leakage of fungal cellular content (Keukens *et al.*, 1995). Fol can overcome the defence response of host through the secretion of tomatinase, a tomatine-degrading enzyme (Lairini *et al.*, 1996). A recent study has examined *Tom*1 gene, which encodes the tomatinase enzyme, Tom1 that detoxifies α-tomatine. Overexpression and gene knockout analysis implicated its role in virulence of Fol (Pareja-Jaime *et al.*, 2008). The study also revealed the existence of four other tomatinase genes in tomato genome. Resistance of tomato against Fol exhibits gene-for-gene relationships (Flor, 1971). Non virulence proteins in Fol that have corresponding counterparts in tomato have been recently recognised. Rep *et al.* (2004) identified a 12 kDa cysteine-rich protein secreted by Fol during colonisation of tomato xylem vessels. Mass spectrometry was used to obtain the peptide sequences which allowed the identification of the coding sequence. The gene was designated Six1 (Secreted In Xylem 1). The results from the studies suggested that in an incompatible reaction, the tomato I-3 gene recognises Fol Six 1 gene. Houterman *et al.*, (2008) later renamed Six1 gene Avr3. More recently, other avirulence (Six) genes matching other tomato resistance I-genes have also been identified (Houterman *et al.*, 2007; 2008).

TOMATO-*FUSARIUM* ARMS RACES

The interactions between plants and pathogens have been in existence for hundreds of millions of years (Takken and Rep, 2010) with each competing partner engaging in a sort of warfare, with series of advances and retreats that eventually lead to novel resistance or virulence traits (Kareiva, 1999). This antagonistic coevolution starts when the pathogen attacks and exploits the host; this exploitation has a negative effect on the plant and results in selection for a novel defence strategy that spreads through the entire host population. This leads to a decrease in the pathogen fitness, which in turn results in selection for a genotype that can overcome the defence that also spreads through the pathogen population (Stahl and Bishop, 2000). This evolutionary dance between plants and pathogens is what Karaive (1999) refers to as "reciprocal adaptive genetic changes within populations of interacting species that act as selective agents for one another."

Similarly, the interactions between tomato and Fol have become an evolutionary arms race, resulting in the formation of several races of the pathogen on the basis of the ability to overcome specific I gene in tomato (Keen, 1990). The I gene for resistance against Fol race 1 had gained wide use before 1945 when Alexander and Tucker (1945) reported another evolving strain that was pathogenic on tomato cultivar having the I resistance gene. The new strain was designated as race 2. It was in the early 1960s that Fol race 2 appeared in Florida (Stall, 1961) and later spread to other parts of the world. Meanwhile, in 1955, Alexander and Hoover (1955) found a source of resistance to both races 1 and 2 and the gene was eventually designated as I-2 (Cirulli and Alexander, 1966). In the interim, race 3 had emerged which was pathogenic on cultivars with the I-2 gene (Jones and Woltz, 1981; Grattidge and O'Brien, 1982; Reis *et al.*, 2005; Holguín-Peña, 2005). An overview of the arms race between tomato and *F. oxysporum* has been comprehensively reviewed (Takken and Rep, 2010).

Quantitative Molecular Diagnostic Tools

Fusarium can be differentiated at the species level using various morphological characteristics on selective media (Burgess, *et al.*, 1994), but the various forms (e.g. f. sp. *lycopersici)* and races (e.g. 1, 2 and 3) cannot be identified using this method. More so, Fol and most vascular pathogens grow in the confinement of the vascular system of plants, hence more sensitive and specific diagnostic assays are required to detect and quantify amount in infected field, seeds, plant or soil sample. More recently, sensitive and accurate quantitative diagnostic tools that have been employed involve nucleic acid sequence amplification-based (NASAB) methods. These rely on the recognition of specific DNA sequences such as the internal transcribed spacer regions and/or the intergenic regions of the rRNA genes for differentiating various species of *Fusarium,* while avirulence gene (van der Does *et al.*, 2008; Houterman *et al.*, 2008), endo- and exo-polygalacturonase gene sequences (Hirano and Arie, 2006) are used to differentiate various races. Possible approaches include the use of the polymerase chain reaction (PCR) (Nicholson, *et al.*, 1998; Hirano and Arie, 2006; Balogun, 2007), real time PCR (Okubara, *et al.*, 2005; Lievens, *et al.*, 2006; Inami, *et al.*, 2010) and PCR enzyme-linked immunosorbent assay, ELISA (Cullen, *et al.*, 2001).

Economic Importance and Disease Management

Fusarium wilt of tomato caused by Fol leads to great losses in warm climates and sandy soils of temperate regions (Wokoma, 2008). Entire plants and fruit (if any) are lost. Several chemical fungicides such as prochloraz and carbendazim (Song *et al.*, 2004), Bavistin (Alam *et al.*, 2010) and salicylic acid (Amel *et al.*,2010), are used to suppress the disease by inducing resistance, but these chemicals have a negative impact on human health and are hazardous to the environment. The use of chemicals as the sole method of control may indeed result in non-target effects on soil microorganism populations. Others may cause the tomato plant to become resistant to these chemicals, over time. One way of overcoming this problem could be to alternate chemicals with different modes of action. A better alternative to chemicals are the soil microbes such as *Trichoderma harzianum, T. asperellum, T. koningii, Penicillium* spp. and *Streptomyces griseoviridis* residing in the rhizosphere of the plants and have the ability to suppress the pathogens and stimulate plant growth by the production of phytohormones and/or degradation of complex substrates (Osuide *et al.*, 2002; Syed *et al.*, 2010; Borrero *et al.,* 2011). Several disadvantages are associated with these biocontrol measures. One is that *in vitro* results do not always translate to successful field trials; because biofungicides are usually composed of living organisms, their efficacy may vary under a wide range of environmental, cultural, and biotic conditions. Besides, several chemical fungicides are effective against soil-borne pathogens and possess competitive advantages not yet available with their biofungicide counterparts. Thirdly, conventional fungicides are often cheaper and easier to use and their efficacy is constant except in very severe conditions. Due to such difficulties, integrated pest management strategies offer bright opportunities of success, provided that the mode of action of the various components is understood. Then the registration procedures for those that have proved effective are cumbersome and difficult in many countries. Examples of registered biocontrol agents that can be used to help control the disease are as follows. *Streptomyces griseoviridis* Strain K61 (Mycostop®) can be applied to seedlings in rock wool after emergence or immediately after transplanting. *Trichoderma harzianum* strain KRL-AG2 wettable powder (Rootshield®) can be applied as a drench or incorporated as granules into the seeding or transplanting medium.

Severity of disease can also be reduced by soil amendment through composting (Cotxarrere *et al*., 2002; Ahmed, 2011) and by heat treatment (Anchis *et al.*, 2008). Care must be taken with respect to exclusion and early quarantine, the best control measures. Cultivation of resistant varieties is commonly practiced as well as maintaining good hygiene (such as avoidance of contamination, during planting and seedling transplanting, irrigation, and clearing of debris from previous year's planting) (Jones and Woltz, 1981).

Symptomless carriers of fungus, for example, certain weeds (Fassihiani, 2000) should be removed from around the tomato field. It is advisable not to plant solanaceous plants in the field if soil is heavily infested because these may act as symptomless carriers (OISAT, 2011). By raising the soil pH to 6.5-7 and using nitrates rather than ammonia for nitrogenous fertilisation, disease development is limited (AVRDC, 2005; Borrero *et al.,* 2011). Crop rotation is generally ineffective due to the effective survival strategies of the pathogen. There is also the potential for the development of new races that may overcome cultivar resistance (AAFC, Canada, 2006). Some laboratory investigations have indicated the possible use of plant extracts for the control of *F. oxysporum* (Okungbowa and Edema, 2006; Okungbowa 2011).

Conclusion

Fusarium wilt still remains one of the major causes of crop loss in tomato production all over the world, despite measures aimed at prevention and control. Use of resistant cultivars is the best method of control, but the most popular cultivars are not resistant. Moreover, new races of the pathogen can develop and fumigants are being lost from the market (mainly due to impact on the environment). Alternative control strategies are being investigated. More research is needed to elucidate certain aspects of the disease, such as actual role of root exudates and genetic control of spore germination and whether it is vascular occlusion or presence of toxin that causes death of the plant, in order to develop alternative control methods of which biocontrol is one.

References

AAFC, Canada (2006). Crop profile for greenhouse tomato in Canada. Agriculture and Agri-Food Canada. http://dsp-psd.pwgsc.gc.ca/collection_2009/agr/A118-10-24-2006E.pdf.

Agrios, G. N. (2005). Plant Pathology. 5th ed. Burlington, USA: Elsevier Academic Press. 432 pp.

Ahmed, M. (2011). Management of *Fusarium* wilt of tomato by soil amendment with *Trichoderma koningii* and a white sterile fungus. *Indian Journal of Research, 5*:35-38.

Alam, S.S., Sakamoto, K., Amemiya, Y. & Inubushi, K. (2010). Biocontrol of soil-borne *Fusarium* wilts of tomato and cabbage with a root colonizing fungus, *Penicillium* sp. EU0013. *Proceedings of the 19th World Congress of Soil Science: Soil Solutions for a Changing World.* 1 – 6 August 2010, Brisbane, Australia. Pp.20-22.

Alexander, L. J. & Tucker, C. M. (1945). Physiologic specialization in the tomato wilt fungus *Fusarium oxysporum* f. sp. *lycopersici. Journal of Agricultural Research, 70*, 303-313.

Alexander, L. J. & Hoover, M. M. (1955). Disease resistance in wild species of tomato. Ohio Agric. Exp. Stn. Res. Bull. 752.

Amel, A.H., Soad, M., A. & Ahmed A.I. (2010). Activation of Tomato Plant Defense Response Against *Fusarium* Wilt Disease Using *Trichoderma Harzianum* and Salicylic Acid under Greenhouse Conditions. Research Journal of Agriculture and Biological Sciences, *6(3)*: 328-338.

Anchisi, M., Gennari, M. & Matta A. (1985). Retardation of *Fusarium* wilt symptoms in tomato by pre- and post-inoculation treatments of the roots and aerial parts of the host in hot water. *Physiological Plant Pathology, 26 (2)* :175-183.

Arif, M., Pani, D. R., Zaidi, N. W. & Singh, U. S. (2011). PCR-based identification and characterization of *Fusarium* sp. associated with mango malformation. *Biotechnology Research International*. doi:10.4061/2011/141649.

Armstrong, G.M., & Armstrong, J.K. (1981). Formae speciales and races of *Fusarium oxysporum* causing wilt disease. In *Fusarium*: Disease, biology, and taxonomy. P.E. Nelson, T.A. Toussoun & R.J. Cook (eds). Pp. 391–399. University Park, PA: Pennsylvania State University Press.

AVRDC, (2005). The World Vegetable Centre. *Fusarium Wilt*. Fact Sheet, Vol. 5,pg. 627. http://www.avrdc.org/pdf/tomato/*Fusarium*.pdf. Accessed 19th August, 2011.10.30pm.

Balali, G.R. & Iranpoor, M. (2006). *Iranian Journal of Science and Technology,Transaction A, 30(1):* 91-102.

Balogun, O. S. (2007). Evaluation of the Efficiency of the Products of a Non-Phenol-Chloroform Dna Extraction Method as Templates for PCR Analysis of Some *Fusarium oxysporum* Isolates *American-Eurasian Journal of Scientific Research 2 (2)*, 141-145.

Beckman, C. H. (1987). *The Nature of Wilt Diseases of Plant*. St. Paul, Minnesota: American Phytopathological Society Press. 175p.

Beckman, C.H. (2000). Phenolic-storing cells: keys to programmed cell death and periderm formation in wilt disease resistance and in general defence responses in plants. Physiological and *Molecular Plant Pathology, 57*, 101–110.

Beckman, C. H., Mueller, W. C., Tessier, B. J. & Harrison, N. A. (1982). Recognition and callose deposition in response to vascular infection in *Fusarium* wilt-resistant or susceptible tomato plants. *Physiology and Plant Pathology, 20*, 1-10.

Beckman, C. H. & Talboys, P. W. (1981). Anatomy of resistance. In: Fungal Wilt Diseases of Plants. M. E. Mace, A. A. Bell, & C. H. Beckman (Eds.) Academic Press New York, 487-521.

Bell, A. A. & Mace, M. E. (1981). Biochemistry and Physiology of resistance. In: *Fungal Wilt Diseases of Plants*. Mace, M. E., Bell, A. A., & Beckman, C. H. (Eds.), New York: Academic Press. pp. 431-486.

Bergeson, G.B. (1972). Concepts of nematode-fungus association in plant disease complexes: A review. *Experimental Parasitology 32(2):* 301-314.

Bishop, C. D. & Cooper, R. M. (1983). An ultrastructure study of root invasion in three vascular wilt diseases. *Physiology and Plant Pathology, 22*: 15-27.

Booth, C. (1971). *Fusarium: laboratory guide to the identification of the major species*. Kew, Surrey: Commonwealth Mycological Institute.

Booth, C. (1981). Perfect states (teleomorphs) of *Fusarium* species. In P. E. Nelson, T. A. Toussoun, & R. J. Cook (eds.). *Fusarium: Diseases, Biology Taxonomy*. Pp. 446-452. University Park, Pennsylvania: Pennsylvania State University Press.

Borrero, C., Trillas, M., Delgado, A. & Avilés, M. (2011). Effect of ammonium/nitrate ratio in nutrient solution on control of *Fusarium* wilt of tomato by *Trichoderma asperellum* T34. *Plant Pathology*. doi: 10.1111/j.1365-3059.2011.02490.x

Burgess, L.W., Summerell, B. A., Bullock, S., Gott, K.P. & Backhouse, D. (1994). Laboratory Manual for *Fusarium* Research 3 Edn. University of Sydney Australia.

Caperton, C.M., Martyn, R.D. & Starr, J.L.(1986). Effects of *Fusarium* innoculum density and root-knot nematodes on wilt resistance in summer squash. *Plant Disease 70(3):* 207-209.

Cirulli, M. & Alexander, L. J. (1966). A comparison of pathogenic isolates of *Fusarium oxysporum* f. *lycopersici* and differet sources of resistance in tomato. *Phytopathology, 56*, 1301-1304.

Cotxarrera, L.,Trillas-Gay, M.I., Steinberg, C.& Alabouvette, C. (2002). Use of sewage sludge compost and *Trichoderma asperellum* isolates to suppress *Fusarium* wilt of tomato. *Soil Biology and Biochemistry 34*:467-476.

Cullen, D. W., Toth, I. K., Pitkin, Y., Boonham, N., Walsh, K., Barker, I. & Lees, A. K. (2005). Use of Quantitative Molecular Diagnostic Assays to Investigate *Fusarium* Dry Rot in Potato Stocks and Soil. *Phytopathology, 95,* 1462-1471.

De Hoog, G. S., Guarro, J., Gene J. & Figueras, M.J. (2000). *Atlas of Clinical Fungi*, 2nd ed. Vol. 1. Centraalbureau voor Schimmelcultures, Utrecht, The Netherlands.

Desjardins, A. E. & Proctor, R. H. (2007). Molecular biology of *Fusarium* mycotoxins. *International Journal of Food Microbiology, 119*, 47–50.

Di Pietro, A., Madrid, M.P., Caracuel, Z., Delgado-Jarana, J. & Roncero, M.I.G. (2003). *Fusarium oxysporum*: exploring the molecular arsenal of a vascular wilt fungus. *Molecular Plant Pathology, 4*, 315–325.

Di Pietro, A.D., García-Maceira, F.I., Méglecz, E. & Roncero, I.G. (2001). A MAP kinase of the vascular wilt fungus *Fusarium oxysporum* is essential for root penetration and pathogenesis. *Molecular Microbiology 39*:1140–1152.

Edreva, A., Velikova, V., Tsonev, T., Dagnon, S. Gürel, A., Aktaş, L. & Gesheva, E. (2008). Stress-Protective Role of Secondary Metabolites: Diversity of Functions and Mechanisms. *General and Applied Plant Physiology*, Special Issue, 34 (1-2), 67-78.

El-Hariry, M.A., Revri, R.& Abdallah, H.M. (2004). *Tomato Manual*. 59p.

Elias, K.S. & Schneider, RW. (1991). Vegetative compatibility groups in *Fusarium oxysporum* f.sp. *lycopersici*. *Phytopathology, 81(2)*:159-162]

El-Kazzaz, M.K., El-Fadly, G.B., Hassan, M.A.A. & El-Kot G.A.N. (2008). Identification of some *Fusarium* spp. using molecular biology techniques. *Egyptian Journal of Phytopathology 36 (1-2)*:57-69.

Fassihiani, A. (2000). Symptomless carriers of the causal agent of tomato wilt pathogen. *Journal of Agriculture,Science & Technology 2*: 27-32.

Flor, H.H. (1971). Current status of the gene-for-gene concept. *Annual Review of Phytopathology, 9*, 275–296.

Francis, R.G. & Burgess, L.W. (1975). Surveys of fusaria and other fungi associated with stalk rot of maize in eastern Australia. *Australian Journal of Agricultural Research*, 26:801-807.

Grattidge, R. & O'Brien, R. G. (1982). Occurrence of a third race of *Fusarium* wilt of tomatoes in Queensland. *Plant Disease, 66*, 165-166.

Green, R. J. (1981). An Overview. In: Fungal Wilt Diseases of Plants. Mace, M. A., Bell, A. A. and Beckman, C. H. (Eds.), New York, USA. Academic Press, Inc. pp. 640.

Guarro, J. & Gene J. (1992). *Fusarium* infections. Criteria for the identification of the responsible species. *Mycoses 35*:109-114

Hardham, A.R. (2001). Cell Biology of fungal infection of plants. In: R. Howard & N. Gow (eds). Biology of the fungal cell. Berlin, Germany. Springer-Verlag.

Harris, S.D. (2005). Morphogenesis in germinating *Fusarium graminearum* macroconidia. *Mycologia 97(4)*: 880-887.

Hennequin, C., Abachin E., Symoens F., Lavarde V., Reboux C., Nolard N. & Berche P. (1999). Identification of *Fusarium* species involved in human infections by 28S rRNA gene sequencing. *Journal of Clinical Microbiology. 37*:3586-3589.

Heuvelin, E. (2005). Tomatoes. E. Heuvelink (ed). CABI Publishing Co., Oxfordshire, UK. 339pp.

Hirano, Y. & Arie, T (2006). PCR-based differentiation of *Fusarium oxysporum* f. sp. *lycopersici* and *radicis-lycopersici* and races of *F. oxysporum* f. sp. *lycopersici. Journal of Genetics and Plant Pathology*, 72, 273–283.

Hogan, L. H., Klein, B. S. & Levitz, S. M. (1996). Virulence Factors of Medically Important Fungi. *Clinical Microbiology Reviews*, 9(4), 469–488.

Holguín-Peña, R. J. (2005). *Fusarium* Wilt of Tomato Caused by *Fusarium oxysporum* f. sp. *lycopersici* Race 3 in Baja California Sur, Mexico, *Plant Disease, 89(12)*, 1360.

Houterman, P.M., Cornelissen, B.J. & Rep, M. (2008). Suppression of plant resistance gene-based immunity by a fungal effector. *PLoS Pathogens, 4*, e1000061.

Huang, C. C. & Lindhout, P. (1997). Screening for resistance in wild *Lycopersicon* species to *Fusarium oxysporum* f. sp. *lycopersici* race 1 and race 2. *Euphytica, 93*, 145–153.

Inami, K., Yoshioka, C., Hirano, Y., Kawabe, M., Tsushima, S., Teraoka, T. & Arie, T. (2010). Real-time PCR for differential determination of the tomato wilt fungus, *Fusarium oxysporum* f. sp. *lycopersici*, and its races. *Journal of Genetics and Plant Pathology*, 76, 116–121.

Inoue, I., Namiki, F. & Tsuge T. (2002). Plant colonization by the vascular wilt fungus *Fusarium oxysporum* requires FOW1, a gene encoding a mitochondrial protein. *The Plant Cell 14*:1869-1883.

Jeschke, N., Nelson, P.E. & Maracas, F.O. (1990). *Fusarium* species isolated from soil samples collected at different altitudes in the Transkei, Southern Africa. *Mycologia 82*: 727-733.

Jones, J. P. & Woltz, S. S. (1981). *Fusarium* incited diseases of tomato and potato and their control. In P. E. Nelson, T. A. Toussoun, & R. J. Cook (Eds.), *Fusarium*: Diseases, Biology and Taxonomy (pp. 157-168). Pennsylvania State University Press, University Park.

Kareiva, P. (1999). Coevolutionary arms races: is victory possible? *Proceedings of the National Academy of Sciences*, USA, *96*, 8-10.

Keen, N.T. (1990). Gene-for-gene complementarity in plant–pathogen interactions. *Annual Review of Genetics, 24*, 447–463.

Keukens, E.A.J., deVrije, T., vandenBoom, C., deWaard, P., Plasman, H.H., Thiel, F., Chupin, V., Jongen, W.M.F. & deKruijff, B. (1995). Molecular basis of glycoalkaloid induced membrane disruption. *Biochiica et Biophysica Acta, Biomembranes, 1240*, 216–228.

Keukens, E.A.J., Devrije, T., Fabrie, C.H.J.P., Demel, R.A., Jongen, W.M.F. & Dekruijff, B. (1992). Dual specificity of sterol-mediated glycoalkaloid induced membrane disruption. *Biochimica et Biophysica Acta, 1110*, 127–136.

Kinet, J. M. and Peet, M. M. (1997). Tomato. In: The Physiology of Vegetable Crops. H.C. Wien (ed.). CAB International, Wallingford, UK. pp. 207-258.

Kover, P. X. & Schaal, B. A. (2002). Genetic variation for disease resistance and tolerance among *Arabidopsis thaliana* accessions. *Proceedings of the National Academy of Sciences,* USA, 99 (17): 11270-11274.

Lacmanova, I., Pazlarova, J., Kostelanska, M.& Hajslova, J. (2009). PCR-based identification of toxigenic *Fusarium* species. *Czech Journal of Food Science, 27(2)*: 90-94.

Lairini, K., PerezEspinosa, A., Pineda, M. & RuizRubio, M. (1996). Purification and characterization of tomatinase from *Fusarium oxysporum* f. sp. *lycopersici*. *Applied Environmental Microbiology, 62*, 1604–1609.

Larone, D. H. (1995). *Medically Important Fungi - A Guide to Identification,* 3rd ed. Washington, D.C: ASM Press.

Lievens, B., Houterman, P.M. & Rep, M. (2009). Effector gene screening allows unambiguous identification of *Fusarium oxysporum* f. sp. lycopersici races and discrimination from other formae speciales. *FEMS Microbiology Letters, 300(2)*:201-215.

Lievens, B., Brouwer, M., Vanachter, A. C. R. C., Cammue, B. P. A. & Thomma. B. P. H. J. (2006). Real-time PCR for detection and quantification of fungal and oomycete tomato pathogens in plant and soil samples. *Plant Science*, *171*, 155–165.

Ludwig, R. A. (1952). Studies on the physiology of hadromycotic wilting in the tomato plant. MacDonald College Technical Bulletin No. 20, McGill University, Montreal, 37:546.

Madrid, M. P., Di Pietro, A. & Roncero, M. I. G. (2003). Class V chitin synthase determines pathogenesis in the vascular wilt fungus *Fusarium oxysporum* and mediates resistance to plant defence compounds. *Molecular Microbiology, 47*, 257–266.

Mandal, S. and Mitra, A. (2008). Accumulation of cell wall-bound phenolic metabolites and their upliftment in hairy root cultures of tomato (*Lycopersicon esculentum* Mill.). *Biotechnology Letters, 30*:1253–1258.

Mandeel, Q.A. (2007). Modeling competition for infection sites on roots by nonpathogenic strains of *Fusarium oxysporum. Mycopathologia 163 (1)*: 9-20.

Martin-Urdiroz, M., Roncero, M.I.G., Gonzalez-Reyes, J.A. & Ruiz-Roldan, C. (2008). ChsVb, a class VII chitin synthase involved in septation, is critical for pathogenicity in *Fusarium oxysporum. Eukaryotic Cell, 7*: 112–121.

Michielse, C. B. & Rep, M. (2009). Pathogen profile update: *Fusarium* oxysporum. *Molecular Plant Pathology, 10*: 311–324.

Michielse, C.B., van Wijk, R., Reijnen, L., Cornelissen, B.J. & Rep., M. (2009). Insight into the molecular requirements for pathogenicity of *Fusarium oxysporum* f. sp. *lycopersici* through large-scale insertional mutagenesis. *Genome Biology, 10*, R4.

Miller, S.A., Rowe, R.C. & Riedel, R.M. (2011). *Fusarium* and *Verticillium wilts of tomato, potato, pepper, and eggplant.* Fact Sheet. Columbus, Ohio State University.

Moreau, M., Catesson, A., Peresse, M., & Czaninski, Y. (1978). Dynamique comparee des reactions cytologiques du xyleme de l'oeillet en presence de parasites vascularies. *Phytopathology, 91*, 289-306.

Mueller, W. C. & Beckman, C. H. (1984). Ultrastructure of the cell wall of vessel contact cells in the xylem of tomato stems. *Annals of Botany, 53*, 107-114.

Morrell, J.J. & Bloom, J.R. (1981). Influence of *Meloidogyne incognita* on *Fusarium* wilt of tomato at or below the minimum temperature for wilt development. *Journal of Nematology, 1(1):* 57-60.

Mueller, W.C. & Beckman C. H. (1988). Correlated light and electron microscope studies of callose deposits in vascular parenchyma cells of tomato plants inoculated with *Fusarium oxysporum* f. sp *lycopersici*. *Physiological and Molecular Plant Pathology 33*: 201-208.

Mui-Yun, W. (2003). *Fusarium oxysporum f. sp. lycopersici (Sacc.): PP728 Soil-borne Plant Pathogen Class Project.* North Carolina State University.

Mycology (2011).Mycology online. http://www.mycology.adelaide.edu.au/Fungal_Descriptions/Hyphomycetes_(hyaline)/*Fusarium*/

Namiki, F., Shiomi, T., Kayamura, T., & Tsuge, T. (1994). Characterization of the formae speciales of *Fusarium oxysporum* causing wilt of cucurbits by DNA fingerprinting with nuclear repetitive DNA sequences. *Applied Environmental Microbiology. 60*: 2684–2691.

Nelson, E. B. (1991). Exudate molecules initiating fungal responses to seed and roots. In D. L. Keister, & P.B. Cregan (Eds.) The rhizosphere and plant growth. Dordrecht: Kluwer Academic Publishers. 197-209.

Nelson, P.E., Dignani M.C. & Anaisse E.J. (1994).Taxonomy, biology, and clinical aspects of *Fusarium* species. *Clinical Microbiology Reviews, 7(4)*: 479-504.

Nelson, P. E., Toussoun, T. A. & Maracas, W. F. O. (1983). *Fusarium species: an illustrated manual for identification.* University Park, Pennsylvania, PA: Pennsylvania State University Press.

Nicholson, P., Simpson, D. R., Weston, G., Rezanoor, H. N., Lees, A. K., Parry, D. W. & Joyce, D. (1998). Detection and quantification of *Fusarium* culmorum and *Fusarium graminearum* in cereals using PCR assays. *Physiological and Molecular Plant Pathology,* 53, 17-37.

Nogués, S., Cotxarrera, L., Alegre, L. & Trillas, M.I. (2002). Limitations to photosynthesis in tomato leaves induced by *Fusarium* wilt. *New Phytologist*,154 (2): 461-470.

OISAT (2011). Online Information Service for Non-Chemical Pest Management in the Tropics. *Fusarium* Wilt. http://www.cd3wd.com/cd3wd_40/Biovision/export/defaultct87$pests.html.

Okubara, P. A., Schroeder, K. L. & Paulitz, T. C. (2005). Real-time polymerase chain reaction: applications to studies on soil borne pathogens. *Canadian Journal of Plant Pathology*, *27*, 300–313.

Okungbowa, F.I. (2011). Screening of four food spices of plant origin for antifungal activity against a *Fusarium oxysporum* soil isolate. *Jurnal Teknologi & Industri Pangan, 22 (1)*:93-96.

Okungbowa, F.I. & Edema N.E. (2006). Antifungal activity of six asteraceous plants extracts against *Fusarium oxysporum*. *Nigerian Journal of Botany, 20 (1)*:45 - 49.

Olivain, C., Humbert, C., Nahalkova, J., Fatehi, J., Floriane H. & Alabouvette C. (2006). Colonization of tomato root by pathogenic and nonpathogenic *Fusarium oxysporum* strains inoculated together and separately into the soil. *Applied Environmental Microbiology*,*72(2)*: 1523–1531.

Osuide, M.I.,Aluya, E.I. & Emoghene, A.O. (2002). Control of *Fusarium* wilt of tomato (*Lycopersicon esculentum* Mill) by *Trichoderma* species. *Acta Phytopathologica et Entomologica Hungarica, 37(1-3)*: 47-55.

PADIL (2011). Pests and Diseases Image Library. *Fusarium* wilt of chicken pea (*Fusarium oxysporum* f.sp. ciceris). http://old.padil.gov.au/pbt/index.php?q=node/15&pbtID=160.

Pareja-Jaime, Y., Roncero, M. I. G. & Ruiz-Roldan, M. C. (2008). Tomatinase from *Fusarium oxysporum* f. sp. *lycopersici* is required for full virulence on tomato plants. *Molecular Plant Microbe Interaction, 21*, 728–736.

Pedley, K. F. & Martin, G. B. (2003). Molecular basis of *Pto*-mediated resistance to bacterial speck disease in tomato. *Annual Review of Phtytopathology, 41*, 215-243.

Pegg, G. F. & Brady, B. L. (2002). *Verticillium wilts*. Wallingford, Oxon, UK: CABI Publishing. 416 pp.

Pitt, J. I., Hocking A.D., Bhudhasamai K., Miscamble B.F., Wheeler, K.A., & Tanboon-Ek, P. (1994). The normal mycoflora of commodities from Thailand. 2. Beans, rice, small grains and other commodities. *International Journal of Food Microbiology, 23*:35-43.

Reis, A., Costa, H., Boiteux, L. S. & Lopes, C. A. (2005). First Report of *Fusarium oxysporum* f. sp. *lycopersici* Race 3 on Tomato in Brazil. *Brazillian Phytopathology, 30(4)*, 426-428.

Rep, M., van der Does, H. C., Meijer, M., van Wijk, R., Houterman, P. M., Dekker, H. L., de Koster, C. G. & Cornelissen, B. J. C. (2004). A small, cysteine-rich protein secreted by *Fusarium oxysporum* during colonization of xylem vessels is required for I-3-mediated resistance in tomato. *Molecular Microbiology, 53*, 1373–1383.

Rodríguez-Gálvez, E. & Mendgen, K. (1995). The infection process of *Fusarium oxysporum* in cotton root tips. *Protoplasma 189*: 61–72.

Rodríguez-Molina, M.C. (2003).Vascular colonisation patterns in susceptible and resistant tomato cultivars inoculated with *Fusarium oxysporum* f.sp. lycopersici races 0 and 1. *Plant Pathology 52*: 199–203.

Shittu, H. O. & Robb, E. J. (2011). Tomato displays characteristics of a model organism for genetic analysis. *Nigerian Journal of Life Sciences* (In press).

Shittu, H. O., Shakir, A. S., Nazar, R. N. & Robb, J. (2009). Endophyte-induced *Verticillium* protection in tomato is range-restricted. *Plant Signaling and Behavior, 4 (2):* 160-161.

Song, W., Zhou, L., Yang, C., Cao, X., Zhang, L. & Liu, X. (2004). Tomato *Fusarium* wilt and its chemical control strategies in a hydroponic system. *Crop Protection 23(3)*: 243-247.

Stall, R. E. (1961). Development of *Fusarium* wilt on resistant varieties of tomato caused by a strain different from race 1 isolates of *Fusarium oxysporum* f. *lycopersici*. *Plant Disease Report*, 45, 12-15.

Steinkellner,S.,Mammerler,R.& Vierheilig, H. (2008). Germination of *Fusarium oxysporum* in root exudates from tomato plants challenged with different *Fusarium oxysporum* strains. *European Journal of Plant Pathology 122*:395-401.

Stukenbrock, E. H. & McDonald, B. A. (2009). Population genetics of fungal and oomycetes effectors involved in gene-for-gene interactions. *Molecular Plant Microbe Interaction, 22*, 371–380.

Simons, G., Groenendijk, J., Wijbrandi, J., Reijans, M., Groenen, J., Diergaarde, P., Van der Lee,T., Bleeker, M., Onstenk, J., De Both, M., Haring, M., Mes, J., Cornelissen, B., Zabeau, M. and Vos, P. (1998). Dissection of the *Fusarium* I-2 gene cluster in tomato reveals six homologs and one active gene copy. *Plant Cell, 10*, 1055–1068.

Stahl, E.A. & Bishop, J.G. (2000). Plant–pathogen arms races at the molecular level. *Current Opinion in Plant Biology, 3*, 299–304.

Stoessl, A. Stothers, J. B. & Ward, E. W. (1976). Sequiterpenoid stress compounds of the Solanaceae. *Phytochemistry, 15*, 855-872.

Syed, S.A., Kazunori. S., Yoshimiki, A. & Kazuyuki I. (2010). Biocontrol of soil-borne *Fusarium* wilts of tomato and cabbage with a root colonizing fungus, *Penicillium* sp. EU0013. Proceedings of the 19th World Congress of Soil Science, Soil Solutions for a Changing World,1 – 6 August 2010, Brisbane, Australia.

Takken, F. & Rep, M. (2010). The arms race between tomato and *Fusarium oxysporum*. *Molecular Plant Pathology, 11(2)*, 309–314.

Togawa, M. (1992). Effect of sterilization methods, plant varieties and leaf stages on conidia and perithecia formation in the genus *Fusarium* in CLA-culture. *Transactions of Mycological Society of Japan, 33*, 385–393.

Toussoun, T.A. & Nelson, P.E. (1976). *A Pictorial Guide to the Identification of Fusarium species* Second Edition. University Park, Pennsylvania State University Press.

Van der Does, H. C., Lievens, B., Claes, L., Houterman, P. M., Cornelissen, B. J. C., Rep, M. (2008). The presence of a virulence locus discriminates *Fusarium oxysporum* isolates causing tomato wilt from other isolates. *Environmental Microbiology, 10*, 1475–1485.

VanderMolen, G. E., Beckman, C. H., & Rodehorst, E. (1977). Vascular gelation: A general response phenomenon following infection. *Physiology and Plant Pathology, 11*, 95-100.

Voigt, C. A., Schafer, W. & Salomon, S. (2005). A secreted lipase of *Fusarium graminearum* is a virulence factor required for infection of cereals. *The Plant Journal 42,* 364–375.

Walter, S., Nicholson, P. & Dooha, F.M. (2009). Action and reaction of host and pathogen during *Fusarium* head blight disease. *New Phytologist, 185*: 54–66.

Wang, H., Xiao, M., Kong, F., Chen, S., Dou, H., Sorrell, T., Li, R. & Xu, Y. (2011). Accurate and Practical Identification of 20 *Fusarium* Species by Seven-Locus Sequence Analysis and Reverse Line Blot Hybridization, and an In Vitro Antifungal Susceptibility Study. *Journal of Clinical Microbiology 49(5)*: 1890-1898.

Wanjiru, W.M., Kang, Z. & Buchenauer H. (2002). Importance of cell wall degrading enzymes produced by *Fusarium graminearum* during infection of wheat heads. *European Journal of Plant Pathology*, *108*:803–810.

Wanjiru, W.M., Zhensheng, K. & Buchenauer, H. (2002) Importance of cell wall degrading enzymes produced by *Fusarium graminearum* during infection of wheat heads. *European Journal of Plant Pathology, 108,* 803–810.

Wokoma, E.C.W. (2008). Preliminary report on diseases of tomato in Choba, Rivers State. *Journal of Applied Sciences and Environmental Management 12(3)*: 117-121.

Xu, J.R. (2000). MAP kinase in fungal pathogens. *Fungal Genetics and Biology 31*:137-152.

In: Fusarium
Editors: Tony Ferarri Rios and Emilia Reyes Ortega

ISBN: 978-1-61942-539-2

Chapter 5

ETIOLOGY AND MANAGEMENT OF ASPARAGUS CROWN AND ROOT ROT

A. Borrego-Benjumea[1,*], M. J. Basallote Ureba[2,*] and J. M. Melero-Vara[1]

[1]Instituto de Agricultura Sostenible, CSIC, Córdoba, Spain
[2]Instituto de Investigación y Formación Agraria y Pesquera (IFAPA) Las Torres-Tomejil, Alcalá del Río (Sevilla), Spain

ABSTRACT

Crown and root rot is the most important disease of asparagus worldwide. Symptoms include rots and vascular discoloration of roots, rhizomes and stems, necrotic lesions of roots and stems, yellowing and senescence of stems and spears and dwarfing and death of plants. This brings about reductions in plant vigour and individual plants yield related to decline of plants and replant problem. The complex etiology of Fusariumcrown and root rot of asparagus (FCRR), with several species of *Fusarium* reported as causal agents, depend on soil and climate conditions whereby prevalence of the species involved differs among crop areas. *Fusarium oxysporum* f. sp. *asparagi*, *F. proliferatum* and *F. verticillioides* (syn. *F. moniliforme*) are the pathogens most commonly associated to FCRR. *F. solani*, less widespread in asparagus fields, include some isolates highly pathogenic to asparagus, at least in Spain. Other *Fusarium* species are often recovered although with much lower frequencies than *F. oxysporum* f. sp. *asparagi* or *F. proliferatum*. Among those, *F. subglutinans*, *F. redolens* f. sp. *asparagi* and *F. culmorum* seem to play a pathogenic role in the disease complex.

Asparagus plants stressed either by biotic or abiotic factors, such as drought or old crops, are more prone to infections by different *Fusarium* spp. in the complex. With regard to optimal temperature for fungal growth and disease development, roughly ranging 20-30ºC, *F. solani* is favoured by higher temperature than *F. oxysporum* f. sp. *asparagi* and *F. proliferatum*.

Within *Liliaceae*, *F. oxysporum* f. sp. *asparagi*and *F. solani* show a certain degree of pathogenic specialisation which contrasts with the lack of specificity of the other species above mentioned. The genetic intra-specific diversity, mainly in *F. oxysporum* f. sp.

* First authors.

asparagi, is related to the host range, as studies of populations have commonly shown these to be composed of many different vegetative compatibility groups (VCG) which frequently include a reduced or single number of isolates; these also differ among studies conducted in different countries.

Disease management can not be achieved by resistant cultivars because of the very narrow genetic diversity of *Asparagus officinalis*. The old practice of applying rock salt to asparagus soils proved effective in reducing populations of *Fusarium* spp. and infections, and increasing plant vigour, possibly because of a shift of microbial populations towards a disease suppressiveness composition. Except for the use of methyl isothiocianate as a soil fumigant and thiabendazol as a fungicide, which costs seem unaffordable to growers, no chemical control measures are available.

However, use of organic soil amendments followed by polyethylene tarping of soil seems to be effective to control FCRR under some conditions. Incorporation of fresh residues of broccoli or grass into infested soils reduces Foa densities due to anaerobiosis and the accumulation of toxic compounds. The latter could also be the main mechanism by which fresh poultry manure application achieves reduction of Foa propagules in soil, disease levels, and increases fresh weight of plants and yields. In addition, promising biological control agents such as some *Trichoderma* isolates or non-pathogenic isolates of *F. oxysporum* that proliferate in the presence of toxic asparagus residues encourage long-term experiments aimed at implementing such control practice, which is compatible with Ecological Agriculture requirements.

INTRODUCTION

Asparagus (*Asparagus officinalis* L.) plant was originated and domesticated in Ancient time in some areas of Asia and/or the seacoast of Greece (Bailey & Bailey, 1976). The crop spread along the Mediterranean basin more than 20 centuries ago. It is a perennial crop able to give profitable yields for 10-15 years after three years of crop establishment. Fusarium crown and root rot of asparagus (FCRR) is probably its most important disease worldwide. Seedling blight was the name originally given to the disease (Graham, 1955). FCRR has been associated with asparagus decline, accounting for the reduction in quantity and size of the spears produced, and the death of plants. Furthermore, early decline or replant disease are denominations of the damage caused to asparagus crops when old asparagus fields are replanted with asparagus; yields are slowly reduced in 5-10 years old crop, and plants are unable to establish productive plantings (Grogan & Kimble, 1959). Although accumulation in the soil of autotoxins released by asparagus roots have been suspected as a possible promoter of replant disease (Hartung et al., 1989), studies in the Netherlands proved this not to be true when root residues were incorporated into unsterilized soil, as antagonistic microorganisms in those counteract the pathogen which has a low competitive saprophytic ability (Blok & Bollen, 1996b). The FCRR has a complex etiology with the occurrence of several soil-borne *Fusarium* spp. as causal agents associated to asparagus roots (Elmer, 2001).

Symptomatology

Disease symptoms include reduced plant size, wilt of aerial plant parts, brown or brown-reddish necrotic lesions in the root apex and at the insertion of secondary and storage roots, as well as necrotic lesions of variable length developed on them, decrease of the number of secondary roots, rhizome rots, vascular discoloration of rhizome and stem bases and necrotic flecks or lesions in stem bases (Cohen & Heald, 1941; Johnston et al, 1979; Schreuder et al., 1995; Elmer, 2001; Corpas-Hervias et al., 2006).

ETIOLOGY

The Causal Agents

For this complex etiology the main fungal species involved are *Fusarium oxysporum* Schlecht. emend. Snyder & Hansen and *F. proliferatum* (Matsush.) Nirenberg (Damicone & Manning, 1985; LaMondia & Elmer, 1989; Elmer, 2001; Schreuder et al., 1995; Corpas-Hervias et al., 2006). *F. redolens* Wollenw. f.sp. *asparagi* Baayen, morphologically similar to *F. oxysporum*, seems to be also an important pathogen in this disease complex (Baayen et al., 2000b). *F. verticillioides* (Saccardo) Nirenberg (syn = *F. moniliforme* Sheld.) and *F. solani* (Martius) Apple & Wollenw. emend. Snyder & Hansen, have also been associated with FCRR (Endo & Burkholder, 1971; Johnston et al., 1979; Tu, 1980; LaMondia & Elmer, 1989; Schreuder et al., 1995; Corpas-Hervias et al., 2006). Furthermore, several other *Fusarium* spp., like *F. subglutinans* (Wollenw. & Reink.) Nelson, Tousson & Marasas, *F. culmorum* (W.G. Smith) Saccardo, *F. equiseti* (Corda) Saccardo and *F. lateritium* Nees, may be minor etiological components of the complex disease as well (Gindrat *et al.*, 1984; Tello *et al.*, 1985; Blok & Bollen, 1995; Durán, 2001; Elmer, 2001; Corpas-Hervias et al., 2006).

Distribution and Host Range

The distribution of *Fusarium* spp. affecting asparagus vary between fields and areas, depending on the environmental conditions that may differentially favour one or several of the species (Damicone & Manning, 1985; Fantino & Fantuz, 1990; Hartung et al., 1990; Elmer, 1995a; Blok & Bollen, 1996a; Elmer et al., 1997; Moretti et al., 1997; Doan & Carris, 1998; Corpas-Hervias et al., 2006; Vujanovic et al., 2006; Wong & Jeffries, 2006). Recovery of *F. oxysporum* and *F. proliferatum* were at similar rates in samples from South Africa and Spain (Schreuder et al., 1995; Wong & Jeffries, 2006), but *F. oxysporum* was prevalent in crowns widely sampled from diverse Spanish asparagus nurseries (Corpas-Hervias et al., 2006). Previous results point to the higher frequency of *F. proliferatum* in warmer climates (Moretti et al., 1997; Elmer, 2001) and the low or nil occurrence in northern European countries (Blok & Bollen, 1996a; Wong & Jeffries, 2006). A biogeographical study carried out in eastern Canada, also with very low temperature conditions, showed *F. proliferatum* to be ubiquitous and abundant, which contrasts with the failure to recover isolates of *F. proliferatum* and *F. verticillioides* from the Netherlands (Van Bakel & Kerstens, 1970; Blok & Bollen, 1996a), and the infrequent occurrence of *F. proliferatum* in UK, where *F. oxysporum* appeared as predominant (Wong & Jeffries, 2006). *F. redolens* f.sp. *asparagi* was rather infrequent in most fields sampled in UK and Spain, it has probably been neglected in early works due to the difficulties to be discriminated from *F. oxysporum* (Baayen et al., 2000b).

A study on the host specificity of *F. oxysporum* f.sp. *asparagi*, conducted in the Netherlands, concluded that it has a narrow host range as a pathogen, restricted to asparagus, although it colonizes root tissues of other 11 plant species, including several leguminous crops (with pea and lupine showing mild symptoms), carrot, maize, sugar beet, wheat and grasses, whereas onion and leek crops were free of root colonization (Blok & Bollen, 1997). In contrast with these results, in a recent work with isolates of *F. oxysporum*, *F. proliferatum* and *F. solani* recovered from symptomatic asparagus plants in Spain, root colonization of onion plants inoculated with one *F. oxysporum* isolate obtained from asparagus, reached a level similar to that of asparagus. Furthermore, root symptoms were observed in garlic inoculated with the three *Fusarium* species studied, and *F. solani* was also pathogenic to onion. *F. oxysporum* and *F. solani* showed a certain degree of pathogenic specialisation within Liliaceae which contrasts with the lack of specificity of *F. proliferatum* (Molinero-Ruiz et al., 2010). In this regard, phylogenetic analysis using the translation elongation factor-1 alpha (EF1α) sequences indicated that isolates of *F. oxysporum* pathogenic to asparagus are spread across a number of clades within the species complex, supporting the hypothesis that pathogenicity to asparagus in this species is a relatively unspecialized trait (Wong & Jeffries, 2006).

Pathogenicity

F. oxysporum and *F. proliferatum* are highly pathogenic to asparagus in pathogenicity tests. It is interesting to mention that only *F. oxysporum* f.sp. *asparagi* appeared as the causal agent of "early decline" and "replant-bound early decline" (Blok & Bollen, 1996a). Although this *forma specialis* was described originally for isolates of *F. oxysporum* attacking asparagus, further studies found this pathogenicity to asparagus also occurred in isolates of *F. oxysporum* obtained from an ample number of hosts and situations in which asparagus was lacking and those from asparagus are sometimes pathogenic also to other crops, and an extremely large genetic diversity of those isolates was found (Elmer & Stephens, 1989; Elmer, 1991; Yergeau et al., 2005; Vujanovic et al., 2006; Wong & Jeffries, 2006). The pathogenicity of *F. solani* to asparagus varies; it was non-pathogenic or slightly virulent in USA and South Africa (LaMondia & Elmer, 1989; Schreuder et al., 1995), in contrast with Taiwan, Spain and Japan, where highly virulent isolates of *F. solani* were found (Tu, 1980; Corpas-Hervias et al., 2006; Nahiyan et al., 2011). It seems clear that a considerable diversity, both genomic and pathogenically, occurs in the populations of *F. solani* (Wong & Jeffries, 2006). *F. redolens* pathogenicity to asparagus seedlings was as high as that of *F. oxysporum* f.sp. *asparagi* and *F. proliferatum*, suggesting the former as a significant component in asparagus decline in particular fields (Wong & Jeffries, 2006).

Inoculum Survival and Conditions that Favour Infections

The *Fusarium* species pathogenic to asparagus survive in the soil as chlamydospores (Nelson et al., 1983), or as mycelium in infected plant debris (Nyvall & Kommendahl, 1968;

Hartung et al., 1990). These pathogens are transmitted with the propagation materials, crowns and seeds (Inglis, 1980; Damicone et al., 1981; Fantino & Fantuz, 1990; Blok & Bollen, 1996c; Doan & Carris, 1998; Corpas et al., 2006), being of much higher epidemiological importance the crown than the seed inoculum, as the crown is the regular organ transplanted to crop fields and, at least *F. oxysporum* f.sp. *asparagi*, is mainly confined to the seed surface and, therefore, can be readily eliminated by regular seed disinfestation (Inglis, 1980).

Asparagus plants stressed either by abiotic or biotic factors, such as drought, viral and insect damage, or old crops, are more prone to infections by different *Fusarium* spp. in the disease complex (Nigh, 1990; Elmer et al., 1996). With regard to optimal temperature for fungal growth and disease development, roughly ranging 20-30ºC, *F. solani* is favoured by higher temperature than *F. oxysporum* and *F. proliferatum.* Under controlled conditions, the incubation at 20ºC and ≥90% relative humidity (RH) was associated with more crown and root rot, increasing the severity of both aboveground and root symptoms, as well as the asparagus dry weight losses, as compared to non-inoculated plants. The temperature effect on stem symptoms was modified by relative humidity, with high severity observed when RH≥90% (Corpas & Basallote, unpublished data).

Genetic Diversity

The genetic diversity of *F. oxysporum* and *F. moniliforme* populations from asparagus fields in Michigan and Connecticut, USA, indicated that the vegetative compatibility groups (VCGs) to which the isolates of these two species could be assigned were very large, at least 43 that contained *F. oxysporum* strains pathogenic to asparagus, and at least 13 for *F. moniliforme* (Elmer & Stephens, 1989; LaMondia & Elmer, 1989). For both species, an ample number of isolates were unique and classed as single members of different VCGs. This indicates a broad genetic diversity of the populations that are, therefore, genetically isolated. This contrasts with the grouping of isolates of other *Fusarium* species into a reduced number of VCGs (mainly different *formae speciales* of *F. oxysporum*), allowing geographical or host-specific virulence to be related with these groups. This has been confirmed in the Netherlands, where 24 isolates of *F. oxysporum* f.sp. *asparagi* were ascribed to 18 different VCGs (Blok & Bollen, 1997). Pathogenic isolates of *F. proliferatum* from seven asparagus growing regions in Australia were placed into 10 VCGs, the largest of these was also previously found to be the largest VCG in USA, suggesting that seed and crowns carried out pathogen dissemination between countries (Elmer et al., 1999).

Diversity within a single field seemed to be as ample as that between two different European countries (Wong & Jeffries, 2006), and diversity within each region of Japan was similar as that from the different regions (Nahiyan et al., 2011). With regard to the genetic diversity revealed by single-stranded conformational polymorphism (SSCP) profiles, it was found one single profile in *F. redolens*, similarly to what happened for UK populations of *F. proliferatum*, but two additional profiles, very rarely found, were recorded in Spain (Wong & Jeffries, 2006). This sharply contrasts with the wide intraspecific variation among isolates of *F. oxysporum* f.sp. *asparagi* determined by VCG tests, as above mentioned. Two SSCP profiles, OXY1 and OXY2, were found both in UK and Spain, but OXY1 was predominant in these countries as well as in all the European and Peruvian isolates tested, in contrast with the isolates from USA, which were predominantly OXY2 type (Wong & Jeffries, 2006).

Genomic sequences of one or several genes or portions of the nuclear or mitochondrial ribosomal RNA coding region are commonly used for analyses of taxonomic relationships between species. A molecular approach using PCR-denaturing gradient gel electrophoresis (DGGE) based on specific primers targeting the EF1α gene, allowed the clear differentiation of 19 species of *Fusarium* and made possible the accurate discrimination of *Fusarium* species most frequently recovered in asparagus fields in North America (mainly *F. oxysporum, F. proliferatum* and *F. solani*). Moreover, this high resolution technique allowed the detection and identification of *Fusarium* spp. *in planta* (Yergeau et al., 2006). When multiple banding patterns were present, due to the presence of several slightly dissimilar EF1α sequences in a single strain, they might be used, in conjunction with band position, to help in the differentiation among species on the gel (Yergeau et al., 2006). Using primers Alfie 1 and Alfie 2 combined, the *Fusarium* spp. were specifically amplified. All sequences of excised bands from asparagus plants sampled in one field related mainly to *F. oxysporum* and *F. proliferatum*, although also to *F. solani*. Band migrating positions and sequences retrieved from *F. oxysporum* were variable, coinciding with the reported variation in genetic diversity, VCGs, and pathogenicity traits (Yergeau et al., 2006). In contrast, *F. proliferatum*-associated bands from different isolates migrated to a single position and sequences showed little variation, as corresponds to the presence of one single mating population of this *Fusarium* species (Elmer, 1995a). Species differentiation is possible for most bands, except when migration position is common to several species (Yergeau et al., 2006). Using the EF1α gene, the sequences of two different isolates of *F. oxysporum* from Québec, Canada were respectively highly similar (>99%) to one *F. oxysporum* f.sp. *asparagi* isolate from northern USA and to one isolate of *F. oxysporum* f.sp. *dianthi*. Likewise, comparing sequences of two isolates of *F. proliferatum* from USA and two from Québec indicated a very high similarity in one case and an identical sequence in the other. In this way, the larger intraspecific variability in *F. oxysporum*, as compared to *F. proliferatum* populations, was confirmed (Vujanovic et al., 2006).

The SSCP PCR analysis of the ITS2 (internal transcribed sequence) region of the rDNA identified the fusarial populations associated with the roots of asparagus in UK and Spain (Wong & Jeffries, 2006). Most of the isolates of the four main species (*F. oxysporum* f.sp. *asparagi*, *F. proliferatum, F. redolens* and *F. solani*) were easily differentiated by SSCP and their frequencies determined. Although intra-regional variation among different fields was considerable, *F. oxysporum* f.sp. *asparagi* was most frequently isolated from the UK, whereas similar proportions were overall found for *F. oxysporum* f.sp. *asparagi* and *F. proliferatum* isolates from Spain (Wong & Jeffries, 2006). Results from Japan surveys, also using PCR-SSCP analysis, indicated the predominance of *F. oxysporum* f.sp. *asparagi* and a lower frequency of *F. proliferatum*, that increased from the north to the south of the country, confirming the trend of higher occurrence in warmer areas (Moretti et al., 1997; Corpas-Hervias et al., 2006; Nahiyan et al., 2011). The four species were found both in UK and Spain although *F. redolens* and *F. solani* were relatively infrequent there (Wong & Jeffries, 2006), whereas *F. redolens* failed to be recovered from Japan (Nahiyan et al., 2011). However, pathogenicity to asparagus seedlings was as high as that of *F. oxysporum* f.sp. *asparagi* and *F. proliferatum*, suggesting that *F. redolens* may be a significant component in asparagus decline in particular fields (Wong & Jeffries, 2006). In contrast, pathogenicity was nil for the Spanish isolates of *F. solani*, except for one population, similarly to what happened in other crop areas of Spain in a different work (Corpas-Hervias et al., 2006). This species complex was also recovered in relatively low proportions from asparagus samples from USA, South Africa, Australia, and, more rarely, from Japan, were highly pathogenic isolates predominated (Schreuder et al., 1995; Elmer, 2001; Nahiyan et al., 2011). Diversity within a single field seemed to be as ample as that between two different European countries, and within each region of Japan diversity was similar as that from the different regions (Wong & Jeffries, 2006).

The phylogenetic analysis of the EF1α sequences of selected *F. oxysporum* f.sp. *asparagi* isolates greatly extended the range of these isolates from that reported on the basis of the eight established VCGs of *F. oxysporum* f.sp. *asparagi* from USA (Baayen et al., 2000a). This degree of para- and polyphyleticism is larger than in other multiclade *F. oxsporum formae speciales*, and support the hypothesis that pathogenicity to asparagus is a rather unspecialized trait (Wong & Jeffries, 2006).

Management of FCRR

Resistant Cultivars

Asparagus cultivars with resistance to crown and root rot are difficult to develop due to the high genetic diversity within *Fusarium* spp. pathogenic to this host and the narrow genetic basis of asparagus. However, all male asparagus hybrids made a significant advancement of the selection of cultivars for field resistance to FCRR that increased germplasm used by crop breeders (Ellison & Kinelski, 1985, Elmer et al., 1996). In a recent study, the reactions of current asparagus cultivars to *F. oxysporum, F. proliferatum, F. moniliforme* and *F. solani* demonstrated that cvs. Dariana, Plasenesp and Verde-Morado had overall the lowest severity values for the pathogenic populations evaluated (Corpas-Hervias *et al.*, 2006). Cultivar Dariana showed the lowest dry weight losses when compared to the non-inoculated control plants. In addition, the agronomical value of the three cultivars would make them suitable for cropping areas with a previous history of FCRR. In contrast, the widely planted cv. Grande was the most susceptible to the isolates of *Fusarium* tested (Corpas-Hervias *et al.*, 2006).

Chemical Control

Strategies for controlling FCRR include eradication of the pathogens from seeds and planting crowns, as well as from the soil. Soil fumigation achieves partial control since pathogens frequently re-colonize the soil after the treatment (Manning & Vardaro, 1977; Lacy, 1979; Damicone et al., 1981). The use of fungicide chemicals offers also possibilities of controlling FCRR of asparagus (Lacy, 1979; Falloon et al., 1989; Reid *et al.*, 2002).

Benomyl was reported to eradicate *F. oxysporum* f.sp. *asparagi* and *F. moniliforme* associated to asparagus seeds, the solution of benomyl in acetone improving the results since the germination of the seeds was improved, and rinsing seeds in a Na hypochlorite solution prior to the seed treatment gave the most satisfactory control (Manning & Vardaro, 1977). Crown treatments with benomyl and captafol reduced disease levels for a short period and increased size and vigor of the plants (Manning & Vardaro, 1977). In contrast, Lacy (1979) found ineffective the treatment of one-year old crowns with benomyl. However, according to Reid *et al.* (2002), asparagus root weight was increased and disease decreased by benomyl and fludioxonil drenching applications to seedlings growing under low levels of *F. proliferatum* and *F. oxysporum*, but fludioxonil limited disease caused by *Fusarium* spp. at both low and high inoculum levels. Application of those fungicides via drip irrigation in fields with a history of asparagus production resulted in lack of significant differences among treated and untreated plots (Counts & Hausbeck, 2008). All these results point to a diversity of outcomes depending on inoculum levels, rates and methods of application of fungicides (mainly benomyl) and crop stage. Soaking crowns with methyl thiophanate or formononetin, a plant isoflavone, as well as the application of the later as soil drench, have also been shown to reduce root lesions caused by *Fusarium* spp. although application of rock salt or sodium chloride (NaCl) to soil were more effective (Elmer, 2002).

Soil treatment with NaCl was frequently used since the late 19th Century as a soil amendment increasing root and crown vigour and yield of asparagus, a salt tolerant crop, although caution was recommended in the rates of application to avoid undesirable effects in future crops other than asparagus, as well as on soil properties. By the mid of the 20th Century, the abandon of the practices of salting asparagus soil, led to increased problems of asparagus decline due to FCRR, responsible for the disappearance of a large part of the asparagus fields in northeastern USA (Elmer, 1992; Elmer et al.,1996). The use of NaCl as a soil amendment showed promise as a FCRR management strategy, resulting in significant reductions of rhizospheric populations of *F. proliferatum* and *F. oxysporum* (Elmer, 1992). Moreover, in growth chamber and greenhouse studies, NaCl was more effective than other forms of chloride salts ($CaCl_2$, NH_4Cl and $MnCl_2$) in controlling disease caused by both pathogens. NaCl applications increased yield and fern health in the research plots, but not in commercial fields (Reid et al., 2001). The use of NaCl may provide immediate benefits in suppressing disease and increasing yields when circumstances force to re-establish asparagus in abandoned fields (Elmer, 2002). Furthermore, this was an economically feasible practice as compared to soil fumigation (Manning & Vardaro, 1977).

Cultural Practices Using Soil Amendments

Soil amendments with fresh broccoli or grass (3.4-4.0 kg m^{-2}) subsequently covered with air-tight plastic significantly reduced the populations of *F. oxysporum* f. sp. *asparagi* at 15 cm depth, after 15 weeks incubation. This was due to the release of toxic volatile compounds generated by glucosinolates present in cruciferous residues, subsequent to the anaerobic conditions simultaneously developed (Blok et al., 2000). This occurs in the absence of thermal inactivation due to the high temperature conditions that operate during soil solarization processes, yet they might contribute to inactivation of pathogens by weakening the survival structures. On the other hand, incorporation of both amendments into the soil did not inactivate *F. oxysporum* f.sp. *asparagi* nor other fungal pathogens tested, suggesting a crop-non-specific effect. In addition, some of the products generated under reducing soil conditions, such as ethylene, organic acids, alcohols and aldehydes, may accumulate in the soil and are potentially fungitoxic. This method of control is applicable to high-value crops (Blok et al., 2000).

Other studies involving animal manures, such as poultry manure and the pellet of poultry manure, and composts derived from agro-industrial wastes as by-products obtained in the process of oil extraction from olives, combined with the incubation of soil at different temperatures (30 and 35°C) and periods (15, 30 and 45 days), showed a reduction of the density of viable inocula of isolates of *F. oxysporum* f.sp. *asparagi*, *F. proliferatum* and *F. solani* (Borrego-Benjumea et al., 2010; Borrego-Benjumea, unpublished data). This reduction was also observed in the infested un-amended soil when it was incubated at 35°C for 30-45 days. However, when organic amendments were applied, the eradication of these pathogens occurred at less temperature and shorter incubation periods (Borrego-Benjumea et al., 2010). Thus, except for *F. proliferatum*, pellet of poultry manure reduced the viability of the propagules over 90% at temperatures at or above 30°C when incubation was carried out for at least 30 days. The poultry manure resulted in similar reductions of viability at the two temperatures tested when incubation was performed for 45 days in almost all cases, while the compost required a temperature of 35ºC for at least 30 days to reach the same levels of effectiveness. In any case, the poultry manure pellet was the most effective amendment in reducing the viability of *F. oxysporum* f.sp. *asparagi*, *F. solani* and *F. proliferatum* (Borrego-Benjumea et al., 2010; Borrego-Benjumea, unpublished data). This amendment has a higher nitrogen content (4.3% N) than poultry manure (2.4% N) and compost (1.5% N) (Borrego-Benjumea, unpublished data), that could be associated to the inhibitory action on the pathogen by relative amounts of toxic compounds (NH_3 and HNO_2) that are generated from nitrogenous organic amendments (Tenuta and Lazarovits, 2002), the soil pH being the crucial factor in determining the relative amounts of these toxic compounds.

Mitigation of climate change through reduction of atmospheric CO_2 by carbon (C) sequestration can be achieved by the thermal degradation (pyrolysis) of organic materials in the nearly complete lack of oxygen, leading to a charcoal-like C-rich mineral product called biochar that can withstand decomposition to CO_2 for centuries. This new compound may be foreseen as widely available in large amounts in the near future, and it will be ready for use as a mineral fertilizer, to manage pathogens and as a carrier of biocontrol agents (Lehman et al., 2011). Its ample variability in composition, adsorptive properties and texture, depending of the type of biomass used, and several other factors, prompted controversy about beneficial effects on the crops to which it is applied, as harmful or neutral effects were reported in other cases (Glaser et al., 2002).

A recent work on the effects of biochar derived from hardwood dust showed the possibility of its interaction with soil properties (Elmer & Pignatello, 2011). Reduction of FCCR was observed following biochar applications at 1.5 and 3.0% (w/w) to infested soil, and associated with root colonization by arbuscular mycorrhizae (AM), which was doubled at the 3% application rate. This may reduce infection by *F. oxysporum* f.sp. *asparagi* and *F. proliferatum*, and the subsequent disease development. Moreover, the effect of biochar on alkalinisation of acidic soils could lead to disease suppression, by decreasing the availability of Fe to the pathogens, and promoting root health and antagonistic microorganisms. Evidence was given that deleterious effects of allelopathic residues in asparagus soil may be overcome by biochar. The size of biochar-treated plants was reduced even though they had previously shown increased roots colonization by AM fungi. As this soil received rainfall greater than average, and biochar increases the water-holding capacity of soils, as it retains much moisture, root rot might have been increased and, thus, plants growth impaired. Therefore, this detriment of root health is more likely to occur following biochar treatment of heavy clay soils (Elmer & Pignatello, 2011). This study needs to be demonstrated in the commercial asparagus fields where FCRR is well established.

Biological and Integrated Control

Asparagus seedlings inoculated under field conditions, with the AM fungus *Glomus fasciculatum* (That. sensu Gerd.) Gerd. & Trappe) showed lower FCRR ratings in high-phosphorous (P) soils than non-mycorrhizal plants grown in low-P soils, but rhizosphere populations of *F. oxysporum* were lowest in high-P soils, regardless of the AM status. In greenhouse, artificial inoculation of well-watered asparagus plants with both *F. oxysporum* and *G. fasciculatum* resulted in lower disease levels and rhizosphere populations of *F. oxysporum* than plants without *G. fasciculatum* inoculation, but these results appeared not to be related to final levels of P in plant tissues (Wacker et al., 1990).

Although asparagus yield increases in NaCl-treated fields was not always related to increases of root-colonizing bacteria, percentages of fluorescent pseudomonads or antagonistic bacteria, shifts in the proportion of rhizobacteria that reduce Mn *in vitro* were found in 70% of the fields amended with NaCl. This treatment also very frequently increases by 19% the Mn levels in roots. The percentage of fluorescent *Pseudomonas* was significantly higher in 50% of the cases, and they were more frequently Mn-reducing strains. Therefore, the interrelationships among this soil treatment, FCRR reduction and increased levels of Mn-reducing rhizobacteria seem to support the hypothesis that Mn availability plays a role in asparagus disease resistance (Elmer, 1995b).

The use of non-pathogenic strains of *F. oxysporum* has been effective, under experimental conditions, to biologically control wilt and rot diseases of many different crops caused by a diversity of *formae speciales* of *F. oxysporum*, the biocontrol capacity of the formers being achieved by mechanisms of action that comprise competition for nutrients, that may deprive chlamydospores of elements required for germination, competition with pathogenic strains for sites of infection on the roots, and systemic acquired resistance (SAR) triggered by plant defence reactions. Biocontrol efficacy depends much of the application conditions that are encountered under commercial field production (Chet, 1990; Fravel et al., 2003). As with many other pathosystems, non-pathogenic strains of *F. oxysporum* become optimal candidates to suppress FCRR of asparagus, as reported in Europe and North America (Blok et al., 1997; He et al., 2002; Reid et al., 2002), and association between activation of defence-related enzymes such as peroxidase and phenylalanine ammonia-lyase, lignin accumulation and increase of antifungal activity of root exudates with SAR induced by non-pathogenic strains of *F. oxysporum* against *F. oxysporum* f.sp. *asparagi* (He et al., 2002). However, long-term effectiveness in control of this disease was not assessed. Screening five non-pathogenic strains of *F. oxysporum* with and without NaCl treatment, resulted in root lesion reductions by all the strains tested as well as by NaCl, which did not hinder the ability of non-pathogenic strains of *F. oxysporum* to suppress disease, allowing the combination of both control practices although there was no interaction between them. Furthermore, one of those strains increased root weight but NaCl did not affect root growth (Elmer, 2004).

Combination of that method with other control practices improves sometimes the results and may give consistency to the disease control. Therefore, a field study was conducted by Elmer (2004) to compare the efficacy of three selected non-pathogenic strains of *F.*

oxysporum applied to asparagus crowns, alone or combined with moderate doses of rock salt amendments (280-560 Kg ha^{-1}) in 4-year experiments conducted in two fields where asparagus declined after nine years cropping. Disease was reduced by one of the non-pathogenic strains of *F. oxysporum* in both sites and increased marketable yield in the silt loam soil field. NaCl increased yield over controls in the field with sandy soil but not in the other. That strain was the only one maintaining root colonization after four years, although another strain initially colonized roots better than the others, and increased root weight as compared to the control. As a perennial plant, asparagus may require the non-pathogenic strains of *F. oxysporum* to have the trait of long survival, along with biocontrol capabilities (Elmer, 2004).

However, in greenhouse experiments conducted with biological control agents to test their ability to control the disease caused by *F. oxysporum* f. sp. *asparagi* and *F. proliferatum* showed non-pathogenic isolates of *Fusarium oxysporum* not to be effective in controlling disease. This contrasted with *Trichoderma harzianum* T-22 treatment that increased root weight and decreased root disease as compared with the infested control, when a low level of the pathogens was used, but it was not effective at high rates of inoculum (Reid *et al.*, 2002). The ability of *Trichoderma* spp. to increase plant growth has been reported in several crops, with most of the effective isolates colonizing and growing in association with the roots of the plants (Harman, 2000; 2006; Harman *et al.,* 2004). This mechanism, designated as "rhizosphere competence" (Harman, 2000), seemed to be operating in the case of isolates *T. asperellum* T108 and *T. harzianum* T2 (both from asparagus rhizosphere), which promoted a greater development of the aboveground parts of the plants simultaneously to the profuse colonization of their roots. The antagonistic effect of the two *Trichoderma* spp. isolates against *F. oxysporum* and *F. proliferatum* from asparagus, observed in dual culture assays, resulted in the control of the disease in pathogenicity tests. However, the effectiveness of the biocontrol exerted by both *Trichoderma* strains depended of the virulence of the *Fusarium* isolate used (Rubio-Pérez et al., 2008).

ACKNOWLEDGMENT

Reviewed by Dr. Pervaiz Abbasi. Southern Crop Protection and Food Research Centre, Agriculture and Agri-Food Canada, 1391 Sandford Street, London, ON, Canada, N5V 4T3.

REFERENCES

Baayen, R.P., O'Donnel, K., Bonants, P.J.M., Cigelnik, E., Kroon, L. P. N., Roebroeck, E. J. A., and Waalwijk, C., 2000a. Gene genealogies and AFLP analysis in the *Fusarium oxysporum* complex identify monophyletic and nonmonophyletic formae speciales causing wilt and rot disease. *Phytopathology* 90:891-900.

Baayen, R.P., van den Boogert, P.H.J.F., Bonants, P.J.M., Poll, J.T.K., Blok, W.J. and Waalwijk, C., 2000b. *Fusarium redolens* f.sp. *asparagi*, causal agent of asparagus root rot, crown rot an spear rot. *Eur. J. Plant Pathol.* 106: 907-912.

Bailey, L. H., and Bailey, E. Z. 1976. *Hortus Third: A Concise Dictionary of Plants Cultivated in the United States and Canada*. MacMillan Publishing Co., New York.

Blok, W. J. & Bollen, G. J. 1995. Fungi on the roots and stem bases of asparagus in the Netherlands: Species and pathogenicity. *Eur. J. Plant Path.* 101:15-24.

Blok, W. J. & Bollen, G. J. 1996a. Etiology of asparagus replant-bound early decline. *Eur. J. Plant Pathol.* 102: 87-98.

Blok, W. J. & Bollen, G. J. 1996b. Interactions of asparagus root tissue with soil microorganisms as a factor in early decline of asparagus. *Plant Pathology* 45:809-822.

Blok, W.J. and Bollen, G.J. 1996c. Inoculum sources de *Fusarium oxysporum* f sp. *asparagi* in asparagus production. *Ann. Appl. Biol.* 128: 219-231.

Blok, W. J. & Bollen, G. J. 1997. Host specificity and vegetative compatibility of Ducth isolates of *Fusarium oxysporum* f. sp. *asparagi*. *Can. J. Bot.* 75:383-393.

Blok, W. J., Lamers, J. G., Termorshuizen, A. J., and Bollen, G. J. 2000. Control of soilborne plant pathogens by incorporating fresh organic amendments followed by tarping. *Phytopathology* 90:253-259.

Borrego-Benjumea, A., Melero-Vara, J. M., and Basallote-Ureba, M. J. 2010. Effects of soil organic amendments and incubation temperature on the control of *Fusarium oxysporum* f.sp. *asparagi*. *Acta Horticulturae* 883:369-375.

Chet, I. (1990) Biological control of soil-borne plant pathogens with fungal antagonists in combination with soil treatments. In *Biological Control of Soil-borne Plant pathogens*, pp: 15-43. Ed. D. Hornby. Oxon, UK: CAB International.

Cohen, S. I., and Heald, F. D. 1941. A wilt and root rot of asparagus caused by *Fusarium oxysporum* Schlecht. *Plant Dis. Reptr.* 25:503-509.

Corpas-Hervias, C., Melero-Vara, J. M., Molinero-Ruiz, M. L., Zurera-Muñoz, C. and Basallote-Ureba, M. J. 2006. Characterization of isolates of *Fusarium* spp. obtained from asparagus in Spain. *Plant Dis.* 90: 1441-1451.

Counts, J. W. and Hausbeck, M. K. 2008. Srategies for managing fusarium crown and root rot on asparagus. *Acta Horticulturae* 776: 167-173.

Damicone, J. P., Cooley, D. R., and Manning, W. J. 1981. Benomyl in acetone eradicates *Fusarium moniliforme* and *F. oxysporum* from asparagus seed. *Plant Dis.* 65:892-893.

Damicone, J. P., and Manning, W. J. 1985. Frequency and pathogenicity of *Fusarium* spp. isolated from first-year asparagus grown from transplants. *Plant Dis.* 69:413-416.

Doan, M. C., and Carris, L. M. 1998. Characterization of *Fusarium* populations in asparagus fields in the Pacific Northwest. *Acta Horticulturae* 479:219-226.

Durán, R.M. 2001. *Etiología de enfermedades de espárrago causadas por patógenos de suelo en Andalucía*. Ph.D. Thesis, University of Córdoba, Spain. 258 pp.

Ellison, J. H., and Kinelski, J. J. 1985. `Jersey Giant´, an all male asparagus hybrid. *HortSci.* 20:1141.

Elmer, W. H. 1991. Vegetative compatibility groups of *Fusarium proliferatum* from asparagus and comparisons of virulence, growth rates, and colonization of asparagus residues among groups. *Phytopathology* 81:852-857.

Elmer, W. H. 1992. Suppression of Fusarium crown and root rot of asparagus with sodium chloride. *Phytopathology* 82:97-104.

Elmer, W. H. 1995a. A single mating population of *Gibberella fujikuroi (Fusarium proliferatum*) predominates in asparagus fields in Connecticut, Massachusetts, and Michigan. *Mycologia* 87:68-71.

Elmer, W. H. 1995b. Association between Mn-reducing root bacteria and NaCl applications in suppression of *Fusarium* crown and root rot of asparagus. Phyto*p*athology 85:1461-1467.

Elmer, W. H. 2001. *Fusarium diseases of asparagus*, pp. 248-262. *En*: *Fusarium* Paul E. Nelson Memorial Symposium Summerell, B. A.., J. F. Leslie, D. Backhouse, W.L. Bryden and L. W. Burgess, (eds.). APS Press, St. Paul Minnesota. 392 pp.

Elmer, W. H. 2002. Influence of formononetin and NaCl on mycorrhizal colonization and Fusarium crown and root rot of asparagus. *Plant Dis.* 86:1318-1324.

Elmer, W. H. 2004. Combining non-pathogenic strains of *Fusarium oxysporum* with sodium chloride to suppress fusarium crown rot of asparagus in replanted fields. *Plant Pathology* 53:751-758.

Elmer, W. H., Johnson, D. A., and Mink, G. I. 1996. Epidemiology and management of the diseases causal to asparagus decline. *Plant Dis.* 80:117-125.

Elmer, W. H., and LaMondia, J. A. 1998. Studies on the suppression of Fusarium crown and root rot with NaCl. *Acta Horticulturae* 479:211-215.

Elmer, W. H., and Pignatello, J.J. 2011. Effect of biochar amendments on mycorrhizal associations and Fusarium crown and root rot of asparagus in replant soils. *Plant Dis.* 95:960-966.

Elmer, W. H., and Stephens, C. T. 1989. Classification of *Fusarium oxysporum* f. sp. *asparagi* into vegetatively compatible groups. *Phytopathology* 79:88-93.

Elmer, W. H., Summerell, B. A., Burgess, L.W., Backhouse, D., and Abubaker, A.A. 1997. *Fusarium* species associated with asparagus crowns and soil in Australia and New Zealand. *Australasian Pl. Path.* 26:255-261.

Elmer, W. H., Summerell, B. A., Burgess, L. W., and Nigh, E. L. 1999. *Vegetative compatibility groups of Fusarium proliferatum from asparagus in Australia Mycologia* 91:650-654.

Endo, R. M., and Burkholder, E. C. 1971. The association of *Fusarium moniliforme* with the crown rot complex of asparagus *Phytopathology* (Abstr.) 61:891.

Falloon, P. G., Fraser, H., and Nikologg, A. S. 1989. *Control of Fusarium rot with Thiabendazole. Final Report*. Crop Research Division, DSIR, Christchurch, NZ.11p.

Fantino, M. G. and Fantuz, F. 1990. The sanitary state of asparagus nurseries in Emilia-Romagna in 1988-1989. *Acta Horticulturae* 271:120-125.

Fravel, D., Olivain, C. and Alabouvette, C. 2003. *Fusarium oxysporum* and its biocontrol. *New Phytologist* 157:493-502.

Gindrat, D.,Varady, C., and Neury, G. 1984. Le dépérissement de l´asperge. Rev. Suisse Vitic. Arboric Hortic. 16:23-26.

Glaser, B., Lehman, J., & Zech, W. 2002. Ameliorating physical and chemical properties of highly weathered soils in the tropics with charcoal--a review. Biol. *Fertil. Soils* 35:219-230.

Graham, K. M. 1955. Seedling blight, a fusarial disease of asparagus. *Can. J. Bot.* 33:374-400.

Grogan, R. G., and Kimble, K. A. 1959. The association of Fusarium wilt with the asparagus decline and replant problem in California. *Phytopathology* 49:122-125.

Harman, G. E. 2000. Myths and dogmas of biocontrol. Changes in perceptions derived from research on *Trichoderma harzianum* T22. *Plant Dis.* 84:337-393.

Harman, G. E. 2006. Overview of mechanism and uses of *Trichoderma* spp. *Phytopatholgy* 96:190-194.

Harman, G. E., Howell, C. R., Viterbo, A., Chet, I., and Lorito, M. 2004. *Trichoderma* species-*opportunistic, avirulent plant simbionts. Nature* Reviews-Microbiology 2:43-56.

Hartung, A. C., Putnam, A. R., and Stephens, C. T. 1989. Inhibitory activity of asparagus root tissue and extracts on asparagus seedlings. *J. Am. Soc. Hortic. Sci.* 114:144-148.

Hartung, A. C., Stephens, C. T., and Elmer, W. H. 1990. Survey of *Fusarium* populations in Michigan's asparagus fields. *Acta Horticulturae* 271:395-401.

He, C. Y., Hsiang, T., and Wolyng, D. J. 2002. Induction of systemic resistance and pathogen defence responses in *Asparagus officinalis* inoculated with non-pathogenic strains of *Fusarium oxysporum. Plant Pathology* 51:225-230.

Inglis, D. A. 1980. Contamination of asparagus seed by *Fusarium oxysporum* f. sp. *asparagi* and *Fusarium moniliforme.* Plant Dis. 64:74-76.

Johnston, S. A., Springer, J. K., and Lewis, G. D. 1979. *Fusarium moniliforme* as a cause of stem and crown rot of asparagus and its association with asparagus decline. *Phytopathology* 69:778-780.

Lacy, M. L. 1979. Effects of chemicals on stand establishment and yields of asparagus. *Plant Dis. Rept.* 63:612-616.

LaMondia, J. A., and Elmer, W. H. 1989. Pathogenicity and vegetative compatibility among isolates of *Fusarium oxysporum* and *F. moniliforme* colonizing asparagus tissues. *Can. J. Bot.* 67:2420-2424.

Lehman, J., Rilling, M.C., Thies, J., Masiello, C.A., Hockaday, W.C., and Crowly, D. 2011. Biochar effects on soil biota—A review. *Soil Biol Biochem.* 43:1812-1836.

Manning, W. J., and Vardaro, P. M. 1977. Soil fumigation and preplant fungicide crown soaks: Effects on plant growth and Fusarium incidence in newly planted asparagus. *Plant Dis. Rept.* 61:355-357.

Molinero-Ruiz, L., Rubio-Pérez, E., González-Domínguez, E., and Basallote-Ureba, M. J. 2010. Alternative hosts for *Fusarium* spp. causing crown and root rot of asparagus in Spain. *J. Phytopathol.* 59: 114-116.

Moretti, A., Logrieco, A., Doko, B., Frisullo, S., Visconti, A., and Bottalico, A. 1997. *Fusarium proliferatum* from asparagus, in Italy: occurrence, fertility and toxigenicity. Proceedings of the Fifth European Fusarium Symposium. *Cereal Research Communication* 25:785-786.

Nahiyan, A.S.M., Boyer, L.R., Jeffries, P., and Matsubara, Y. 2011. PCR-SSCP analysis of *Fusarium* diversity in asparagus decline in Japan. *Eur. J. Plant Pathol.* 130:197-203.

Nelson, P. E., Toussoun, T. A. and Marasas, W. F. O. 1983. Fusarium species: An illustrative manual for identification. Pennsylvania State. University Press, University Park. 193 pp

Nigh, E. L.1990. Stress factors influencing *Fusarium* infection in asparagus. *Acta Horticulturae* 271:315-322.

Nyvall, R. & Kommendahl, T. 1968. Individual thickened hyphae as survival structures of *Fusarium moniliforme* in corn. *Phytopathology* 58:1704-1707.

Reid, T. C., Hausbeck, M. K. and K. Kizilkaya. 2001. Effects of sodium chloride on commercial asparagus and of alternative forms of chloride salt on Fusarium crown and root rot. *Plant Dis.* 85:1271-1275.

Reid, T.C., Hausbeck, M.K., and Kizilkaya, K. 2002. Use of fungicides and biological control in the suppression of Fusarium crown and root rot of asparagus under greenhouse and growth chamber conditions. *Plant Dis.* 86:493-498.

Rubio-Pérez, E., Molinero-Ruiz, M. L., Melero-Vara, J. M., and M. J. Basallote-Ureba. 2008. Selection of potential antagonists against asparagus crown and root rot caused by *Fusarium* spp. *Communications in Agricultural and Applied Biological Sciences* 73: 203-206.

Schreuder, W., Lamprecht, S. C., Marasas W. F. O, and Calitz, F. J. 1995. Pathogenicity of three *Fusarium* species associated with asparagus decline in South Africa. *Plant Disease* 79:177-181.

Tello, J.C., González, M. L. and Lacasa, A. 1985. The "Fusariosis" (Disease produced by *Fusarium* spp.) of asparagus in Spain. Pp. 126-135 In: Proc. Six Int. Asparagus Symposium. E.C. Loungheed and H. Tiessen, (eds.). University of Guelph, Ontario, Canada.

Tenuta, M., Lazarovits, G. 2002. Ammonia and nitrous acid from nitrogenous amendments kill the microsclerotia of *Verticillium dahliae*. *Phytopathology* 92:255-264.

Tu, C. C. 1980. Studies on fusarial wilt and root disease complexes of asparagus in Taiwan. Pp. 180-186 *In*: *Proc. 5th Int. Asparagus Symposium*. G. Reuter, ed. Wageningen, The Netherlands.

Van Bakel, J. M. M. & Kertsen; J. J. K. 1970. Foot rot in asparagus caused by *Fusarium oxysporum* f. sp. *asparagi*. *Neth. J. Phytopathol.* 76:320-325.

Vujanovic, V., Hamel, C., Yergeau, E., and St-Arnaud, M. 2006. Biodiversity and biogeography of *Fusarium* species from Northeastern North American asparagus fields based on microbiological and molecular approaches. *Microbial Ecology* 51:242-255.

Wacker, T. L., Safir, G. R. and Stephens, C. T. 1990. Effect of *Glomus fasciculatum* on the growth of asparagus and the incidence of Fusarium root rot. *J. Amer. Soc. Hort. Sci.* 115:550-554.

Wong, J.Y. and Jeffries, P. 2006. Diversity of pathogenic Fusarium populations associated with asparagus roots in decline soils in Spain and the UK. *Plant Pathol.* 55:331-342.

Yergeau, E., Filion, M., Vujanovic, V., and St-Arnaud, M. 2005. A PCR-denaturing gradient gel electrophoresis approach to assess Fusarium diversity in asparagus. *Journal of Microbiological Methods* 60:143-154.

In: Fusarium
Editors: Tony Ferarri Rios and Emilia Reyes Ortega
ISBN: 978-1-61942-539-2

Chapter 6

LECTINS: IMPORTANT TOOLS FOR BIOCONTROL OF *FUSARIUM* SPECIES

***Andréa F. S. Santos*[2], *Thiago H. Napoleão*[1], *Patrícia M. G. Paiva*[1] *and Luana C. B. B. Coelho*[1,*]**

[1]Departamento de Bioquímica, Centro de Ciências Biológicas, Universidade Federal de Pernambuco, Cidade Universitária, Recife, Pernambuco, Brazil

[2]University of Minho, IBB-Institute for Biotechnology and Bioengineering, Centre of Biological Engineering
Campus de Gualtar, Braga, Portugal

ABSTRACT

Lectins are proteins from non-immune origin, containing two or more binding sites to mono or oligosaccharides. Plant lectins, found in different plant tissues, show a marked heterogeneity with respect to their molecular structure and sugar-binding specificity. Fungi are frequently pathogens of cultivated plant species representing one of the main causes of crop losses. Growth of fungi has been considered a major factor in biodeterioration of materials such as wood, stored grains, paintings, sculptures, leather, and oil. *Fusarium* is a *filamentous fungi* genus widely distributed in soil and in association with plants. *Fusarium* is a pathogen of bananas, corn, rice, wheat, tomatoes, peanuts, among other crops. Some *Fusarium* species produce mycotoxins that have been found infecting breakfast and infant cereals. *Fusarium* species can affect human health causing cutaneous infections and keratitis, for example. The intensive use of synthetic fungicides for plant disease control is cause of public concern because of the presence of residues in water and crops for animal and human consumption. Biodegradable fungicides derived from plants have been searched aiming to minimize environmental damage. Antifungal effects on *Fusarium* species have been reported for lectins isolated from different plant tissues. The role of lectins in fungi plant defense mechanisms may have evolved from its ability to recognize and agglutinate cells or glycoconjugates in sites for potential invasion by infectious agents, with ability to inhibit fungal growth and

*E-mail address: lcbbcoelho@gmail.com.

germination. Antifungal activity has been reported mainly for chitin-binding lectins which are able to bind chitin present in fungal cell wall and interfere in spore germination. In summary, this chapter discusses the optimistic perspectives for lectin preparations that can be applied in agriculture as cheap and biodegradable antifungal agents preventing pollution of the environment without damages to the biodiversity.

1. Introduction

1.1. Lectins: Definition and Applications

Lectins constitute a protein group of non-immune origin, containing two or more binding sites to mono or oligosaccharides; these proteins interact with glycoconjugates present on cell surfaces inducing the agglutination phenomenon (Figure 1). Lectins are widespread in nature and have been isolated from animals, bacteria, viruses, fungi, and plants. Plant lectins show a marked heterogeneity with respect to their molecular structure, sugar-binding specificity as well as temporal and spatial regulation. Most plant lectins are able to recognize glycans from other organisms, acting in recognition and defense-related mechanisms (Van Damme et al., 2004). These proteins can be found in different plant tissues such as bark, bulb, flower, fruit, heartwood, leaf, root and seeds (Correia et al., 2008; Sá et al., 2009b; Santos et al., 2009; Napoleão et al., 2011; Souza et al., 2011).

In plants, several physiological roles have been ascribed to lectins such as defense against pathogenic microorganisms and insects, participation in the root nodulation mechanism, regulation of the storage and mobilization of reserve carbohydrates and proteins, elongation of the wall cell, and activation of some enzymes (Hirsch, 1999; Aoyama et al., 2001; Limpens and Bisseling, 2003, Kestwal et al., 2007; Sá et al., 2008).

Lectins can be applied for several biotechnological purposes. Currently, the lectins appear as highlighting tools in the field of biorecognition technology. Lectin microarray techniques have been developed as efficient tools to decipher the glycocode reflecting the cell physiological state (Gemeiner et al., 2009). Lectins are used for cytochemical localization of glycoconjugates and for histochemical analysis of structures of normal and transformed tissues (Franceschini et al., 2000; Lima et al., 2010). Lectins are also important tools in isolation and structural studies of oligosaccharides and glycoconjugates (Xie et al. 2009); chromatographic columns using lectins immobilized on inert supports were effective in separating and isolate glycoproteins (Lima et al., 1997; Paiva et al., 2003; Silva et al., 2011). Lectins from *M. oleifera* seeds have coagulant properties and can be used for water treatment (Santos et al., 2009, Ferreira et al., 2011) and humic acid removal (Santos et al., 2011).

Lectins have shown several biological activities, such as inducing of proliferation of lymphocytes, macrophages and neutrophils (Timoshenko et al., 1995; Andrade et al., 1999; Maciel et al., 2004), stimulation of the production of interferon γ by lymphocytes (Melo et al., 2010) as well as anti-inflammatory (Melo et al., 2010), antitumor (Fang et al., 2010), hypotensive (Wang et al., 1996), antinociceptive (Figueiredo et al., 2009), antiviral (Sato et al., 2011), insecticidal, antibacterial and antifungal activities.

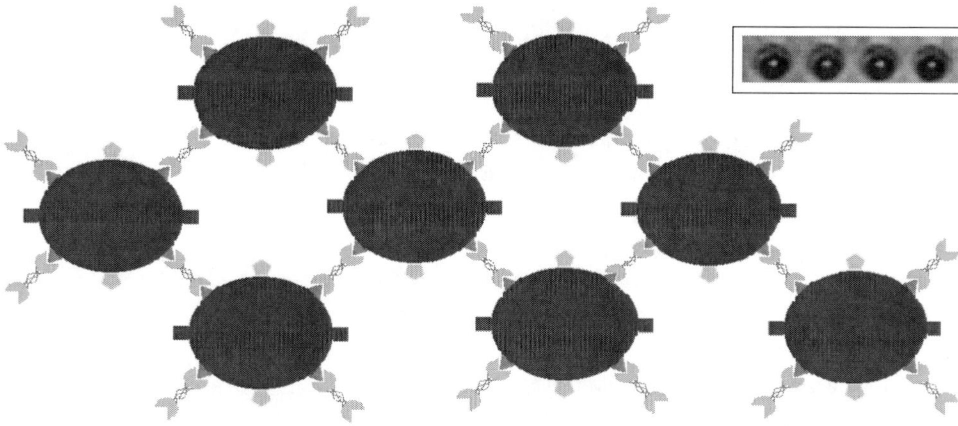

Figure 1. Schematic representation of cell agglutination promoted by lectins. Each lectin molecule forms a bind between two cells (a lectin–cell–lectin matrix) and these bounds constitute a network that prevents the cells from sedimenting. A sample containing lectin has the ability to maintain the cells in suspension (inset).

The insecticidal activity of plant lectins has been well documented against a large variety of insect species (Macedo et al., 2002; Coelho et al., 2007; Macedo et al., 2007; Sá et al., 2008; Coelho et al., 2009; Sá et al., 2009a, Silva et al., 2009; Napoleão et al., 2011;; Oliveira et al., 2011; Souza et al., 2011; Napoleão et al., 2012). This biological property is a potential way of using plant lectins as biodegradable insecticide agents against pests and vectors of human diseases (Vasconcelos and Oliveira, 2004; Paiva et al., 2011a). Oliveira et al. (2011) showed that a lectin from *Moringa oleifera* seeds (cMoL) exerted a significant negative impact on larvae and pupae of the flour moth *Anagasta kuehniella*, a polyphagous pest that feeds on a wide variety of stored products. The *Bauhinia monandra* leaf lectin (BmoLL) was insecticidal against *Callosobruchus maculatus* (cowpea weevil), *Zabrotes subfasciatus* (Mexican been weevil) and *A. kuehniella* (Macedo et al., 2007). Another lectin from *M. oleifera* seeds (WSMoL) promoted the delay of larval development and mortality of *Aedes aegypti* (dengue mosquito) larvae (Coelho et al., 2009); lectins from *Myracrodruon urundeuva* bark (MuBL), heartwood (MuHL) and leaf (MuLL) also showed larvicidal activity on this mosquito species (Sá et al., 2009b; Napoleão et al., 2012). Termiticidal activity on *Nasutitermes coniger* was detected for MuBL, MuHL and MuLL as well as for lectins isolated from *Opuntia ficus indica* cladodes (OfiL) and *B. monandra* secondary roots (BmoRoL) (Paiva et al., 2011b; Souza et al., 2011).

Lectins purified from plant tissues have shown antimicrobial activity. The lectins from *Archidendron jiringa* seeds, *Eugenia uniflora* seeds, *M. urundeuva* heartwood, and *Phthirusa pyrifolia* leaf were active against different pathogenic bacteria such as *Bacillus subtilis, Corynebacterium* spp., *Escherichia coli*, *Klebsiella pneumoniae*, *Pseudomonas aeruginosa*, *Staphylococcus aureus* and *Staphylococcus epidermidis* (Oliveira et al., 2008; Sá et al., 2009a; Costa et al., 2010; Charungchitrak et al., 2011). WSMoL was highly active against *S. aureus* being able to promote the coagulation and inactivation of this bacterium; this lectin was also effective against lake water bacteria, indicating that it can contribute to disinfection of water (Ferreira et al., 2011).

The antifungal activity of plant lectins have been reported against human and plant pathogens such as *Candida albicans*, *Colletrotrichum gloesporioides*, *Rhizoctonia* and *Fusarium* species (Sá et al., 2009a; Santana et al., 2009; Costa et al., 2010; Vaz et al., 2010; Souza et al., 2011).

1.2. *Fusarium* Species and Environmental Effects

Fungi are widely distributed in nature and several species are pathogens of different plant species representing one of the main causes of economic crop losses (Freire et al., 2002). Also, growth of fungi has been implicated in the biodeterioration of wood, stored grains, paintings, sculptures, leather, oil, and several other materials (Sterflinger, 2010).

The species belonging to *Fusarium* genus are filamentous fungi widely distributed in soil and can be found on subterranean and aerial plants, plant debris, and other organic substrates. They can be found in tropical, temperate regions, desert, alpine, and even in arctic areas where harsh climatic conditions prevail. Many species are abundant in fertile cultivated and rangeland soils but are relatively uncommon in forest soils (Nelson et al., 1994).

The damages produced by these fungi have several harmful effects on agriculture. *Fusarium* species are important pathogens of bananas, corn, rice, wheat, tomatoes, peanuts and other crops (Agrios, 2005; Rojo et al., 2007; Wagacha and Muthomi, 2007). Some *Fusarium* species produce mycotoxins and this genus includes the most important toxigenic fungi in northern temperate regions (Gutleb et al., 2002). Worldwide, the major mycotoxins produced by *Fusarium* are trichothecenes, zearalenone, and fumonisins which can be detected in cereal grains, animal feeds; other important *Fusarium* mycotoxins include moniliformin and fusaric acid (D'Mello et al., 1999). In addition, *Fusarium* toxins have been found infecting breakfast and infant cereals (Mahnine et al., 2011). These toxins can affect human and animal health; for example, fumonisins are epidemiologically associated with human esophageal cancer and neural tube defect, and cause leukoencephalomalacia in equine, pulmonary edema in swine as well as nephrotoxicity and hepatotoxicity in several animals (Hesseltine, 1974; Curtis et al., 2011).

Ears of maize are affected by two distinct diseases caused by *Fusarium*: the *Fusarium* ear rot (or pink ear rot) and *Gibberella* ear rot (or red ear rot), both of which can result in mycotoxin contamination of maize grains. The primary causal agent for *Fusarium* ear rot is *Fusarium verticilloides*, but this disease is also occasioned by *Fusarium subglutinans* and *Fusarium proliferatum*. Gibberella ear rot is caused primarily by *Fusarium graminearum* (also known as *Gibberella zeae*) but can also occur due to infection by *Fusarium culmorum* (Munkvold, 2003). The most common species, *Fusarium oxysporum*, causes vascular wilt disease in a wide variety of economically important crops (Roncero et al., 2003). These and other plant diseases caused by different *Fusarium* species are summarized in Table 1.

Fusarium causes superficial, locally invasive, and disseminated infections in humans. Among the diseases caused in humans are cutaneous infections by *F. oxysporum* (Nucci and Anaissie, 2002) and keratitis by *Fusarium solani* (Touvron et al., 2009). *Fusarium* nail infections are relatively common and characterized by the combination of proximal subungual onychomycosis with subacute or acute paronychia involving fingernails or toenails

(Baran et al., 1997). Some other infections caused by *Fusarium* are sinusitis, pneumonia, organ fungemia, septic arthritis and osteomyelitis (Nucci and Anaissie, 2007). The most common species associated with infections are *Fusarium moniliforme*, *F. solani*, *F. oxysporum*, and *F. verticillioides* and (Nucci and Anaissie, 2007).

Fusarium species are also important deteriorative agents, attacking wood (Apetorgbor et al., 2004) concrete samples (Gu et al., 1998), and stored grains (Bhattacharya and Raha, 2002).

Table 1. Some plant diseases caused by *Fusarium* species

Disease	Species	Damaged plant
***Fusarium* ear rot**	*F. proliferatum* *F. verticillioides* *F. subglutinana*	Maize grain
***Fusarium* head blight**	*F. avenaceum* *F. culmorum* *F. graminearum* *F. poae*	Small grain cereals
Gibberella ear rot	*F. graminearum* *F. culmorum*	Maize grain
Melon Fusarium wilt	*F. oxysporum* f. sp. *melonis*	Melon
Watermelon *Fusarium* wilt	*F. oxysporum* f. sp. *niveum*	Watermelon
***Fusarium* sheath rot**	*F. proliferatum*	Rice
Tomato vascular wilt	*F. oxysporum* f. sp. *lycopersici*	Tomato
Tomato crown and root rot	*F. oxysporum* f. sp. *radicis-lycopersici*	Tomato
Brinjal crop wilt	*F. solani* f. sp. melongenae	Brinjal

References: Abbas et al., 1998; Larkin and Fravel, 1998; Freeman et al., 2001; Munkvold, 2003; Pirgozliev et al., 2003; Suárez-Estrella et al., 2004; Joseph et al., 2008; Lori et al., 2009; Curtis et al., 2011; Zhao et al., 2011.

2. Synthetic Antifungal Agents for Fusarium Control

The intensive cultivation of monoculture cereals has required the use of pesticides and antifungal agents aiming to maximize the yields (Magan et al., 2002). Synthetic antifungal agents (such as phenylamides, dithiocarbamates, benzamidazoles, copper oxychlorides, pentachlorophenol, tributyltin oxide and zinc carboxylate) have been used for control of plant diseases. Studies have focused on the development of antifungal agents aiming to prevent both infection and mycotoxin production. Antifungal agents such as tebuconazole and metconazole give good control of both *Fusarium* infection of ears and control of deoxynivalenol production (Magan et al., 2002).

Synthetic fungicides exert their deleterious effects acting on mitosis and cell division (benzimidazoles, benzamides, phenylureas and thiophanates), cell respiration (iprodione, vinclozolin, methoxyacrylates, methoxy-carbamates, oximino acetates, oxazolidine-diones, dihydro-dioxazines, imidazolinones), sterol synthesis (imidazoles, piperazines, pyrimidines and triazoles), nucleic acid synthesis (acylalanines, phenylamides, heteroaromatics, carboxylic acids), signal transduction (dicarboximides), protein synthesis (cyprodinil), lipid and membrane synthesis (carboxylic acid amides), and host-plant defense induction (benzo-thiadiazole) (Fungicide Resistance Action Comitee, 2009).

2.1. Why Look for Alternatives to Synthetic Antifungal Agents?

The exhaustive application of these fungicides is cause for public concern since they have several disadvantages such as high cost, handling hazards and threaten human health and the environment because of their toxic effects and persistence, being detected in water and crops for animal and human consumption (Bajpai et al., 2009). The synthetic antifungal agents have been removed from markets and their use to control post-harvest deterioration has been restricted due to their carcinogenicity and teratogenicity. The residual effects due to a long degradation period results in environmental pollution, contamination of food and water as well as other side-effects on humans (Tripathi and Dubey, 2004, Bajpai et al., 2009; Verma et al., 2008). In addition, development of resistance to commonly used fungicides within populations of pathogens has also become a significant problem (Tripathi and Dubey, 2004). In this sense, several non-chemical treatments or natural chemical alternatives have been investigated for fungal decay control.

3. Natural Antifungal Agents

The control of plant pathogens using natural compounds is a potential alternative to the use of synthetic antifungal agents. Medicinal plants are potential sources of a wide variety of biologically active compounds including biodegradable fungicides. Microbial fungicides are other alternatives for fungi control. A compound produced by *Burkholeria cepacia* showed antifungal effect on *F. solani*, promoting abnormal chitin deposition and swelling of hyphae (Li et al., 2009). The studies on fungicides derived from bioconversion of fatty acids by microorganism have been intensified; an eicosapentaenoic acid produced from bioconversion of fatty acids by *Pseudomonas aeruginosa* PR3 inhibited the growth of *F. oxysporum* and *F. solani* in 58% and 57%, respectively (Bajpai et al., 2008).

Plant-derived fungicides are the most promising because they are less toxic and more environmentally-friendly. Alkaloids, coumarins, essential oils, flavonoids, phenols, phenolic acids, quinones, saponins, terpenoids, xanthones, and plant proteins are potential antifungal agents (Arif et al., 2011).

3.1. Plant-derived Compounds with Antifungal Activity against *Fusarium*

Studies have evaluated plant extracts as natural antifungal agents to control pathogenic fungi, reducing the dependence on the synthetic antifungal agents. Aqueous extract from five plants (*Cinnamomum zeylanicum*, *Cymbopogon proximus*, *Laurus nobilis*, *Persea americana* and *Zingiber officinale*) showed antifungal activity against *F. oxysporum* (Fawzi et al., 2009). The efficacy of plant extracts from *Azadirachta indica*, *Artemisia annua*, *Eucalyptus globulus*, *Ocimum sanctum* and *Rheum emodi* was evaluated against *F. solani* f. sp. *melongenae* and all the extracts promoted significant reduction in fungus growth (Joseph et al., 2008).

Organic extracts and essential oil from *Nandina domestica* showed antifungal activity against *F. oxysporum* and *F. solani* (Bajpai et al., 2009). The essential oil extracted from mature seeds of *Trachyspermum ammi* has potential use as a biofungicide against *F. oxysporum* f.sp. *lycopersici*, *F. oxysporum* f. sp. *cubense* and *F. oxysporum* f. sp. *capsici*, the wilt pathogens of tomato, banana and chili, respectively (Siripornvisal, 2010). *F. oxysporum* f. sp *cicer* (the pathogen of chickpea, *Cicer arietinum*) was sensitive to several essential oils from lemongrass, clove, cinnamon bark, cinnamon leaf, cassia, fennel, basil and evening primrose; the fungitoxic effect of essential oils may be due to the synergistic activity of their different compounds (Pawar and Thaker, 2007).

Some plant proteins with antifungal activity are basic cysteine-rich proteins, chitin-binding proteins, chitinases, chitinase-like proteins, defensins, glucanases, protease inhibitors, ribosome-inactivating proteins, thaumatins, thionins, and lectins (Wong et al., 2010). The defensin from the flowers of *Nicotiana alata* showed antifungal activity against *F. oxysporum* f. sp. *vasinfectum* and its action mechanism includes the permeabilization of the fungal plasma membranes by formation of an aperture, and entering several hyphae promoting granulation of the cytoplasm and cell death (van der Weerden et al., 2008).

4. Antifungal Activity of Plant Lectins on Fusarium

Antifungal effects on *Fusarium* species have been reported for lectins isolated from different plants tissues. Lectin from *Sebastiania jacobinensis* bark inhibited the mycelial growth of *F. moniliforme* and *F. oxysporum* and showed a higher degree of selectivity in comparison with reference chemical fungicides; in addition, this lectin showed low ecotoxicity when evaluated by artemicidal and *Biomphalara glabatra* embriotoxicity assays (Vaz et al., 2010). Lectins isolated from *Archidendron jiringa* seeds, *Astragalus mongholicus* roots, *Amaranthus viridis* seeds, *Phaseolus vulgaris* seeds, and *Pisum sativum* seeds, inhibited the growth of *F. oxysporum* (Ye et al., 2001; Yan et al., 2005; Kaur et al., 2006; Sitohy et al., 2007; Charungchitrak et al., 2011). The lectin from *B. monandra* secondary roots (BmoRoL) purified in milligram quantities showed antifungal activity against the phytopathogenic species *F. oxysporum* and *F. solani* (Souza et al., 2011). The lectin from *Capsicum frutescens* seeds exhibited a strong inhibitory effect on growth and spore germination of *F. moniliforme* (Ngai and Ng, 2007).

Antifungal activity has been reported mainly for chitin-binding lectins which are able to bind chitin present in fungal cell wall (Figure 2). Chitin-binding lectins are able to weak the developing of fungi by binding to the cell wall polymers; in addition, lectins can penetrate the fungal cell wall and disrupt the balance between hydrolysis and synthesis of wall polymers by blocking the enzymes involved in cell wall production (Ciopraga et al., 1999; Baker et al., 2009).

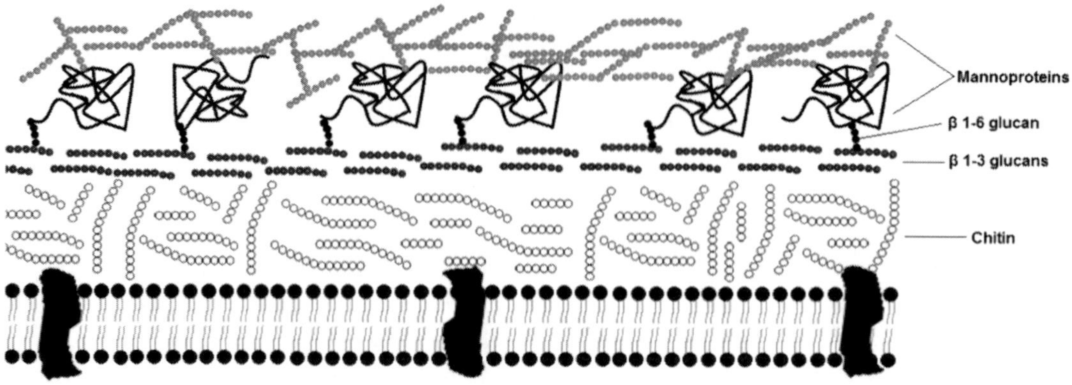

Figure 2. Schematic representation of a fungal cell wall highlighting the potential molecules that can be targets for binding of lectins. The most antifungal lectins are chitin-binding proteins. Lectins able to interact with mannoproteins and other saccharides may also exert antifungal activity.

Analysis by photomicrograph of the mycelia of *F. oxysporum* after 60 h of growth in the presence of *Talisia esculenta* seed lectin revealed inhibition of hyphal development; the authors suggested that the antifungal activity could be due to the interaction of lectin with chitin present in the cell wall and other fungal structures (Freire et al., 2002). The chitin-binding lectin from *M. urundeuva* heartwood showed antifungal activity against *F. decemcellulare*, *F. moniliforme*, *F. oxysporum*, and *F. solani* and little effect against *F. lateritium* (Sá et al., 2009a).

The chitin-binding lectin (wheat germ agglutinin) isolated from a Romanian dihaploid of wheat demonstrated antifungal effect on *F. graminearum* and *F. oxysporum*. Morphological effects detected after 8 h of incubation with the lectin included lysis of hyphal tips, swelling and vacuolation of cell content, affecting the growth of the germ tubes. In addition, conidial suspensions treated with this lectin lost their infecting potential on potato tubers (Ciopraga et al., 1999).

Mannan present in mannoproteins and other saccharides are other important components of most fungal cell walls (Figure 2) and *T. esculenta* seed lectin may inhibit fungal growth also by interacting with these carbohydrates (Freire et al., 2002).

The biodegradability of lectins stimulates their use as natural compounds for fungi control safer than synthetic agrochemicals currently applied. In addition, these proteins may be used to treat *Fusarium* infections in humans and animals. The toxicity of lectins has been studied aiming to assure their safety for environment and human health. Further studies are needed aiming to evaluate the toxicity of antifungal lectins on non-target organisms as well as to develop strategies for its production with the low cost possible and economic viability.

CONCLUSION

In conclusion, this chapter supports the optimistic prospects of lectins from plants, which can be applied in agriculture as a biodegradable alternative for biocontrol of *Fusarium* species.

ACKNOWLEDGMENTS

The authors express their gratitude to the *Conselho Nacional de Desenvolvimento Científico e Tecnológico* (CNPq) for research grants and fellowships (PMGP and LCBBC). We are also grateful to the *Fundação de Amparo à Ciência e Tecnologia do Estado de Pernambuco* (FACEPE) and the *Coordenação de Aperfeiçoamento de Pessoal de Nível Superior* (CAPES) for financial support. It is also acknowledged the Portuguese *Fundação para Ciência e a Tecnologia* (FCT) through the Post-doctoral grant SFRH/BPD/37349/2007 (AFSS).

REFERENCES

Abbas, HK; Cartwright, RD; Shier, WT; Abouzied, MM; Bird, CB; Rice, LG; Ross, PF; Sciumbato, LG; Meredith, FI. Natural occurrence of fumonisins in rice with *Fusarium* sheath rot disease. *Plant Disease*, 1998, 82, 22-25.

Agrios, GN. Plant Pathology. 5th ed. New York: Elsevier Academic Press; 2005.

Andrade JL; Arruda S; Barbosa, T; Paim, L; Ramos, MV; Cavada, BS; Barral-Neto, M. Lectin-induced nitric oxide production. *Cellular Immunology*, 1999, 194, 98-102.

Aoyama, H; Cavagis, ADM; Taga, EM; Ferreira CV. Endogenous lectin as a possible regulator of the hydrolysis of physiological substrates by soybean seed acid phosphatase. *Phytochemistry*, 2001, 58, 221-225.

Apetorgbor, MM; Darkwa, NA; Frimpong, O; Agyeman, VK. Biodeteriorating agents associated with three tropical timber species. *Forest Ecology and Management*, 2004, 195, 311-323.

Arif, T; Mandal, TK; Dabur, R. Natural products: Anti-fungal agents derived from plants. In: Tiwari, VK; Mishra BB, editors. *Opportunity, Challenge and Scope of Natural Products in Medicinal Chemistry*. Kerala: Research Signpost; 2011; 283-311.

Bajpai, VK; Lee, TJ; Kang, SC. Chemical composition and *in vitro* control of agricultural plant pathogens by the essential oil and various extracts of *Nandina domestica* Thunb. *Journal of the Science of Food and Agriculture*, 2009, 89, 109-116.

Bajpai, VK; Shin, SY; Kim, HR; Kang SC. Anti-fungal action of bioconverted eicosapentaenoic acid (bEPA) against plant pathogens. *Industrial Crops and Products*, 2008, 27, 136-141.

Baker, RL; Brown, RL; Chen, Z-Y; Cleveland, TE; Fakhoury, AM. A maize lectin-like protein with antifungal activity against *Aspergilus flavus*. *Journal of Food Protection*, 2009, 72, 120-127.

Baran, R; Tosti, A; Piraccini, MB. Uncommon clinical patterns of Fusarium nail infection: report of three cases. The British Journal of Dermatology, 1997, 136, 424-427.

Bhattacharya, K; Raha, S. Deteriorative changes of maize, groundnut and soybean seeds by fungi in storage. *Mycopathologia*, 2002, 155, 135-141.

Charungchitrak, S; Petsom, A; Sangvanich, P; Karnchanatat, A. Antifungal and antibacterial activities of lectin from the seeds of *Archidendron jiringa* Nielsen. *Food Chemistry*, 2011, 126, 1025-1032.

Ciopraga, J; Gozia, O; Tudor, R; Brezuica, L; Doyle, RJ. *Fusarium* sp. growth inhibition by wheat germ agglutinin. *Biochimica et Biophysica Acta*, 1999, 1428, 424-432.

Coelho, JS; Santos, NDL; Napoleão, TH; Gomes, FS; Ferreira, RS; Zingali, RB; Coelho, LCBB; Leite, SP; Navarro, DMAF; Paiva, PMG. Effect of *Moringa oleifera* lectin on development and mortality of *Aedes aegypti* larvae. *Chemosphere*, 2009, 77, 934-938.

Coelho, MB; Marangoni, S; Macedo MLR. Insecticidal action of *Annona coriacea* lectin against the flour moth *Anagasta kuehniella* and the rice moth *Corcyra cephalonica* (Lepidoptera: Pyralidae). *Comparative Biochemistry and Physiology Part C*, 2007, 146, 406-414.

Correia, MTS; Coelho, LCBB; Paiva, PMG. Lectins, carbohydrate recognition molecules: Are they toxic? In: Siddique, YH, editor. *Recent Trends in Toxicology vol. 37*, Kerala: Transworld Research Network; 2008, 47-59.

Costa, RMPB; Vaz, AFM; Oliva, MLV; Coelho, LCBB; Correia, MTS; Carneiro-da-Cunha, MG. A new mistletoe *Phthirusa pyrifolia* leaf lectin with antimicrobial properties. *Process Biochemistry*, 2010, 45, 526-533.

Curtis, FD; Cicco, VD; Haidukowski, M; Pascale, M; Somma, S; Moretti, A. Effects of agrochemical treatments on the occurrence of *Fusarium* ear rot and fumonisin contamination of maize in Southern Italy. *Field Crops Research*, 2011, 123, 161-169.

D'Mello, JPF; Placinta, CM; Macdonald, AMC. *Fusarium* mycotoxins: a review of global implications for animal health, welfare and productivity. *Animal Feed Science and Technology*, 1999, 80, 183-205.

Fang, EF; Lin, P; Wong, JH; Tsao, SW; Ng, TB. A lectin with anti-HIV-1 reverse transcriptase, antitumor, and nitric oxide inducing activities from seeds of *Phaseolus vulgaris* cv. extralong autumn purple bean. *Journal of Agricultural and Food Chemistry*, 2010, 58, 2221-2229.

Fawzi, EM; Khalil, AA; Afifi, AF. Antifungal effect of some plant extracts on *Alternaria alternata* and *Fusarium oxysporum. African Journal of Biotechnology*, 2009, 8, 2590-2597.

Ferreira, RS; Napoleão, TH; Santos, AFS; Sá, RA; Carneiro-da-Cunha, MG; Morais, MMC; Silva-Lucca, RA; Oliva, MLV; Coelho, LCBB; Paiva, PMG. Coagulant and antibacterial activities of the water-soluble seed lectin from *Moringa oleifera*. Letters in Applied Microbiology, 2011, 53, 186-192.

Figueiredo, JG; Bitencourt, FS; Beserra, IG; Teixeira, CS; Luz, PB; Bezerra, EH; Mota, MR; Assreuy, AM; Cunha, FQ; Cavada, BS; Alencar, NM. Antinociceptive activity and toxicology of the lectin from *Canavalia boliviana* seeds in mice. *Naunyn-Schmiedeberg's Archives of Pharmacology*, 2009, 380, 407-414.

Franceschini, V; Lazzari, M; Ciani, F. Lectin cytochemical localization of glycoconjugates in the olfactory system of the lizards *Lacerta viridis* and *Podarcis sicula*. *Anatomy and Embryology*, 2000, 202, 49-54.

Freeman, S; Zveibil, A; Vintal, H; Maymon, M. Isolation of nonpathogenic mutants of *Fusarium oxysporum* f. sp. *melonis* for biological control of *Fusarium* wilt cucurtits. *Phytopathology*, 2002, 92, 164-168.

Freire, MGM; Gomes, VM; Corsini, RE; Machado, OLT; De Simone, SG; Novello, JC; Marangoni, S; Macedo, MLR. Isolation and partial characterization of a novel lectin from *Talisia esculenta* seeds that interferes with fungal growth. *Plant Physiology and Biochemistry*, 2002, 40, 61-68.

Fungicide Resistance Action Comitee. FRAC Code List©: Fungicides sorted by mode of action (including FRAC Code numbering), 2009.

Gemeiner, P; Mislovičová, D; Tkáč, J; Švitel, J; Pätoprstý, V; Hrabárová, E; Kogan, G; Kožár, T. Lectinomics: II. A highway to biomedical/clinical diagnostics. *Biotechnology Advances*, 2009, 27, 1-15.

Gu, JD; Ford, TE; Berke, NS; Mitchell, R. Biodeterioration of concrete by the fungus *Fusarium*. *International Biodeterioration and Biodegradation*, 1998, 41, 101-109.

Gutleb, AC; Morrison, E; Murk, AJ. Cytotoxicity assays for mycotoxins produced by *Fusarium* strains: a review. *Environmental Toxicology and Pharmacology*, 2002, 11, 309-320.

Hesseltine, CW. Natural occurrence of mycotoxins in cereals. *Mycopathology*, 1974, 53, 141-153.

Hirsch, AM. Role of lectins (and rhizobial exopolysaccharides) in legume nodulation. *Current Opinion in Plant Biology*, 1999, 2, 320–326.

Joseph, B; Dar, MA; Kumar, V. Bioefficacy of plant extracts to control *Fusarium solani* f. sp. *melongenae* incitant of Brinjal Wilt. *Global Journal of Biotechnology & Biochemistry*, 2008, 3, 56-59.

Kaur, N; Dhuna, V; Kamboj, SS; Agrewala, JN; Singh J. A novel antiproliferative and antifungal lectin from *Amaranthus viridis* Linn seeds. *Protein and Peptide Letters*, 2006, 13, 897-905.

Kestwal, RM; Konozy, EME; Hsiao, C; Roque-Barreira, MC; Bhide, SV. Characterization of α-mannosidase from *Erythrina indica* seeds and influence of endogenous lectin on its activity. *Biochimica et Biophysica Acta – General Subjects*, 2007, 1770, 24-28.

Larkin, RP; Fravel, DR. Efficacy of various fungal and bacterial biocontrol organisms for control of *Fusarium* wilt of tomato. *Plant Disease*, 1998, 82, 1022-1028.

Li, X; Quan, C-S; Yu, H-Y; Wang, J-H; Fan, S-D. Assessment of antifungal effects of a novel compound from *Burkholderia cepacia* against *Fusarium solani* by fluorescent staining. *World Journal of Microbiology and Biotechnology*, 2009, 25, 151-154.

Lima, ALR; Cavalcanti, CCB; Silva, MCC; Paiva, PMG; Coelho, LCBB; Beltrão, EIC; Correia, MTS. Histochemical Evaluation of human prostatic tissues with *Cratylia mollis* seed lectin. *Journal of Biomedicine and Biotechnology*, 2010, 2010, ID 179817.

Lima, VLM; Correia, MTS; Cechinel, YMN; Sampaio, CAM; Owen, JS; Coelho, LCBB. Immobilized *Cratylia mollis* lectin as potential matrix to isolate plasma glycoproteins including lecithin cholesterol acyltransferase. *Carbohydrate Polymers*, 1997, 33, 27-32.

Limpens, E; Bisseling, T. Signaling in symbiosis. *Current Opinion in Plant Biology* 2003, 6, 343-350.

Lori, GA; Sisterna, MN; Sarandón, SJ; Rizzo, I; Chidichimo, H. *Fusarium* head blight in wheat: Impact of tillage and other agronomic practices under natural infection. *Crop Protection*, 2009, 28, 495-502.

Macedo, MLR; Freire, MGM; Novello, JC; Marangoni, S. *Talisia esculenta* lectin and larval development of *Callosobruchus maculatus* and *Zabrotes subfasciatus* (Coleoptera: Bruchidae). *Biochimica et Biophysica Acta – General Subjects*, 2002, 1571, 83–88.

Macedo, MLR; Freire, MGM; Silva, MBR, Coelho, LCBB, 2007. Insecticidal action of *Bauhinia monandra* leaf lectin (BmoLL) against *Anagasta kuehniella* (Lepidoptera: Pyralidae), *Zabrotes subfasciatus* and *Callosobruchus maculatus* (Coleoptera: Bruchidae). *Comparative Biochemistry and Physiology Part A*, 2007, 146, 486-498.

Maciel, EVM; Araújo-Filho, VS; Nakazawa, M; Gomes, YM; Coelho, LCBB; Correia, MTS. Mitogenic activity of *Cratylia mollis* lectin on human lymphocytes. *Biologicals*, 2004, 32, 57-60.

Magan, N; Hope, R; Colleate, A; Baxter, ES. Relationship between growth and mycotoxin production by *Fusarium* species, biocides and environment. *European Journal of Plant Pathology*, 2002, 108, 685-690.

Mahnine, N; Meca, G; Elabidi, A; Fekhaoui, M; Saoiabi, A; Font, G; Mañes, J; Zinedine, A. Further data on the levels of emerging *Fusarium* mycotoxins enniatins (A, A1, B, B1) beauvericin and fusaproliferin in breakfast and infant cereals from Morocco. *Food Chemistry*, 2011, 124, 481-485.

Melo, CML; Castro, MC; Oliveira, AP; Gomes, FO; Pereira, VR; Correia, MTS; Coelho LCBB; Paiva PMG. Immunomodulatory response of Cramoll 1,4 lectin on experimental lymphocytes. *Phytotherapy Research*, 2010, 24, 1631-1636.

Munkvold, GP. Epidemiology of *Fusarium* diseases and their mycotoxins in maize ears. *European Journal of Plant Pathology*, 2003, 109, 705-713.

Napoleão, TH; Gomes, FS; Lima, TA; Santos, NDL; Sá, RA; Albuquerque, AC; Coelho, LCBB; Paiva, PMG. Termiticidal activity of lectins from *Myracrodruon urundeuva* against *Nasutitermes corniger* and its mechanisms. *International Biodeterioration and Biodegradation*, 2011, 65, 52-59.

Napoleão, TH; Pontual EV; Lima TA; Santos NDL; Sá RA; Coelho LCBB; Navarro DMAF; Paiva PMG. Effect of *Myracrodruon urundeuva* leaf lectin on survival and digestive enzymes of *Aedes aegypti* larvae. *Parasitology Research*, 2012, doi:10.1007/s00436-011-2529-7.

Nelson, P; Dignan, MC; Anaissie, E. Taxonomy, biology and clinical aspects of *Fusarium* species. *Clinical Microbiology Reviews*, 1994, 7, 479-504.

Ngai, PHK, Ng TB. A lectin with antifungal and mitogenic activities from red cluster peppers (*Capsicum frutescens*) seeds. *Applied Microbiology and Biotechnology*, 2007, 74, 366-371.

Nucci, M; Anaissie, E. Cutaneous infection by *Fusarium* species in healthly and immunocompromised hosts: Implications for diagnosis and management. *Clinical Infection Diseases*, 2002, 35, 909-920.

Nucci, M, Anaissie, E. *Fusarium* infections in immunocompromised patients. *Clinical Microbiology Reviews*, 2007, 20, 695-704.

Oliveira, MDL; Andrade, CAS; Santos-Magalhães, NS; Coelho, LCBB; Teixiera, JA; Carneiro-da-Cunha, MG; Correia, MTS. Purification of a lectin from *Eugenia uniflora* L. seeds and its potential antibacterial activity. *Letters in Applied Microbiology*, 2008, 46, 371-376.

Oliveira, CFR; Luz, LA; Paiva, PMG; Coelho, LCBB; Marangoni, S; Macedo, MLR. Evaluation of seed coagulant *Moringa oleifera* lectin (cMoL) as a bioinsecticidal tool with potential for the control of insects. *Process Biochemistry*, 2011, 46, 498-504.

Paiva, PMG; Souza, AF; Oliva MLV; Kennedy JF; Cavalcanti MSM; Coelho LCBB; Sampaio CAM. Isolation of a trypsin inhibitor from *Echinodorus paniculatus* seeds by affinity chromatography on immobilized *Cratylia mollis* isolectins. *Bioresource Technology*, 2003, 88 75-79.

Paiva, PMG; Napoleão, TH; Santos, NDL; Correia, MTS; Navarro, DMAF; Coelho, LCBB. Plant compounds with *Aedes aegypti* larvicidal activity and other biological properties, in: Liong, M-T., editor. *Bioprocess Sciences and Technology*. New York: Nova Publishers Inc.; 2011; 271-296.

Paiva, PMG; Santana, GMS; Souza, IFAC; Albuquerque, LP; Agra-Neto, AC; Albuquerque, AC; Luz, LA; Napoleão, TH; Coelho, LCBB. Effect of lectins from *Opuntia ficus indica* cladodes and *Moringa oleifera* seeds on survival of *Nasutitermes corniger*. *International Biodeterioration and Biodegradation*, 2011, 65, 988-995.

Pawar, VC; Thaker, VS. Evaluation of the anti-*Fusarium oxysporum* f.sp. *cicer* and anti-*Alternaria porri* effects of some essential oils. *World Journal of Microbiology and Biotechnology*, 2007, 23, 1099-1106.

Pirgozliev, SR; Edwards, SG; Hare, MC; Jenkinson, P. Strategies for the control of *Fusarium* head blight in cereals. *European Journal of Plant Pathology*, 2003, 109,731-742.

Rojo, FG, Reynoso, MM, Ferez, M, Chulze, SN, Torres, AM. Biological control by *Trichoderma* species of *Fusarium solani* causing peanut brown root rot under field conditions. *Crop Protection*, 2005, 26, 549-555.

Roncero, MIG; Hera, C; Ruiz-Rubio, M; Maceira, FIG; Madrid, MP; Caracuel, Z; Calero, F; Delgado-Jarana, J; Roldán-Rodríguez, R; Martínez-Rocha, AL; Velasco, C; Roa, J; Martín-Urdiroz, M; Córdoba, D; Di Pietro, A. *Fusarium* as a model for studying virulence in soilborne plant pathogens. *Physiological and Molecular Plant Pathology*, 2003, 62, 87-98.

Sá, RA; Gomes, FS; Napoleão, TH; Santos, NDL; Melo, CML.; Gusmão, NB; Coelho, LCBB; Paiva, PMG; Bieber, LW. Antibacterial and antifungal activities of *Myracrodruon urundeuva* heartwood. *Wood Science and Technology*, 2009a, 43, 85-95.

Sá, RA; Napoleão, TH; Santos, NDL; Gomes, FS; Albuquerque, AC; Xavier, HS; Coelho, LCBB; Bieber, LW; Paiva, PMG. Induction of mortality on *Nasutitermes corniger* (Isoptera, Termitidae) by *Myracrodruon urundeuva* heartwood lectin. *International Biodeterioration and Biodegradation*, 2008, 62, 460-464.

Sá, RA; Santos, NDL; Silva, CSB; Napoleão, TH; Gomes, FS; Cavada, BS; Coelho, LCBB; Navarro, DMAF; Bieber, LW; Paiva PMG. Larvicidal activity of lectins from *Myracrodruon urundeuva* on *Aedes aegypti*. Comparative Biochemistry and Physiology Part C, 2009b, 149, 300-306.

Santana, GMS; Albuquerque, LP; Simões, DA; Gusmão, NB; Coelho, LCBB; Paiva, PMG. Isolation of lectin from *Opuntia ficus indica* cladodes. *Acta Horticulturae*, 2009, 811, 281-286

Santos, AFS, Luz, LA, Argolo, ACC, Teixeira, JA, Paiva, PMG; Coelho, LCBB. Isolation of a seed coagulant *Moringa oleifera* lectin. *Process Biochemistry*, 2009, 44, 504-508.

Santos, AFS; Paiva, PMG; Teixeira, JAC; Brito, AG; Coelho, LCBB; Nogueira, R. Coagulant properties of *Moringa oleifera* protein preparations: application to humic acids removal. *Environmental Technology*, 2011, doi: 10.1080/09593330.2010.550323

Sato, Y; Morimoto, K; Hirayama, M; Hori K. High mannose-specific lectin (KAA-2) from the red alga *Kappaphycus alvarezii* potently inhibits influenza virus infection in a strain-independent manner. *Biochemical and Biophysical Research Communications*, 2011, 405, 291-296.

Silva, MCC; Santana, LA; Silva-Lucca, RA; Lima, ALR; Ferreira, JG; Paiva, PMG; Coelho, LCBB; Oliva, MLV; Zingali, RB; Correia, MTS. Immobilized *Cratylia mollis* lectin: An affinity matrix to purify a soybean (*Glycine max*) seed protein with *in vitro* platelet antiaggregation and anticoagulant activities. *Process Biochemistry* 2011, 46, 74-80.

Silva, MDC; Sá, RA; Napoleão, TH; Gomes, FS; Santos, NDL; Albuquerque, AC; Xavier, HS; Paiva, PMG, Correia, MTS; Coelho, LCBB. Purified *Cladonia verticillaris* lichen lectin: Insecticidal activity on *Nasutitermes corniger* (Isoptera: Termitidae). *International Biodeterioration and Biodegradation*, 2009, 63, 334-340.

Siripornvisal, S. Antifungal activity of Ajowan oil against *Fusarium oxysporum*. KMITL *Science and Technology Journal*, 2010, 10, 45-51.

Sitohy, M; Doheim, M; Badr H. Isolation and characterization of a lectin with antifungal activity from Egyptian *Pisum sativum* seeds. *Food Chemistry*, 2007, 104, 971-979.

Souza, JD; Silva, MBR; Argolo, ACC; Napoleão, TH; Sá, RA; Correia, MTS; Paiva, PMG; Silva, MDC; Coelho, LCBB. A new *Bauhinia monandra* galactose-specific lectin purified in milligram quantities from secondary roots with antifungal and termiticidal activities. *International Biodeterioration and Biodegradation*, 2011, 65, 696-702.

Sterflinger, K. Fungi: Their role in deterioration of cultural heritage. *Fungal Biology Reviews*, 2010, 24, 47-55.

Suárez-Estrella, F; Vargas-García, MC; López, MJ; Moreno, J. Survival of *Fusarium oxysporum* f.sp. *melonis* on plant waste. *Crop Protection*, 2004, 23, 127-133.

Timoshenko, AV; Cherenkevich, SN; Gabius, HJ. *Viscum album* agglutinin-induced aggregation of blood cells and the lectin effects on neutrophil function. *Biomedicine & Pharmacotherapy*, 1995, 49, 153-158.

Touvron, G; Denis, D; Doat, M; Girard, A; Brandely, M-L; Chast, F; Renard, G. Efficacité de l'amphotéricine B liposomale dans le traitement d'une kératite à *Fusarium solani* résistante. Journal Français d'Ophtalmologie, 2009, 32, 721-726.

Tripathi, P; Dubey, NK. Exploitation of natural products as an alternative strategy to control postharvest fungal rotting of fruit and vegetables. *Postharvest Biology and Technology*, 2004, 32, 235-245.

Van Damme, EJM; Barre, A; Rougé, P; Peumans, WJ. Cytoplasmic/nuclear plant lectins: a new story. *Trends in Plant Science*, 2004, 9, 484-489.

van der Weerden, NL; Lay, FT; Anderson, MA. The plant defensin, NaD1, enters the cytoplasm of *Fusarium oxysporum* hyphae. *The Journal of Biological Chemistry*, 2008, 283, 14445-14452.

Vasconcelos, IM; Oliveira, JTA. Antinutritional properties of plant lectins. *Toxicon*, 2004, 4, 385-403.

Vaz, AFM; Costa, RMPB; Melo, AMMA; Oliva, MLV; Santana, LA; Silva-Lucca, RA; Coelho, LCBB; Correia, MTS. Biocontrol of *Fusarium* species by a novel lectin with low ecotoxicity isolated from *Sebastiania jacobinensis*. *Food Chemistry*, 2010, 119, 1507-1513.

Verma, RK; Chaurasia, L; Katiyar, S. Potential antifungal plants for controlling building fungi. *Natural Product Radiance*, 2008, 7, 374-387.

Wagacha, JM, Muthomi, JW. *Fusarium culmorum*: Infection process, mechanisms of mycotoxin production and their role in pathogenesis in wheat. *Crop Protection*, 2007, 26, 877-885.

Wong, JH; Ng, TB; Cheung, RCF; Ye, XJ; Wang, HX; Lam, SK; Lin, P; Chan, YS; Fang, EF; Ngai; PHK; Xia, LX; Ye, XY; Jiang, Y.; Liu, F. Proteins with antifungal properties and other medicinal applications from plants and mushrooms. *Applied Microbiology and Biotechnology*, 2010, 87, 1221-1235.

Xie, M.; Hu, J; Long, YM; Zhang, ZL; Xie, HY; Pang, DW. Lectin-modified trifunctional nanobiosensors for mapping cell surface glycoconjugates. *Biosensors and Bioelectronics*, 2009, 24, 1311-1317.

Yan, Q; Jiang, Z; Yang, S; Deng, W; Han, L. A novel homodimeric lectin from *Astragalus mongholicus* with antifungal activity. *Archives of Biochemistry and Biophysics*, 2005, 442, 72-81.

Ye, XY; Ng, TB; Tsang, PWK; Wang, J. Isolation of a homodimeric lectin with antifungal and antiviral activities from red kidney bean (*Phaseolus vulgaris*) seeds. *Journal of Protein Chemistry*, 2001, 20, 367-375.

Zhao, Q; Dong, C; Yang, X; Mei, X; Ran, W; Shen, Q; Xu, Y. Biocontrol of *Fusarium* wilt disease for *Cucumis melo* melon using bio-organic fertilizer. *Applied Soil Ecology*, 2011, 47, 67-75.

In: Fusarium
Editors: Tony Ferarri Rios and Emilia Reyes Ortega
ISBN: 978-1-61942-539-2

Chapter 7

ANTIFUNGAL PROTEINS FOR CONTROL OF *FUSARIUM* SPECIES

Tzi Bun Ng and Jack Ho Wong
School of Biomedical Sciences, Faculty of Medicine,
the Chinese University of Hong Kong, Hong Kong, China

ABSTRACT

Some *Fusarium* species are human pathogens while other *Fusarium* species are phytopathogens.A diversity of antifungal proteins displays inhibitory activity toward *Fusarium* species. They include chitin binding proteins, chitinases, cyclophilin-like proteins, defensins, embryo-abundant protein-like proteins, glucanases, lectins, lipid transfer proteins, osmotins, peroxidases, proteases, ribonucleases, storage 2S albumins, thaumatin-like proteins,and other antifungal proteins. The antifugal mechanisms of chitinases, defensins, glucanases, and thaumatin-like proteins are well known but those of the other aforementioned proteins await elucidation.

Keywords: *Fusarium*, antifungal proteins, antifungal mechanisms

INTRODUCTION

Pathogenic fungi bring about devastating damage to agricultural crops and cause diseases in humans and animals. Antifungal proteins are produced by a diversity of organisms including humans, other animals, plants, fungi and bacteria to combat invading bacteria. Research on antifungal proteins may disclose some with potent fungistatic and/or fungicidal activities which can be developed into therapeutic agents for use in humans and animals and which can be exploited in some strategies such as transgenic plants to resist fungal attack. The various types of antifungal proteins capable of inhibiting *Fusarium* species are described in the following.

Chitin Binding Proteins

The chitin binding protein Cbp50 from the insecticidal bacterium *Bacillus thuringiensis serovar konkukian* S4 antifungal activity against the phytopathogenic fungus *F.oxysporum* and the opportunistic human pathogen *Aspergillus niger*. It possesses a modular structure comprised of an N-terminal CBM33 domain, two copies of a fibronectin-like domain and a C-terminal chitin binding domain classified as CBM5. Both the N-terminal and C-terminal domains are crucial for chitin binding (Mehmood et al., 2011).

Chitinases

A 29-kDa chitinase from the bulbs of the plant *Urginea indica* (Indian squill) inhibited the fungal pathogens *F.oxysporum* and *Rhizoctonia solani in vitro* (Shenoy et al., 2006).

Two 30-kDa proteins with N-terminal sequence homology to chitinases have been isolated from fruits of the emperor banana inhibited mycelial growth in *F. oxysporum* but not in *Mycosphaerella arachidicola*. The chitinase-like protein more strongly bound on Mono S was obtained with a slightly lower yield and exhibited a higher antifungal potency toward *F. oxysporum* when compared with the less strongly bound chitinase-like protein (Ho and Ng, 2007).

A 30.6-kDa chitinase with antifungal activity was isolated from the Canadian cranberry beans (*Phaseolus vulgaris*) exerted a potent inhibitory action toward fungal species including *F.oxysporum,Botrytis cinerea, Physalospora piricola,* and *Pythium aphanidermatum.* (Wang et al., 2009).

The *Bacillus pumilus* SG2 chitinase gene (ChiS) and its truncated form lacking chitin binding (ChBD) and fibronectin type III (FnIII) domains were transformed to *Arabidopsis* plants and the expression, functionality and antifungal activity of the recombinant proteins were investigated. Although the intact and truncated forms showed equipotent chitinolytic activity, they differed in antifungal activity. Recombinant ChiS in plant protein extracts displayed a high inhibitory effect on spore germination and radial growth of hyphae in *F. graminearum, Alternaria brassicicola,* and *Botrytis cinerea*. The truncated enzyme had no activity. The results reveal that ChBD and FnIII domains are essential in hydrolysis of chitin-glucan complex of fungal cell walls but not in hydrolysis of colloidal chitin (Dehestani et al., 2010).

Cyclophilin-like Proteins

An antifungal protein from Chinese cabbage (*Brassica campestris* L. ssp. pekinensis)with an N-terminal amino acid highly homologous to that of plant cyclophilins inhibited the growth of filamentous fungal pathogens. It also inhibited the growth of several fungal pathogens (Lee et al.,2007).

Defensins

An 7 kDa antifungal defensin peptide from ground beans (*Vigna sesquipedalis* cv. 'Ground Bean') exhibited an antifungal activity toward *F.oxysporum, Botrytis cinerea,* and *Mycosphaerella arachidicola*; and an antibacterial action on *Escherichia coli B, Proteus vulgaris, Mycobacterium phlei* and *Bacillus megaterium.* It exerted antiproliferative activity toward breast cancer (MCF-7) cells and leukemia M1 cells. It also exhibited some inhibitory activity toward human immunodeficiency virus-type 1 reverse transcriptase (Wong and Ng, 2005).

A 6.5-kDa antifungal peptide from the seeds of the shelf bean potently suppressed mycelial growth in *Fusarium oxysporum, Botrytis cinerea,* and *Mycosphaerella arachidicola* with an IC(50) of 2.1,2.9, and 0.34 microM, respectively. [Methyl-(3)H]-thymidine incorporation into isolated mouse splenocytes was stimulated but incorporation into M1 (myeloma) and L1210 (leukemia) cells was inhibited. The peptide reduced the activity of HIV-1 reverse transcriptase and also inhibited translation in a cell-free rabbit reticulocyte lysate system (Wong and Ng, 2006).

A 6-kDa antifungal peptide with inhibitory activity on mycelial growth in *F.oxysporum, Mycosphaerella arachidicola,* and *Physalospora piricola* was isolated from baby lima beans. The peptide suppressed growth in *M. arachidicola* with an IC(50) of 0.87 muM and inhibited activity of HIV-1 reverse transcriptase with an IC(50) of 4 muM. The peptide exhibited an N-terminal amino acid sequence similar to those of leguminous defensins (Wang and Ng, 2006).

A 7.3-kDa antifungal peptide with an N-terminal sequence exhibiting remarkable homology to defensins from other leguminous plants was isolated from *Phaseolus vulgaris* cv. 'Spotted Bean'. It exerted an antifungal action on *F.oxysporum* and *Mycosphaerella arachidicola.* It inhibited mycelial growth in *F. oxysporum* with an IC(50) value of 1.8 microM. It suppressed methyl-(3)H]-thymidine incorporation by leukemia L1210 cells and MBL2 cells with an IC(50) value of 4.0 and 9.0 microM, respectively (Wang and Ng,2007).

A 5.495-kDa heat-stable antimicrobial peptide (Vv-AMP11) isolated from *Vitis vinifera* exhibits sequence homology to the family of plant defensins. It is only expressed in berry tissue at the onset of berry ripening and onward.It demonstrates potent antifungal activity against *F.oxysporum* and *Verticillium dahliae* by inhibiting hyphal elongation. Its inhibitory activity might also be related to changing the fungal membrane permeability (de Beer and Vivier, 2008).

An 7.1-kDa antifungal defensin-like peptide was isolated from dried Nepalese large red beans (*Phaseolus angularis*). The antifungal peptide inhibited mycelial growth in *F. oxysporum* and *Mycosphaerella arachidicola* with an IC(50) value of 1.4 and 1.8 microM, respectively. It did not inhibit HIV-1 reverse transcriptase when tested up to 200 microM. It exerted an antiproliferative action on L1210 leukemia cells and MBL2 lymphoma cells with an IC(50) of 15 and 60 microM, respectively(Ma et al.,2009).

Plant defensin corn 1 (Pdc1) expressed in *P. pastoris* was more efficient than the protein expressed in *E. coli* in inhibiting growth of *F. graminearum* (Kant et al., 2009).

A 5443-Da peptide with sequence homology to defensins from purple pole beans (*Phaseolus vulgaris* cv. 'Extra-long Purple Pole bean') inhibited mycelial growth in *F. oxysporum, Mycosphaerella arachidicola, Helminthosporium maydis, Verticillium dahliae, Rhizoctonia solani, Candida albicans* and *Setosphaeria turcica.* Its antifungal potency was higher than that of the plant defensin coccinin (IC50>50 microM). The antifungal activity was pH stable and thermostable. The peptide inhibited the proliferation of hepatoma (HepG2), breast cancer (MCF7), colon cancer (HT29) and cervical cancer (SiHa) cells but not that of human embryonic liver (WRL68) cells. Its anti-HepG2 activity (IC50=4.1+/-0.8 microM, n=3) was higher than that of another plant defensin, gymnin (IC50>50 microM). Its anti-MCF7 activity (IC50=8.3+/-0.3 microM, n=3) was similar to that of other plant defensins. It reduced the activity of HIV-1 reverse transcriptase with an IC50 of 0.5+/-0.1 microM, n=3, much more potently than other plant defensins (IC50>40 microM). There is the possibility of using the purple pole bean defensin for producing antifungal drugs and/or transgenic plants with fungal resistance (Lin et al., 2009).

A 6.8-kDa peptide from *Phaseolus limensis* legumes designated Limyin, suppressed mycelial growth in *F. solani, Alternaria alternata,*and *Botrytis cinerea.* It showed antiproliferative activity towards tumor cells including human liver hepatoma cells Bel-7402 and neuroblastoma cells SHSY5Y (Wang et al., 2009).

Plant defensin NaD1 appeared to permeabilize cells of *F. oxysporum* f. sp. vasinfectum via a novel mechanism that required the presence of the fungal cell wall. NaD1 entered the cytoplasm of treated hyphae, indicating that cell death is accelerated by interaction with intracellular targets (van der Weerden et al., 2010).

Plant defensins MsDef1 and MtDef4 vary drastically in their antifungal properties as well as modes of antifungal action toward *F. graminearum.* MsDef1 but not MtDef4 induces prolific hyperbranching of fungal hyphae. The γ-core motifs of the two defensins differ in amino acid sequenceand in net charge. The major determinants of the antifungal activity and morphogenicity of these defensins reside in theγ-core motifs. TheMsDef1-γ4 variant with the γ-core motif replaced by that of MtDef4 resembled MtDef4 in antifungal potency and inability to induce hyphal hyperbranching. The γ-core motif of MtDef4 alone but not MsDef1 was capable of inhibiting fungal growth. The cationic and hydrophobic amino acids were important for antifungal activity. MtDef4 was more efficient than MsDef1 in permeabilizing fungal plasma membrane (Sagaram et al., 2011).

Defensin PvD(1)from Phaseolus vulgaris seeds induced membrane permeabilization in the filamentous fungi *F.oxysporum, F.solani,* and *F. laterithium* and in the yeasts *Candida albicans, C. parapsilosis, C. tropicalis, Kluyveromyces marxiannus, Pichia membranifaciens,* and *Saccharomyces cerevisiae.* Defensin-treated *C. albicans* and *C. guilliermondii* cells showed disorganization of both cytoplasmic content and plasma membrane. PvD(1) inhibited glucose-stimulated acidification of the medium by yeast cells and filamentous fungi,and induced the production of reactive oxygen species and nitric oxide in *F. oxysporum* and *C. albicans* cells(Mello et al.,2011).

An antifungal peptide with a defensin-like sequence and exhibiting a molecular mass of 7.3kDa was purified from dried seeds of *Phaseolus vulgaris* 'Cloud Bean'. The antifungal peptide exerted antifungal activity *F. oxysporum* with an IC (50) value of 2.2 μM and against *Mycosphaerella arachidicola* with an IC(50)value of 1.8 μM. Proliferation of L1210 mouse leukemia cells and MBL2 lymphoma cells was inhibited by the antifungal peptide with an IC(50) of 10 μM and 40 μM, respectively (Wu et al., 2011).

The defensin RsAFP2 from radish (*Raphanus sativus*) inhibited growth of *F. graminearum* and *R. cerealis.* The RsAFP2 gene was transformed into Chinese wheat variety Yangmai 12 via biolistic bombardment to assess the effectiveness of the defensin in protecting wheat from the fungal pathogens. Disease tests disclosed that four RsAFP2 transgenic lines (RA1-RA4) exhibited augmented resistance to *F. graminearum* compared to untransformed Yangmai 12 and null-segregated plants. The expression level of RsAFP2 was associated with the degree of resistance. Two of these transgenic lines (RA1 and RA2) also displayed increased resistance to *R. cerealis* (Li et al., 2011).

Embryo-abundant Protein-like Proteins

A novel basic protein with antifungal activity isolated from the seeds of *Ginkgo biloba* inhibited the growth of *F. oxysporum, Trichoderma reesei,* and *Candida albicans* but did not exhibit antibacterial action against *Escherichia coli.* The antifungal protein shows approximately 85% identity with embryo-abundant proteins from *Picea abies* and *Picea glauca* (Sawano et al., 2007).

Glucanases

A 42.7-kDa endo-1,3-beta-D: -glucanase from *Streptomyces* sp. S27 enzyme(BglS27) inhibited the growth of phytopathogenic fungi *F.oxysporum* and *Rhizoctonia solani* and some mycotoxin-producing fungi *F. crookwellense* and *Paecilomyces variotii.* These favorable properties make BglS27 a good candidate for utilization in biotechnological applications such as plant protection, feed and food preservation (Shi et al.,2010).

Lectins

A 66-kDa dimeric D-galactose binding lectin from the roots of *Astragalus mongholicus* exerted antifungalactivity against *Botrytis cincerea, Fusarium oxysporum, Colletorichum* sp., and *Drechslera turcia* but not against *Rhizoctonia solani* and *Mycosphaerella arachidicola.* (Yan et al., 2005).

A monomeric 29.5-kDa mannose/glucose-binding lectin from he seeds of red cluster pepper *Capsium frutescens* L. var. fasciculatum inhibited the germination of *Aspergillus flavus* and *F. moniliforme* spores and hyphal growth in the two fungi (Ngai and Ng,2007).

A 32.4-kDa lectin from *Curcuma amarissima* rhizomes inhibited the growth of *F.oxysporum, Colectrotrichum cassiicola,* and *Exserohilum turicicum,* and showed cytotoxicity against BT474 breast cancer cells *in vitro* (Kheeree et al., 2010).

Banerjee et al.(2011a)converted homodimeric 25-kDa insecticidal, mannose- binding *Allium sativum* leaf agglutinin (ASAL) into a monomeric form. Using SWISS-MODEL program a stable monomer was designed by changing five amino acid residues near the C-terminus of ASAL. By introducing 5 site-specific mutations (-DNSNN-), a β turn was constructed between the 11(th) and 12(th) β strands of subunits of ASAL, leading to a stable monomeric mutant ASAL (mASAL).Although mannose binds efficiently to both mASAL and ASAL, the hemagglutinating activity of purified mASAL was abolished. Furthermore, mASAL had negligible insecticidal activity compared to ASAL. mASAL exhibited strong antifungal activity against the pathogenic fungi *F.oxysporum, Alternaria brassicicola* and *Rhizoctonia solani* associated with changes in fungal membrane permeability.

Lipid Transfer Proteins

An antifungal peptide with a molecular mass of 9412 and an N-terminal sequence exhibiting notable homology to those of lipid transfer proteins was isolated from seeds of the vegetable *Brassica campestris* (Lin et al., 2005) inhibited mycelial growth in *F. oxysporum* and *Mycosphaerella arachidicola* with an IC(50) value of 8.3 microM and 4.5 microM, respectively(Lin et al., 2007).

An 9.4-kDa thermostable and pH-stable antifungal peptide, designated as campesin, was isolated from seeds of the cabbage *Brassica campestris*. It inhibited mycelial growth in *F.oxysporum* and *Mycosphaerella arachidicola*, with an IC(50) of 5.1 microM and 4.4 microM, respectively. The peptide was characterized by remarkable. It inhibited proliferation of HepG2 and MCF cancer cells with an IC(50) of 6.4 microM and 1.8 microM, and the activity of HIV-1 reverse transcriptase with anIC(50) of 3.2 microM. It demonstrated lysolecithin binding activity (Lin et al.,2009).

Osmotins

Solanum nigrum osmotin-like protein exerted antifungal action toward *F. solani* f. sp.glycines, *Colletotrichum* spp. and *Macrophomina phaseolina* (Campos et al., 2008).

CpOsm, a stable osmotin from *Calotropis procera* latex, inhibited the spore germination of *Fusarium solani* in all pH ranges tested. It induced the membrane permeabilization of spores and hyphae from *F. solani* (de Freitas et al., 2011).

Peroxidases

A 32-kDa class III haem peroxidase from leaves of *Acorus calamus* inhibited hyphal growth *Macrophomina phaseolina, Fusarium moniliforme* and *Trichosporium vesiculosum.* Microscopic observations disclosed distortion in the hyphal structure (Ghosh, 2006).

A peroxidase from the large lima bean (*Phaseolus limensis*) seeds suppressed mycelial growth in *Fusarium solani, Mycosphaerella arachidicola* and *Pythium aphanidermatum* with an IC50 of 76, 103, and 119 microM, respectively (Wang et al., 2009).

Proteases

A 12-kDa cytotoxic antifungal protease was isolated from dried fruiting bodies of the medicinal mushroom *Cordyceps militaris*. Its optimum temperature and pH were 37C and pH 7.0-9.0. The enzyme activity was specifically inhibited by the serine protease inhibitor phenylmethylsulfonyl fluoride. It exerted a potent antifungal action against the growth of *F. oxysporum*, and exhibited cytotoxicity against human breast and bladder cancer cells (Park et al., 2009).

The 45-kDa subtilisin-like protease from *Trichoderma harzianum* which parasitizes a large variety of phytopathogenic fungi.exerted broad-spectrum antifungal activity against *F.*

oxysporum,Alternaria alternata, Rhizoctonia solani, Sclerotinia sclerotiorum and *Cytospora chrysosperma* (Yan et al., 2009).

Banerjee et al. (2011b) isolated a glycoprotein cysteine protease from *Zingiber ottensii* rhizomes. It manifested inhibitory effect against *Collectotrichum cassiicola*, while weaker effects against *F. oxysporum* and *Exserohilum turicicum* were observed.

The 43-kDa cysteine protease from turmeric (*Curcuma domestica*) rhizome showed antifungal activity at 10 microg mL(-1) towards pathogens *Fusarium* sp., *Pythium aphanidermatum,* and *Trichoderma viride* (Nagarathnam et al., 2010).

Ribonucleases

A 47-kDa ribonuclease from fresh saffron (*Crocus sativus)* stigmas inhibited *F. oxysporum* growth but not *Penicillium sp.* and *Verticillium dahliae.* It appears to be involved in the active defense response through activation of the jasmonic acid pathway (Gómez-Gómez et al., 2011).

Storage 2S Albumins

A 5-kDa peptide, Pe-AFP1 purified from the seeds of passion fruit (*Passiflora edulis*). inhibited development of the filamentous fungi *Fusarium oxysporum, Trichoderma harzianum,*and *Aspergillus fumigatus* (Pelegrini et al.,2006).

Thaumatin-like Proteins

A 20-kDa protein with substantial N-terminal sequence homology to thaumatin-like proteins was isolated from ripe fruits of the emperor banana, Musa basjoo cv. 'emperor banana'. The thaumatin-like protein inhibited mycelial growth in *F. oxysporum* and *Mycosphaerella arachidicola*. (Ho et al., 2007).

An antifungal protein with sequence homology to thaumatin-like proteins was isolated from seeds of the desert plant *Cynanchum komarovii*. The protein showed antifungal activity against fungal growth of *F. oxysporum, Botrytis cinerea, Rhizoctonia solani, Valsa mali* and *Verticillium dahlia.* Its transcription level was elevated under the stress of abscisic acid, salicylic acid, methyl jasmonate, NaCl and drought, which indicates that CkTLP may play an important role in response to abiotic stresses (Wang et al., 2011).

Other Antifungal Proteins

A novel1370-Da peptide with high inhibitory activity against fungi growth have been isolated from the venom of the Brazilian snake *Bothrops jararaca.* inhibited the growth of different phytopathogenic fungi (*F.oxysporum* and *Colletotrichum lindemuthianum*) and yeast

(*Candida albicans* and *Saccharomyces cerevisiae*). Its effect is related to membrane permeabilization (Gomes et al., 2005).

A10-kDa antifungal peptidefrom coconut flesh displayed antifungal activity against *F. oxysporum, Mycosphaerella arachidicola* and *Physalospora piricola*. The IC50 values of its inhibitory activities on mycelial growth in *M. arachidicola* and HIV-1 reverse transcriptase activity were respectively 1.2 and 52.5 microM (Wang and Ng 2005).

An 32-kDa antifungal protein with a novel N-terminal sequence was isolated from ginger rhizomes exhibited antifungal activity toward various fungi including *F.oxysporum,Botrytis cinerea, Mycosphaerella arachidicola,* and *Physalospora piricola* (Wang and Ng, 2005).

A 15-kDa antifungal protein, designated ganodermin, from the medical mushroom *Ganoderma lucidum* inhibited the mycelial growth of *F. oxysporum Botrytis cinerea*, and *Physalospora piricola* with an IC50 value of 15.2 microM, 12.4 microM and 18.1microM, respectively(Wang Ng,2006). \

An 11-kDa antifungal protein with a lysine-rich N-terminal sequence was isolated from the seeds of the pea *Pisum sativum* var. arvense Poir. It exerted antifungal activity against *Physalospora piricola* with an IC50 of 0.62 microM, and also antifungal activity against *F. oxysporum* and *Mycosphaerella arachidicola*. It inhibited human immunodeficiency virus type 1 reverse transcriptase with an IC50 of 4.7microM (Wang and Ng, 2006).

An 11-kDa antifungal peptide from dry seeds of the red lentil (*Lens culinaris*) inhibited mycelial growth in *Mycosphaerella arachidicola* with an IC50 of 36 microM. It also exhibited antifungal activity against *F. oxysporum* (Wang and Ng, 2007).

The 20-kDa recombinant AhPR10 from peanut *Arachis hypogaea* exhibits ribonuclease activity as well as antifungal activity against the peanut pathogens *F.oxysporum* and *Rhizoctonia solani*. Mutant protein AhPR10-K54N in which lys54 is mutated to asn54 loses ribonuclease and antifungal activities. FITC-labeled AhPR10 and AhPR10-K54N are internalized by hyphae of *F. oxysporum* and *R. solani* but the latter protein does not inhibit fungal growth. This suggests that the ribonuclease function of AhPR10 is essential for its antifungal activity (Chadha and Das, 2006).

An 11-kDa antifungal peptide from dry seeds of the red lentil (*Lens culinaris*) inhibited mycelial growth in *Mycosphaerella arachidicola* with an IC50 of 36 microM. It also exhibited antifungal activity against *F. oxysporum* (Wang and Ng, 2007).

A 4-kDa antifungal peptide isolated from buckwheat seeds, with relatively high thermostability and pH stability, inhibited mycelial growth in *F. oxysporum* and *Mycosphaerella arachidicola* with an IC(50) of 35 and 40 microM, respectively. It inhibited proliferation of, L1210 (leukemia) cells, breast cancer (MCF-7) cells, Hep G2 (hepatoma) cells and liver embryonic WRL 68 cells with an IC(50) of 4, 25, 33,and 37 microM, respectively. It did not elicit a mitogenic response from splenocytes or stimulate nitric oxide production by macrophages. It inhibited HIV-1 reverse transcriptase with an IC(50) of 5.5 microM (Leung and Ng, 2007).

A novel 5773-Da antifungal peptide produced by an indigenous fungal strain (VR) of *Aspergillus clavatus* inhibited growth of *F.oxysporum, F. solani, Aspergillus niger, Botrytis cinerea,* and *Alternaria solani* (Skouri-Gargouri and Gargouri, 2008).

A 17-kDa antimicrobial protein devoid of hemolytic activity and isolated from growth medium containing the filamentous fungus *Aspergillus oryzae* strongly inhibited pathogenic microbial strains, including pathogenic fungi, *F. moniliform* var. subglutinans and *Colletotrichum coccodes*, and showed antibacterial activity against bacteria, including E. coli

O157 and Staphylococcus aureus(Park et al.,2008).Martín-Urdiroz et al.(2009) assayed antifungal protein (AFP) from *Aspergillus giganteus* for toxicity against the *F.oxysporum* wild-type strain and mutants in genes regulating cell signaling (DeltapacC, pacCc Deltafmk1) or cell-wall biogenesis (DeltachsV Deltachs7, Deltagas1). The mutants were divided into two groups in accordance with their sensitivity to AFP: DeltapacC, Deltagas1 and Deltachs7, which were less sensitive to AFP compared with the wild-type, and pacCC, Deltafmk1 and DeltachsV,. Western blot analysis revealed increased binding of AFP to the three resistant mutants, DeltapacC, Deltagas1 andDeltachs7, but also to DeltachsV. Hence differential binding may not be a key determinant for sensitivity. Addition of Ca2+ or K+ markedly curtailed antifungal activity and binding of AFP, suggesting that these cations compete for the same targets as AFP at the fungal cell membrane.

A 5907-Da antifungal peptide was isolated from kale seeds inhibited mycelial growth in a number of fungal species including *F. oxysporum, Helminthosporium maydis, Mycosphaerella arachidicola* and *Valsa mali*, with an IC(50) of 4.3microM, 2.1microM, 2.4microM, and 0.15microM,respectively and exhibited pronounced thermostability and pH stability. Itinhibited proliferation of hepatoma (HepG2) and breast cancer (MCF7) cells withan IC(50) of 2.7microM and 3.4microM, and the activity of HIV-1 reverse transcriptase with an IC(50) of 4.9microM (Lin and Ng, 2008).

An 18.9 kDa antifungal protein designated juncin from seeds of the Japanese takana (*Brassica juncea* var. integrifolia) exhibited antifungal activity toward *F.oxysporum, Helminthosporium maydis,* and *Mycosphaerella arachidicola* with IC(50) values of13.5, 27, and 10 μM, respectively. It was devoid of mitogenic activity toward murine splenocytes and nitric oxide inducing activity toward murine macrophages. It inhibited the proliferation of hepatoma (HepG2) and breast cancer (MCF7) cells with IC(50) values of 5.6 and 6.4 μM, respectively, and the activity of HIV-1 reverse transcriptase with an IC(50) of 4.5 μM. Compared with *B. campestris* and *B. alboglabra* antifungal peptides, juncin exhibits a different molecular mass and N-terminal amino acid sequence but similar biological activities (Ye et al., 2009).

A 42.3-kDa antifungal protein isolated from a culture of *Bacillus subtilis* strain B29 impeded mycelial growth in *F. oxysporum, Fusarium moniliforme, Rhizoctonia solani,*and *Sclerotinia sclerotiorum.* It inhibited conidial spore germination in *F. oxysporum*, suppressed germ-tube elongation, and induced distortion, tumescence, and rupture of a portion of the germinated spores (Li et al., 2009).

A15-kDa antifungal protein, amaryllin, from *Amaryllis belladonna* bulbs, exhibited antifungal activity against *F. oxysporum* and *Aspergillus flavus(*Kumar et al., 2009).

A new antifungal peptide designated as pomegranin, with an N-terminal sequence resembling that of rice disease resistance NB-S-LRR-like protein, inhibited mycelial growth in the fungi *F. oxysporum* and *Botrytis cinerea* with an IC(50) of 6.1 microM and 2 microM, respectively (Guo et al.,2009).

A 5716-Da antifungal peptide with pronounced thermostability and pH stability from *Brassica parachinensis* seeds potently inhibited mycelial growth in a number of fungal species including *F. oxysporum, Helminthosporium maydis, Mycosphaerella arachidicola* and *Valsa mali.* It inhibited proliferation of hepatoma (HepG2) and breast cancer (MCF7) cells and the activity of HIV-1 reverse transcriptase (Lin and Ng, 2009).

Rolland et al. (2010)reported Ls-Stylicin1, a novel antimicrobial peptide from the penaeid shrimp, *Litopenaeus stylirostris.* The predicted mature peptide of 82 residues is negatively charged (theoreticalpI=5.0) and possesses a proline-rich N-terminal region and a cysteine-rich C-terminal region. Recombinant Ls-Stylicin1 in both monomeric and dimeric forms demonstrate potent antifungal activity against *F. oxysporum*, a pathogenic fungus of shrimp, but lower antimicrobial activity against Gram (-) bacteria, *Vibrio* sp.

Barakat et al. (2010)demonstrated that the antifungal protein AFP from *Aspergillus giganteus* inhibited secondary growth of different *Fusarium* species on barley.

A novel noncytotoxic and thermpstable antifungal protein (Pr-2) from pumpkin rinds potently inhibited growth of *F. solani, F.oxysporum, Botrytis cinerea, Colletotrichum coccodes,and Trichoderma harzianum* at 10-20 microM. The results of confocal laser scanning microscopy and SYTOX Green uptake demonstrated that it acted on the fungal cell membrane (Park et al.,2009).A 40 kDa antifungal protein Pr-1from pumpkin rind with relatively high thermostability and no hemolytic activity inhibited growth of *F. oxysporum, F.solani, Botrytis cinerea,*and *Rhizoctonia solani,* as well as the yeast, *Candida albicans*. It did not inhibit growth of *Escherichia coli* or *Staphylococcus aureus*. The protein exerted a membrane permeabilization activity on the fungi (Park et al., 2010).

A 26.9-kDa antifungal peptide isolated from dry seeds of the foxtail millet (*Setaria italica* (L.) Beauv.) exhibited antifungal activity against *F. oxysporum, A. alternate, B. cinerea* and *Trichoderma viride* (Xu et al.,2011).

A 7.3-kDa antifungal peptide from dried red kidney beans exerted an antifungal action on *F. oxysporum* with an IC_{50} of 3.8±0.4 μM (mean±SD, n=3). It also inhibited mycelial growth in *Mycosphaerella arachidicola*. It suppressed growth of lymphoma MBL2 cells and leukemia L1210 cells with an IC_{50} of 5.2±0.4 μM and 7.6±0.6 μM, respectively. HIV-1 reverse transcriptase was inhibited with an IC_{50} of 40±3.2 μM. However, no activity was demonstrated toward other viral enzymes (Li et al., 2011).

The purified recombinant antimicrobial peptide from Argopecten purpuratus hemocytes rApS (3085.80Da) inhibited growth of both biotrophic (*Fusarium oxysporum, Trichoderma harzianum*) and necrotrophic (*Alternaria* spp.,*Botrytis cinerea*) fungi. The high yield obtained and its successful use for controlling pathogenic fungi suggest that this molecule could be applied to agricultural crops using various management strategies (Tapia et al., 2011).

Discussion

In order to protect themselves from attack of fungal pathogens including *Fusarium* species, organisms produce a consortium of antifungal proteins which differ in structure and mechanism of antifungal action. Chitin and glucan, which are constituents of the fungal envelope, are digested by chitinases and glucanases. Defensins alter fungal membrane permeability. Ribonucleases degrade fungal RNA. The mechanisms of antifungal action of cyclophilin-like proteins, embryo abundant protein-like proteins, lipid transfer proteins, peroxidases and storage albumins await full elucidation.

ACKNOWLEDGMENTS

A grant (No.2010SKLAB06-7) from State Key Laboratory for Agrobiotechnology is gratefully acknowledged.

REFERENCES

Banerjee N, Sengupta S, Roy A, Ghosh P, Das K, Das S. Functional alteration of a dimeric insecticidal lectin to a monomeric antifungal protein correlated to its oligomeric status. *PLoS One*. 2011; 6(4):e18593.

Barakat H, Spielvogel A, Hassan M, El-Desouky A, El-Mansy H, Rath F, Meyer V, Stahl U. The antifungal protein AFP from *Aspergillus giganteus* prevents secondary growth of different *Fusarium* species on barley. *Appl Microbiol Biotechnol*. 2010; 87(2):617-24.

Campos Mde A, Silva MS, Magalhães CP, Ribeiro SG, Sarto RP, Vieira EA, Grossi de Sá MF. Expression in *Escherichia coli,* purification, refolding and antifungal activity of an osmotin from *Solanum nigrum*. *Microb Cell Fact*. 2008; 7:7.

de Beer A, Vivier MA. Vv-AMP1, a ripening induced peptide from *Vitis vinifera* shows strong antifungal activity *BMC Plant Biol*. 2008; 8:75.

de Freitas CD, Lopes JL, Beltramini LM, de Oliveira RS, Oliveira JT, Ramos. MV. Osmotin from *Calotropis procera* latex: New insights into structure and antifungal properties. *Biochim Biophys Acta*. 2011; 1808(10):2501-7.

Dehestani A, Kazemitabar K, Ahmadian G, Jelodar NB, Salmanian AH, Seyedi M, Rahimian H, Ghasemi S. Chitinolytic and antifungal activity of a *Bacillus pumilus* chitinase expressed in *Arabidopsis*. *Biotechnol Lett*. 2010; 32(4):539-46.

Gomes VM, Carvalho AO, Da Cunha M, Keller MN, Bloch C Jr, Deolindo P, Alves EW. Purification and characterization of a novel peptide with antifungal activity from *Bothrops jararaca* venom. *Toxicon*. 2005; 45(7): 817-27.

Gómez-Gómez L, Rubio-Moraga A, Ahrazem O. Molecular cloning and characterisation of a pathogenesis-related protein CsPR10 from *Crocus sativus*. *Plant Biol (Stuttg)*. 2011; 13(2):297-303.

Ghosh M. Antifungal properties of haem peroxidase from *Acorus calamus*. *Ann Bot*. 2006; 98(6):1145-53.

Guo G, Wang HX, Ng TB. Pomegranin, an antifungal peptide from pomegranate peels. *Protein Pept Lett*. 2009; 16(1):82-5.

Ho VS, Ng TB. Chitinase-like proteins with antifungal activity from emperor banana fruits. *Protein Pept Lett*. 2007; 14(8):828-31.

Ho VS, Wong JH, Ng TB. A thaumatin-like antifungal protein from the emperor banana. *Peptides*. 2007; 28(4):760-6.

Kamchanatat A, Tiengburanatam N, Boonmee A, Puthong S, Sangvanich P. Zingipain, A cysteine protease from *Zingiber ottensii* Valeton rhizomes with antiproliferative activities against fungi and human malignant cell lines. *Prep Biochem Biotechnol*. 2011; 41(2):138-53.

Kant P, Liu WZ, Pauls KP. PDC1, a corn defensin peptide expressed in *Escherichia coli* and *Pichia pastoris* inhibits growth of *Fusarium graminearum. Peptides*. 2009; 30(9):1593-9.

Kheeree N, Sangvanich P, Puthong S, Karnchanatat A. Antifungal and antiproliferative activities of lectin from the rhizomes of *Curcuma amarissima* Roscoe. *Appl Biochem Biotechnol*. 2010; 162(3):912-25.

Kumar S, Singh N, Sinha M, Kaur P, Srinivasan A, Sharma S, Singh TP. Isolation, purification, crystallization and preliminary crystallographic studies of amaryllin, a plant pathogenesis-related protein from *Amaryllis belladonna*. *Acta Crystallogr Sect F Struct Biol Cryst Commun*. 2009; 65(Pt 6): 635-7.

Li M, Wang H, Ng TB. An antifungal peptide with antiproliferative activity toward tumor cells from red kidney beans. *Protein Pept Lett*. 2011; 18(6): 594-600.

Leung EH, Ng TB. A relatively stable antifungal peptide from buckwheat seeds with antiproliferative activity toward cancer cells. *J Pept Sci*. 2007; 13(11): 762-7.

Li J, Yang Q, Zhao LH, Zhang SM, Wang YX, Zhao XY. Purification and characterization of a novel antifungal protein from *Bacillus subtilis* strain B29. *J Zhejiang Univ Sci B*. 2009; 10(4):264-72.

Li Z, Zhou M, Zhang Z, Ren L, Du L, Zhang B, Xu H, Xin Z. Expression of a radish defensin in transgenic wheat confers increased resistance to *Fusarium graminearum* and *Rhizoctonia cerealis*. *Funct Integr Genomics*. 2011; 11(1):63-70.

Lee JR, Park SC, Kim JY, Lee SS, Park Y, Cheong GW, Hahm KS, Lee SY. Molecular and functional characterization of a cyclophilin with antifungal activity from Chinese cabbage. *Biochem Biophys Res Commun*. 2007; 353(3): 672-8.

Lin P, Ng TB. A novel and exploitable antifungal peptide from kale (*Brassica alboglabra*) seeds. *Peptides*. 2008; 29(10):1664-71.

Lin P, Ng TB.Brassiparin, an antifungal peptide from *Brassica parachinensis* seeds. *J Appl Microbiol*. 2009; 106(2):554-63.

Lin P, Wong JH, Ng TB. A defensin with highly potent antipathogenic ctivities from the seeds of purple pole bean. *Biosci Rep*. 2009; 30(2):101-9.

Lin P, Wong JH, Xia L, Ng TB. Campesin, a thermostable antifungal peptide with highly potent antipathogenic activities. *J Biosci Bioeng*. 2009; 108(3):259-65.

Lin P, Xia L, Ng TB. First isolation of an antifungal lipid transfer peptide from seeds of a Brassica species. *Peptides*. 2007; 28(8):1514-9.

Ma DZ, Wang HX, Ng TB. A peptide with potent antifungal and antiproliferative activities from Nepalese large red beans. *Peptides*. 2009; 30(12):2089-94.

Martín-Urdiroz M, Martínez-Rocha AL, Di Pietro A, Martínez-del-Pozo A, Roncero MI. Differential toxicity of antifungal protein AFP against mutants of *Fusarium oxysporum*. *Int Microbiol*. 2009; 12(2):115-21.

Ngai PH, Ng TB. A lectin with antifungal and mitogenic activities from red cluster pepper (*Capsicum frutescens*) seeds. *Appl Microbiol Biotechnol*. 2007; 74(2):366-71.

Shi P, Yao G, Yang P, Li N, Luo H, Bai Y, Wang Y, Yao B. Cloning, characterization, and antifungal activity of an endo-1,3-beta-D:-glucanase from *Streptomyces* sp. S27. *Appl Microbiol Biotechnol*. 2010; 85(5):1483-90.

Wang S, Shao B, Fu H, Rao P. Isolation of a thermostable legume chitinase and study on the antifungal activity. *Appl Microbiol Biotechnol*. 2009; 85(2): 313-21.

Mehmood MA, Xiao X, Hafeez FY, Gai Y, Wang F. Molecular characterization of the modular chitin binding protein Cbp50 from *Bacillus thuringiensis* serovar konkukian. *Antonie Van Leeuwenhoek*. 2011 Jun 7. [Epub ahead of print].

Mello EO, Ribeiro SF, Carvalho AO, Santos IS, Da Cunha M, Santa-Catarina C, Gomes VM.Antifungal activity of PvD1 defensin involves plasma membrane permeabilization, inhibition of medium acidification, and induction of ROS in fungi cells. Curr Microbiol. 2011;62(4):1209-17.

Nagarathnam R, Rengasamy A, Balasubramanian R. Purification and properties of cysteine protease from rhizomes of *Curcuma longa (*Linn.). *J Sci Food Agric*. 2010; 90(1):97-105.

Park BT, Na KH, Jung EC, Park JW, Kim HH. Antifungal and Anticancer Activities of a Protein from the Mushroom *Cordyceps militaris*. *Korean J Physiol Pharmacol*. 2009; 13(1):49-54.

Park SC, Kim JY, Lee JK, Hwang I, Cheong H, Nah JW, Hahm KS, Park Y. Antifungal mechanism of a novel antifungal protein from pumpkin rinds against various fungal pathogens. *J Agric Food Chem*. 2009; 57(19):9299-304.

Park SC, Lee JR, Kim JY, Hwang I, Nah JW, Cheong H, Park Y, Hahm KS Pr-1, a novel antifungal protein from pumpkin rinds. *Biotechnol Lett*. 2010; 32(1):125-30.

Park SC, Yoo NC, Kim JY, Park HK, Chae BJ, Shin SY, Cheong H, Park Y, Hahm KS. Isolation and characterization of an extracellular antimicrobial protein from *Aspergillus oryzae*. *J Agric Food Chem*. 2008; 56(20):9647-52.

Qi X, Zhou C, Li P, Xu W, Cao Y, Ling H, Ning Chen W, Ming Li C, Xu R, Lamrani M,Mu Y, Leong SS, Wook Chang M, Chan-Park MB. Novel short antibacterial and antifungal peptides with low cytotoxicity: Efficacy and action mechanisms. *Biochem Biophys Res Commun*. 2010; 398(3):594-600.

Rolland JL, Abdelouahab M, Dupont J, Lefevre F, Bachère E, Romestand B.

Stylicins, A new family of antimicrobial peptides from the Pacific blue shrimp *Litopenaeus stylirostris*. *Mol Immunol*. 2010; 47(6):1269-77.

Sagaram US, Pandurangi R, Kaur J, Smith TJ, Shah DM. Structure-activity determinants in antifungal plant defensins MsDef1 and MtDef4 with different modes of action against *Fusarium graminearum*. *PLoS One*. 2011 Apr 13; 6(4):e18550.

Sawano Y, Miyakawa T, Yamazaki H, Tanokura M, Hatano K. Purification, characterization, and molecular gene cloning of an antifungal protein from *Ginkgo biloba* seeds. *Biol Chem*. 2007; 388(3):273-80.

Selim S, Negrel J, Wendehenne D, Ochatt S, Gianinazzi S, van Tuinen D. Stimulation of defense reactions in *Medicago truncatula* by antagonistic lipopeptides from *Paenibacillus* sp. strain B2. *Appl Environ Microbiol*. 2010; 76(22):7420-8.

Shenoy SR, Kameshwari MN, Swaminathan S, Gupta MN. Major antifungal activity from the bulbs of Indian squill *Urginea indica* is a chitinase. *Biotechnol Prog*. 2006; 22(3):631-7.

Shi P, Yao G, Yang P, Li N, Luo H, Bai Y, Wang Y, Yao B. Cloning, characterization, and antifungal activity of an endo-1,3-beta-D:-glucanase from *Streptomyces* sp. S27. *Appl Microbiol Biotechnol*. 2010 Feb; 85(5): 1483-90.

Skouri-Gargouri H, Gargouri A. First isolation of a novel thermostable antifungal peptide secreted by *Aspergillus clavatus*. *Peptides*. 2008; 29(11):1871-7.

Tapia E, Montes C, Rebufel P, Paradela A, Prieto H, Arenas G. Expression of an optimized *Argopecten purpuratus* antimicrobial peptide in *E. coli* and evaluation of the purified recombinant protein by *in vitro* challenges against important plant fungi. *Peptides*. 2011 Jul 20. [Epub ahead of print]

van der Weerden NL, Hancock RE, Anderson MA. Permeabilization of fungal hyphae by the plant defensin NaD1 occurs through a cell wall-dependent process. *J Biol Chem*. 2010; 285(48):37513-20.

Wang H, Ng TB. An antifungal protein from ginger rhizomes. *Biochem Biophys Res Commun*. 2005; 336(1):100-4.

Wang H, Ng TB. Ganodermin, an antifungal protein from fruiting bodies of the medicinal mushroom *Ganoderma lucidum*. *Peptides*. 2006; 27(1):27-30.

Wang HX, Ng TB. An antifungal peptide from the coconut. *Peptides*. 2005; 26(12): 2392-6.

Wang HX, Ng TB. An antifungal peptide from baby lima bean. *Appl Microbiol Biotechnol*. 2006; 73(3):576-81.

Wang HX, Ng TB.Isolation and characterization of an antifungal peptide with antiproliferative activity from seeds of *Phaseolus vulgaris* cv. 'Spotted Bean'. *Appl Microbiol Biotechnol*. 2007;74(1):125-30.

Wang HX, Ng TB. An antifungal peptide from red lentil seeds. *Peptides*. 2007; 28(3):547-52.

Wang S, Rao P, Ye X. Isolation and biochemical characterization of a novel leguminous defense peptide with antifungal and antiproliferative potency. *Appl Microbiol Biotechnol*. 2009; 82(1):79-86.

Wang S, Shao B, Fu H, Rao P. Isolation of a thermostable legume chitinase and study on the antifungal activity. *Appl Microbiol Biotechnol*. 2009; 85(2): 313-21.

Wang SY, Gong YS, Zhou JJ. Chromatographic isolation and characterization of a novel peroxidase from large lima legumes. *J Food Sci*. 2009; 74(3):C193-8.

Wang Q, Li F, Zhang X, Zhang Y, Hou Y, Zhang S, Wu Z. Purification and characterization of a CkTLP protein from *Cynanchum komarovii* seeds that confers antifungal activity. *PLoS One*. 2011; 6(2):e16930.

Wong JH, Ng TB. Sesquin, a potent defensin-like antimicrobial peptide from ground beans with inhibitory activities toward tumor cells and HIV-1 reverse transcriptase *Peptides*. 2005 26(7):1120-6.

Wong JH, Ng TB. Limenin, a defensin-like peptide with multiple exploitable activities from shelf beans. *J Pept Sci*. 2006; 12(5):341-6.

Wu X, Sun J, Zhang G, Wang H, Ng TB. An antifungal defensin from *Phaseolus vulgaris* cv. 'Cloud Bean'. *Phytomedicine*. 2011; 18(2-3):104-9.

Xu W, Wei L, Qu W, Liang Z, Wang J, Peng X, Zhang Y, Huang K. A novel antifungal peptide from foxtail millet seeds. *J Sci Food Agric*. 2011; 91(9): 1630-7.

Yan L, Qian Y. Cloning and heterologous expression of SS10, a subtilisin-like protease displaying antifungal activity from *Trichoderma harzianum*. *FEMS Microbiol Lett*. 2009; 290(1):54-61.

Yan Q, Jiang Z, Yang S, Deng W, Han L. A novel homodimeric lectin from *Astragalus mongholicus* with antifungal activity. *Arch Biochem Biophys*. 2005; 442(1):72-81.

Ye X, Ng TB. Isolation and characterization of juncin, an antifungal protein from seeds of Japanese Takana (*Brassica juncea* Var. integrifolia). *J Agric Food Chem*. 2009; 57(10):4366-71.

In: Fusarium
Editors: Tony Ferarri Rios and Emilia Reyes Ortega
ISBN: 978-1-61942-539-2

Chapter 8

RELEVANCE OF GENOMICS-BASED STUDIES ON MANAGEMENT OF DISEASES CAUSED BY *FUSARIUM* SPECIES

***William Yajima*[1]*, Kishore Chittem*[1] *and Rubella S. Goswami*[2,*]**

[1]Department of Plant Pathology, 306 Walster Hall, Dept. 7660, North Dakota State University, Fargo, ND, US
[2]DuPont Crop Protection, Stine Haskell Research Center, 1090 Elkton Road, Newark, DE, US

ABSTRACT

The annotation of complete genome sequences of phytopathogenic fungi is an ongoing initiative for plant pathologists and different *Fusarium* species, such as *F. solani, F. verticillioides,* and *F. oxysporum* have been the focus of a number of these studies. *F. graminearum* was one of the initial filamentous fungi to have its whole genome sequenced and the resulting data have since served as an invaluable tool to study fungal development, gene expression and function, and population genetics. Techniques involving expressed sequence tags, DNA microarrays, site directed mutagenesis, gene replacement, and large scale DNA sequencing, among others, have enabled a more comprehensive characterization of genes intimately involved in pathogenesis, including virulence factors, such as mycotoxin-encoding genes. Genomics has facilitated the study of relatively large gene sets such as polyketide synthases (PKS) among *Fusarium* species and the determination of the functions of a number of PKS genes in *F. graminearum* was made possible using gene replacement techniques. Additionally, DNA microarrays have been used to identify differentially expressed genes in *F. graminearum* and *F. verticillioides* under diverse conditions while site directed mutagenesis has aided in the characterization of fungicide resistant *F. graminearum*. These as well as other examples of how genomics has contributed to deciphering the pathogenicity of *Fusarium* species are examined. This chapter provides an overview of the techniques commonly utilized in

* E-mail: Rubella.S.Goswami@usa.dupont.com.

genomics-based studies as well as a review of published research highlighting key genomic resources relevant to effective management of diseases caused by different *Fusarium* species. Potential future applications of the burgeoning field of genomics to *Fusarium* research as well as possible consequences of these studies are also discussed.

1. INTRODUCTION

One of the most important factors affecting crop yield and quality is disease and phytopathogenic fungi belonging to the *Fusarium* genus can be particularly problematic for producers of crops such as wheat, corn, and barley. Perhaps the greatest issue concerning infection by *Fusarium* species is the associated mycotoxin contamination by compounds including deoxynivalenol, nivalenol, zearalenone, T2, fumonisins, and diacetoxyscrirpenol, among many others. Harvested crops that have been rendered unusable due to infection by *Fusarium* pathogens can result in considerable economic losses to the agricultural industry. It is paramount that viable management strategies to control outbreaks are continually developed and implemented to deal with the possibility that genetic mutation of pathogenic *Fusarium* isolates may generate strains that can circumvent currently used disease control measures.

As the name suggests, genomics is the study of genes that encode cellular components like proteins, which directly or indirectly affect virtually all cellular activities. By applying genomics-based experimental techniques, researchers have successfully annotated the complete genomes of a number of important plant and fungal species, with the promise that additional genomes will be characterized in the future. The wealth of information contained within these genome sequences has been exploited by others, leading to novel discoveries with relevance to plant pathology. Deciphering the interactions between plant pathogens and susceptible hosts as well as fully characterizing virulence or resistance factors allow researchers to develop new crop varieties and agronomic methods to increase yield and quality.

This chapter explores how genomics has contributed to agriculture by providing background information regarding some of the techniques commonly used in genomics-based studies. How some of these elements, such as DNA microarrays and molecular markers, have been applied to identify phytopathogenic *Fusarium* species is explored. Additionally, the utility of gene expression analyses in terms of identifying virulence and pathogenicity factors that may serve as targets for future disease management strategies is discussed. Gene expression studies have also contributed to a more thorough characterization of sequenced *Fusarium* genomes and the resulting whole genome sequence databases and resources that are publically available to interested parties are examined. Gene manipulation, such as gene disruption and insertional mutagenesis, of selected targets based on genomics-based experiments has aided plant pathologists by enabling the characterization of the phenotypic consequences of the elimination or over-expression of specific genes. Since research on various phytopathogenic *Fusarium* species is extensive, this chapter is not an exhaustive exploration of this agriculturally-important fungal genus. Rather, the studies described in this chapter serve as examples of the work that has been performed on *Fusarium* species that can infect cereals to demonstrate the value of genomics in phytopathology.

2. Pathogen Identification

2.1. Application of Molecular Marker Technology

The identification of the pathogen responsible for a particular disease is, perhaps, one of the most important requirements as it can dictate the selection of the most effective disease management strategy to employ. The value of molecular marker technology can be exemplified by its utility in differentiating between different *Fusarium* species and determining relatedness among the various strains and *formae speciales*. Within the *Fusarium* genus, DNA sequences that have been used to identify different species include β-tubulin, internally transcribed spacer (ITS) regions, calmodulin, and translation elongation factor 1α (TEF-1α) (Waalwijk et al., 1996; O'Donnell et al., 1998; O'Donnell et al., 2000; Geiser et al., 2004; Wulff et al., 2010). While different DNA sequences may be optimal to identify specific *Fusarium* species, TEF-1α may be the most appropriate to discriminate between members of the *Fusarium* genus (Seifert and Levesque, 2004; Kristensen et al., 2005). As whole genome sequencing projects continue to provide more complete genetic information regarding the *Fusarium* genus, the application of DNA molecular markers to accurately and reliably classify pathogenic species as well as identify newly evolved strains will concomitantly improve. Adopting molecular techniques for fungal classification will have considerable implications in differentiating between related *Fusarium* species with no apparent differences in phenotype or morphology.

Polymerase chain reaction (PCR) is a commonly used molecular tool allowing for the rapid amplification of specific DNA sequences and it has led to the development of numerous molecular marker-based techniques including, but not exclusively, random amplified polymorphic DNA (RAPD), amplified and restriction fragment length polymorphism (AFLP and RFLP), single nucleotide polymorphism (SNP), simple sequence repeats (SSR), sequence characterized amplified regions (SCAR), and large-scale pyrosequencing. A thorough review of how these molecular tools have been effectively used in studies of various *Fusarium* species has been written and will, therefore, not be discussed in further detail here (Nayaka, et al., 2011 and references therein).

2.2. DNA Microarrays

In addition to the use of molecular markers, DNA microarrays have also been used to differentiate among phytopathogenic fungal species. An example of such a microarray with significance to disease diagnostics is a sensitive assay to differentiate among a variety of mycotoxin-producing *Fusarium* species (Nicolaisen et al., 2005). Due to the potential presence of multiple fungal species, including numerous *Fusarium* species capable of causing fusarium head blight (FHB), in fields of important cereals such as barley and wheat, the need for accurate identification of pathogenic, mycotoxin producing species is crucial. Included among the species responsible for FHB in cereals are *F. graminearum*, *F. avenaceum*, *F. poae*, *F. sporotrichioides*, *F. culmorum*, *F. equiseti*, *F. langsethiae*, and *F. tricinctum* (Parry et al., 1995; Torp and Nirenberg, 2004). The difference in the suite of mycotoxins produced by the various FHB-causing *Fusarium* species can have a profound effect on the toxicity of infected cereal grains, which would have obvious implications on their use as animal feed and in human consumption. Furthermore, variations in virulence and sensitivity to fungicides observed in different *Fusarium* species have relevance in the determination of effective disease management strategies that cereal growers might implement to protect crops during the growing season. Limitations of traditional methods of fungal identification relying on morphological characteristics include problems differentiating between closely related species with few morphological differences as well as the fact that they can be time-consuming and require a high level of expertise in fungal taxonomy.

Previous efforts using PCR-based techniques to correctly classify *Fusarium* species have yielded viable assays (Parry and Nicholson, 1996; Schilling et al., 1996; Turner et al., 1998). However, the implementation of a DNA microarray-based assay may be a refinement of established techniques by allowing for fast, relatively simple, sensitive, and accurate differentiation of multiple *Fusarium* species in a single sample using one PCR. A microarray consisting of over 50 capture oligonucleotides generated from ITS2 ribosomal DNA sequences was developed and could correctly differentiate between type A and type B trichothecene and non-trichothecene producing *Fusarium* species from infected fields (Nicolaisen et al., 2005). Only minimal cross species hybridization with some of the capture oligonucleotides was observed. Interestingly, although the microarray was designed to detect different *Fusarium* species, a positive correlation between deoxynivalenol content in cereal samples and *Fusarium* DNA abundance has been described, which implies that a DNA microarray might also have applications in quantifying mycotoxin abundance in contaminated cereals (Schnerr et al., 2002; Waalwijk et al., 2004).

Another DNA microarray-based assay to facilitate the detection of diverse *Fusarium* species from infected plant material has been developed and was validated with 14 different species (Kristensen et al., 2007). Capture probes, which were designed to distinguish between different mycotoxin (i.e., trichothecene and moniliformin) producing groups of *Fusarium*, were based on the TEF-1α sequence. The multiplexed *Fusarium* chip contained five consensus and 16 species-specific capture probes to enable the detection of *F. graminearum*, *F. poae*, *F. avenaceum*, *F. sporotrichioides*, *F. langsethiae*, *F. cerealis*, *F. kyushuense*, *F. equiseti*, *F. pseudograminearum*, *F. flocciferum*, *F. sambucinum*, *F. tricinctum*, *F. venenatum*, and *F. torulosum* in one hybridization experiment. In addition to faster detection of *Fusarium* species compared to traditional methods, it is noteworthy that the microarray was able to detect species on infected cereal grains that were not identified through morphological analysis, thereby providing evidence of its utility (Kristensen et al., 2007). Due to the selection of TEF-1α, which is a single copy gene, in this microarray, it is believed that it is more suited for providing quantitative data compared to the previously described microarray based on ribosomal ITS2 sequences, which have a higher copy number (Kristensen et al., 2007). Moreover, because of a greater diversity in TEF-1α sequences among fungal species relative to ITS sequences, it is anticipated that the addition of more *Fusarium* species to the multiplex microarray chip will be more likely compared to the aforementioned ITS2-based microarray (Nicolaisen et al., 2005; Kristensen et al., 2007). For both microarrays, automation of the assay would likely increase throughput and potentially simplify the procedure further, which would provide additional advantages over traditional diagnostic methods based on morphological differences. The continued development of DNA microarrays and generation of new *Fusarium* chips may have profound implications for diagnosticians as well as mycologists.

3. Gene Expression Analysis

3.1. Expressed Sequence Tags

An alternative to whole genome DNA microarrays to study gene expression in phytopathogenic *Fusarium* species uses expressed sequence tags (EST). In addition to a cDNA library made from developing perithecia, researchers generated libraries from *F. graminearum* grown in nutrient (i.e., carbon and nitrogen) starvation conditions to stimulate the expression of genes that may be involved in pathogenicity or virulence (Trail et al., 2003). Between the three *F. graminearum* cDNA libraries, almost 8000 ESTs, corresponding to approximately 2000 identified genes, were generated during this study and submitted to the dbEST database (ID: 13375065-13384819). Based on the assignment of predicted gene functions, gene expression patterns for mycelia cultured in carbon and nitrogen starvation conditions were similar while the ESTs generated from the perithecia cDNA library were different from those identified in the nutrient starvation conditions in terms of genes related to protein destination and synthesis (Trail et al., 2003). The observed differences in gene expression associated with protein destination were suggested to perhaps be due to their involvement in asci maturation occurring within the perithecia (Trail et al., 2003). Additionally, almost half of the ESTs generated from the perithecia cDNA library could not be classified into a specific functional category, which may have been because of an incomplete understanding or compilation of genes regulating this particular developmental stage (Trail et al., 2003). The most prominent EST identified from the perithecia cDNA library (contig 278) was homologous to an *Aspergillus nidulans* DigA, which is responsible for vesicle sorting, presumably during spore membrane maturation (Trail et al., 2003). Of the 16 genes mediating pathogenicity identified in the EST study, nine were expressed only in mycelia grown in nitrogen starvation conditions, three were observed only in the library produced from mycelia cultured in carbon starvation conditions, and three (*RPT4*, *PHR1*, and *cpCPC1*) were common to both starvation conditions (Trail et al., 2003). The remaining pathogenicity-related gene was found only in the perithecia cDNA library.

Another *Fusarium* species that has been analyzed using ESTs is the maize pathogenic *F. verticillioides*, in which fumonisin regulation was examined (Brown et al., 2005). cDNA libraries were prepared from fungus grown under nine different conditions, including in a variety of artificial cultural media for various time points and on maize seedlings (shoots and roots) and developing kernels. From these cDNA libraries, more than 87,000 ESTs representing over 11,000 sequences were generated (Brown et al., 2005). Comparing the results from the prepared libraries demonstrated differential expression of genes in the fumonisin gene cluster at different time points following inoculation of artificial growth medium, including *FUM1*, which showed the highest increase in EST levels (Brown et al., 2005). *FUM1* encodes the enzyme responsible for catalysis of the primary reaction during fumonisin biosynthesis. Furthermore, numerous mycotoxin regulatory genes and genes expected to be linked to pathogen–maize host interactions were characterized. Of particular significance was the discovery of a novel fumonisin gene (*FUM20*) situated between *FUM12* and *FUM13* within the fumonisin gene cluster (Brown et al., 2005). *FUM20* does not share any significant sequence similarity to any genes in the *F. graminearum* genome. A subsequent study identified another novel member of the fumonisin gene cluster, which was designated *FUM21* and was found adjacent to *FUM1* (Brown et al., 2007). The presence and diverse expression patterns of a number of alternative splice forms of *FUM* genes in *F. verticillioides* has been revealed and may indicate that intron splicing plays an important role during fumonisin biosynthesis (Brown et al., 2005; Brown et al., 2007).

Fumonisin biosynthesis in *F. verticillioides* has also been studied using a combination of DNA microarray and EST technology to identify genes possibly involved in secondary metabolism (Pirttila et al., 2004). In this instance, over 700 *F. verticillioides* ESTs were used to create a DNA microarray to assess gene expression patterns in wild-type (WT) and *fcc1* mutant strains cultured on maize as well as in an artificial culture medium set at different pHs. The *FCC1* gene is thought to have a role in fumonisin biosynthesis as well as in conidiation and there are observed amino acid sequence similarities to type-C cyclin Ssn8p, which is involved in phosphorylation of an RNA polymerase in yeast (Liao et al., 1995; Shim and Woloshuk, 2001). A comparison of gene expression during growth on maize revealed 116 ESTs that were differentially expressed between the WT and *fcc1* mutant *F. verticillioides* (Pirttila et al., 2004). Slightly more than half of those ESTs were up-regulated in the WT strain and the mutation in *FCC1* appeared to have an effect on such disparate activities as production of amino acids and proteases and secondary metabolism (mycotoxin production). Additionally, there were 23 ESTs (when grown on maize) and 16 ESTs (when cultured at different pHs) that were expressed only in the wild type *F. verticillioides*, which offers further evidence of the effect of the mutation in *FCC1*. Culturing the fungal strains at different pHs (3 and 8) resulted in the identification of 166 ESTs whose expression patterns differed between WT and the *fcc1* mutant, including 19 whose expression levels were similar to those observed for ESTs representing the *FUM* genes (Pirttila et al., 2004).

3.2. DNA Microarrays

Following the release of the *F. graminearum* whole genome sequence in 2003, it was apparent that the next logical step in genomics-based studies involved the large-scale analysis of the transcription profile of this important phytopathogenic fungus under different conditions using whole genome DNA microarrays. Other phytopathogenic fungi that had undergone gene expression analysis include *Blumeria graminis*, *Cryphonectria parasitica*, *Metarhizium anisopliae*, and *Histoplasma capsulatum* (Allen et al., 2003; Hwang et al., 2003; Both et al., 2005; Wang et al., 2005). Efforts to use the sequenced *F. graminearum* genome to develop a viable DNA microarray for transcription profiling ultimately resulted in the generation of the *F. graminearum* Affymetrix GeneChip (Guldener et al., 2006). The GeneChip microarray consisted of over 18,000 probe sets that were made up of 14 to 15 pairs of short oligonucleotides (25 mers) representing individual genes. To increase the chances that all predicted *F. graminearum* genes would be incorporated into the GeneChip, two sets of putative gene calls, one from the Broad Institute and one from the Munich Information Center for Protein Sequences (MIPS), were used along with gene sets generated through manual annotation efforts. The performance of the generated GeneChip microarray was validated using purified genomic DNA, which confirmed hybridization of DNA from *F. graminearum* as well as from other *Fusarium* species such as *F. boothii*, *F. pseudograminearum*, *F.culmorum*, and *F. asiaticum* (Guldener et al., 2006). The observed cross hybridization with species closely related to *F. graminearum* demonstrated the potential utility of the GeneChip microarray beyond just *F. graminearum*. However, the microarray does not appear to be a viable option to study more distantly related *Fusarium* species such as *F. verticillioides* and *F. oxysporum* due to poor hybridization. Based on hybridization experiments, it was determined that the predicted rate of false negatives for *F. graminearum* using the GeneChip microarray was only 1.3%.

Differences in gene expression patterns were identified in *F. graminearum* grown in different artificial cultural conditions (i.e., nitrogen deficiency or carbon deficiency) and fungus grown on infected barley material at various time points. It was observed that 431 probe sets, including those for genes involved in metabolism, deoxynivalenol biosynthesis, plant cell wall degradation, defense, and homeostasis, were identified in fungus from infected barley and not from fungus cultured in nitrogen or carbon starvation conditions (Guldener et al., 2006). Interestingly, even in the absence of methods for enrichment of *Fusarium* transcripts, the GeneChip microarray could differentiate between fungal and host plant genes with high sensitivity (Guldener et al., 2006).

Another high-throughput whole genome DNA microarray to analyze comprehensive gene expression in *F. graminearum* has been developed to study the effects of azole fungicides, which damage fungal membranes by interfering with an enzyme (CYP51) required for ergosterol synthesis (Becher et al., 2011). Reliance on these fungicides has increased the likelihood that resistant *F. graminearum* strains will evolve and a greater understanding of the gene expression patterns occurring during azole fungicide treatments may offer a more complete idea of how fungicide resistance develops. Unlike the Affymetrix GeneChip microarray, which uses 25 mer oligonucleotides, a DNA microarray developed by Agilent uses longer oligonucleotides (60 mer) that are believed to improve detection sensitivity (Hughes et al., 2001; Relogio et al., 2002). Furthermore, Agilent offers a number of multiplex array formats such as the 8 × 15 k, which consists of eight microarrays, each with 15,000 oligonucleotides, on one slide (Becher et al., 2011). Researchers have used an Agilent 8 × 15 k DNA microarray format (design ID 021301) to detect the effects of tebuconazole on *F. graminearum* transcript accumulation and have identified many genes involved in sterol biosynthesis (FGSG_02771, FGSG_03686, FGSG_04092, FGSG_09764, and FGSG_13888) as well as those encoding transcription factors and ABC transporters whose expression was significantly affected by azole fungicide treatment (Becher et al., 2011). Interestingly, real-time quantitative reverse transcription PCR (qRT-PCR) was performed to verify transcript abundance of 31 genes and there was a strong positive correlation between microarray and real-time qRT-PCR results, thereby validating the accuracy of the microarray data (Becher et al., 2011). In total, expression of 596 genes was significantly up-regulated while expression was down-regulated for 462 genes upon tebuconazole treatment (Becher et al., 2011). Some of the down-regulated genes were those mediating energy generation and amino acid transport or metabolism. Functional classification of genes whose expression was affected by azole fungicide treatment indicated that among the up-regulated genes, many could be categorized as being involved in one of sterol biosynthesis, isoprenoid metabolism, heme binding, or tetra- and pentacyclic triterpenes metabolism (Becher et al., 2011).

Comparison of the DNA microarray results using the Agilent format with those generated with the previously described Affymetrix microarray was performed to ascertain the effects of different stress conditions on *F. graminearum* gene expression. Approximately three quarters of the 596 genes (i.e., 457 genes) whose expression was up-regulated during tebuconazole treatment were up-regulated only during fungicide treatment and not during growth in either nitrogen or carbon starvation conditions while about one quarter (i.e., 139 genes) of the 596 genes were also up-regulated in one or both of the starvation conditions (Becher et al., 2011). In terms of down-regulated gene expression, of the 462 genes that responded to tebuconazole treatment, approximately half (i.e., 232 genes) were also down-regulated during nitrogen and/or carbon starvation, including numerous transcription factors. The fact that the expression of many genes was up- or down-regulated by unrelated stress conditions implies that these genes may be affected by stress in a non-specific manner. Further study of the genes specifically influenced by azole fungicide treatment may offer new insights into the molecular basis of *F. graminearum* response to fungicides as well as possible clues regarding fungicide resistance development.

4. Genome Databases and Resources

Among the various *Fusarium* species, *F. graminearum* (teleomorph: *Gibberella zeae*) is perhaps the most studied and well characterized in terms of genomics. The whole genome of this fungal pathogen was sequenced by the Broad Institute as a result of the collaborations between researchers from several different institutions and published in 2007 (Cuomo et al., 2007). Since then, significant effort has been targeted towards annotation and characterization of this sequence and the latest version is currently available through the Munich Information Center for Protein Sequences *Fusarium graminearum* Genome Database (MIPS-FGDB) at http://mips.helmholtz-muenchen.de/genre/proj/fusarium (Guldener et al., 2006). The genome sequence resulted from efforts to end sequence fosmids, plasmids, and BAC clones. The MIPS-FGDB, which presently contains in excess of 14,000 genes, is intended to be a comprehensive resource available to researchers that provides detailed information concerning the molecular structure and functions of *F. graminearum* genes. Proteins encoded by the provided genes are further classified based on similarity to known proteins and ESTs. Additionally, structural and functional characteristics of proteins are also provided. Browsing of the database and retrieval of gene information can be accomplished using user-friendly online techniques. Furthermore, complete *F. graminearum* gene sequences can be downloaded from the MIPS-FGDB website. Since it is continually updated as newer information becomes available, the MIPS-FGDB is constantly evolving.

The *F. graminearum* genetic map has been studied by various groups and the molecular markers that have been generated have been observed to anchor almost the entirety of the *F. graminearum* (strain PH-1) genome (Jurgenson et al., 2002; O'Donnell et al., 2004; Brown et al., 2006). In addition to the *F. graminearum* genome, other phytopathogenic *Fusarium* species with completely annotated genomes include *F. verticillioides* and *F. oxysporum*. The *Fusarium* comparative genomics database maintained by the Broad Institute allows researchers to simultaneously compare the genome sequences of these related yet phenotypically distinct *Fusarium* species. This comparative analysis offers a method to comprehensively study *Fusarium* pathogenicity and virulence factors as well as the evolutionary relationship among the *Fusarium* species.

Of the three sequenced *Fusarium* genomes, *F. oxysporum* (15 chromosomes) was the largest at 60 Mb, *F. graminearum* (4 chromosomes) was the smallest at 36 Mb, and *F. verticillioides* (11 chromosomes) was in between at 42 Mb (Ma et al., 2010). The greater genomic content of *F. oxysporum* is contained in unique sequences of the additional chromosomes. There was an observed 91% nucleotide sequence identity between orthologs of *F. verticillioides* and *F. oxysporum* and an 85% sequence identity between these orthologs and those of *F. graminearum*. Overall, within the three genomes, there were more than 9000 conserved syntenic orthologs, including numerous transmembrane transporters, transcription factors, and lytic enzymes (Ma et al., 2010). In addition to these genetic similarities among the various *Fusarium* species, there are thousands of unique genes, with many of them located in the telomeric regions of chromosomes (Cuomo et al., 2007). Single nucleotide polymorphisms (SNP) were observed to be fairly prevalent in these telomeric chromosomal segments, which were also associated with increased frequency of recombination events (Gale et al., 2005). It is likely that many of these unique genes are responsible for mediating activities relevant to plant host infections. Another apparent commonality among the different *Fusarium* species is the fact that regions of high genetic diversity were enriched with genes believed to be involved in plant host infections, including carbohydrate-active enzymes, secreted proteins, and genes whose expression was up-regulated during infection (Guldener et al., 2006; Cuomo et al., 2007; Ma et al., 2010).

Upon analyzing the genome sequences of *F. verticillioides*, *F. graminearum*, and *F. oxysporum* f. sp. *lycopersici* (tomato pathogen), researchers uncovered the existence of *F. oxysporum* lineage-specific genomic segments spanning the entirety of chromosomes 3, 6, 14, and 15 as well as segments of chromosomes 1 and 2 (Ma et al., 2010). The lineage-specific domains consisted of 19 Mb, which corresponded to the approximate difference in genome size between *F. oxysporum* and *F. verticillioides*. These lineage-specific regions contained numerous pathogenicity-related effector genes and transposons and it was reported that the horizontal transfer of chromosome 14 from a pathogenic strain of *F. oxysporum* to a genetically-distinct, non-pathogenic strain resulted in the introduction of host-specific pathogenicity (Ma et al., 2010). Interestingly, the effector gene sequences of chromosome 14 have been observed to be highly similar, if not identical, among unrelated tomato-pathogenic *F. oxysporum* strains (Lievens et al., 2009). Apparently, within the *F. oxysporum* genome, the lineage-specific regions vary between fungal strains with different plant host specificities, thereby suggesting that these differences might have a crucial effect on mediating plant host compatibility (Ma et al., 2010). Analysis of the *F. solani* genome, which has also undergone whole genome sequencing, revealed that *F. solani* possesses three distinct lineage-specific chromosomes, which are different from those found in *F. oxysporum* (Coleman et al., 2009; Ma et al., 2010). *F. solani* strains also possess transposon-rich chromosomes containing genes, including those encoding products involved in nutrient use and detoxification, that dictate plant host specificity (Temporini and VanEtten, 2004; Rodriguez-Carres et al., 2008). These findings may have implications in understanding plant-pathogen interactions as well as how pathogenic strains of *Fusarium* species develop or evolve.

In addition to annotating genomes, it is important to delineate how the coordination of gene expression occurs if the full value of genomics is to be realized. Among key regulatory factors are the various transcription factors and there remains much about their activities, as well as their targets, that is still unclear in phytopathogenic *Fusarium* species. With four sequenced *Fusarium* genomes, researchers have undertaken efforts to identify possible regulatory elements through the implementation of comparative genomics techniques to analyze conserved non-coding genomic segments. It is noteworthy that many identified regulatory motifs present in upstream regions or introns among distantly related fungi were highly conserved, which appears to imply that they have a significant evolutionary and biological value (Kumar et al., 2010). Examples of regulatory motifs found within the promoter regions were those with sequence similarities to transcription factor binding sites for carbon catabolite (GCGGG) and nitrogen metabolism (YGATAAG) repression as well as for regulation of responses to environmental pH (GCCARG). Other transcription factor binding sites that were observed to be conserved in the promoter regions of sequenced *Fusarium* species included those involved in phosphate metabolism and environmental stress response (Kumar et al., 2010). In total, 73 potential regulatory motifs predominantly upstream of genes mediating particular cellular activities were identified in promoter regions, while 39 and 69 candidate motifs were detected in the introns and downstream genomic regions, respectively. One of the highly conserved regulatory sequences found in introns (GTNAGT) was a splice donor site (Kumar et al., 2010). Interestingly, known mammalian and yeast polyadenylation sequences were not present in the downstream regions of *Fusarium* genomes, indicating that a different polyadenylation signal is likely used in *Fusarium* species. Better characterization of many of the identified motifs as well as the detection of new ones will likely result as more fungal genomes become fully sequenced and comparative genomics-based studies become more advanced and refined.

The accumulation of data generated from genome sequencing and gene expression endeavors is merely the first step in genomics-based studies of important phytopathogenic *Fusarium* species. Comprehensive data analysis using a myriad of bioinformatics options is crucial to make optimal use of *Fusarium* genomic resources. Compiling lists of genes and expression patterns of identified genes provides relatively limited, albeit important, information. Functional determination of proteins encoded by genes and characterization of their biochemical and physiological roles in fungal development, reproduction, and pathogenicity are somewhat daunting tasks facing scientists studying *Fusarium* species.

As more diverse fungal species undergo whole genome sequencing, the importance of comparative genomics, which can be viewed as the comparison of genomes of different fungi, becomes more apparent. Uncovering differences and similarities among disparate fungal species may offer clues as to the commonality of specific virulence or pathogenicity factors as well as providing indications regarding evolutionary relatedness between species. The aforementioned comparison of the genomes of *F. graminearum*, *F. verticillioides*, *F. oxysporum*, and *F. solani* is but one example of the effective application of comparative genomics.

5. Targeted Gene Disruption and Insertional Mutagenesis

Data generated by DNA microarrays or EST studies identifying candidate genes involved in potentially important activities influencing pathogenicity can be verified by manipulating expression of selected genes of interest and assessing the phenotypic changes that result. One method to accomplish this is through targeted gene disruption or replacement. Characterization of specific genes possibly mediating virulence, including mycotoxin biosynthesis or fungicide resistance, in various *Fusarium* species can provide novel information regarding disease initiation and progression, which may offer clues relevant to the development of newer and more effective disease management strategies. An example of gene characterization through the implementation of gene mutation techniques was the determination that *F. graminearum* gene *Rrr1* was involved in pathogenicity and sporulation (Goswami et al., 2006). Gene disruption mutants were observed to produce significantly fewer macroconidia on artificial growth medium as well as delayed growth or spread on susceptible wheat heads. The *Rrr1* gene was selected for further study based on an EST study involving multiple cDNA libraries produced through subtractive hybridization to examine the initial stages of FHB development on infected wheat (Goswami et al., 2006).

One of the consequences of whole genome sequencing of *Fusarium* species is that it has allowed for a more complete characterization of polyketides, which are important secondary metabolites that exhibit a relatively high degree of functional diversity, including serving as mycotoxins (Bhatnagar et al., 2003; Gaffoor et al., 2005). Polyketides are synthesized with the assistance of a group of enzymes called polyketide synthases (PKSs) that consist of at least three domains and are either reducing or non-reducing (Bingle et al., 1999). Successful gene disruption of the 15 PKS genes present in the *F. graminearum* genome (i.e., *PKS1*, *PKS2*, *PKS5*, *PKS6*, *PKS7*, *PKS9*, *PKS11*, *PKS17*, *ZEA1*, *ZEA2*, *GRS1*, *PLSP1*, *GzFUS1*, *PGL1*, and *AUR1*) has been achieved and analysis of the phenotypes of resultant mutants as well as PKS gene expression in diverse conditions has generated valuable information regarding these important enzymes (Gaffoor et al., 2005). Wild type *F. graminearum* protoplasts were transformed using DNA constructs that allowed for the disruption of the target PKS genes and subsequent insertion of the hygromycin B phosphotransferase gene, which served as a selectable marker. In this instance, transformation was achieved through co-incubation of protoplasts and disruption vector; however, other transformation techniques that have been used for insertional mutagenesis in *Fusarium* species include *Agrobacterium tumefaciens*-mediated transformation (Mullins et al., 2001). While the disruption of PKS genes did not have an appreciable effect on pathogenicity on wheat heads or on perithecia and ascospore production for any of the individual mutants, there were observed differences in terms of fungal growth. Disruption of *ZEA2*, *PKS1*, *PKS6*, *PKS7*, *PKS9*, *PKS11*, and *PGL1* resulted in increased growth compared to wild type while disruption of *GRS1* and *PKS2* led to decreased growth rates (Gaffoor et al., 2005). Insertional mutagenesis of all other PKS genes did not have any significant effect on growth rates of mutants. Production of specific mycotoxins was also inhibited in mutants unable to express *ZEA1* or *ZEA2* (zearalenone), *GzFUS1* (fusarin C), or *AUR1* (aurofusarin) (Gaffoor et al., 2005). Absence of a dark pigment in perithecia produced by mutants unable to express *PGL1* suggested a role for this particular PKS gene.

The significance of enniatin, which is a mycotoxin, on *F. avenaceum* virulence was investigated by generating enniatin non-producing mutants through gene disruption of the enzyme responsible for catalyzing its biosynthesis (Herrmann et al., 1996a). Enniatin synthetases have also been identified in other *Fusarium* species such as *F. sambucinum*, *F. scirpi*, and *F. lateritium* with high sequence similarities between the gene from *F. avenaceum* and *F. scirpi* (Pieper et al., 1992; Herrmann et al., 1996b). Infection of susceptible potato tubers with enniatin synthetase non-expressing mutant *F. avenaceum* and wild type controls revealed a significant reduction in potato tuber decomposition along with absence of enniatin on mutant-infected potato as compared to wild type inoculated material (Herrmann et al., 1996a).

Targeted gene disruption also led to improved characterization of the deoxynivalenol and nivalenol chemotypes of *F. graminearum* by providing evidence for the roles of two genes in trichothecene mycotoxin production (Ichinoe et al., 1983; Lee et al., 2002). Genetic manipulation resulted in the generation of transgenic mutants in which a deoxynivalenol chemotype and a nivalenol chemotype could produce nivalenol and deoxynivalenol, respectively, which was contrary to the expected mycotoxin production pattern for the two chemotypes. Similar to the previously described fumonisin gene cluster, the genes involved in trichothecene biosynthesis are also clustered together. Between the deoxynivalenol and nivalenol *F. graminearum* chemotypes, nine of the 10 *Tri* genes were conserved, with *Tri7* present and expressed in nivalenol chemotypes, but non-functional in deoxynivalenol producers (Lee et al., 2001). Additionally, *Tri13* was observed to be considerably different between the deoxynivalenol and nivalenol chemotypes and, as for *Tri7*, it was not viable in deoxynivalenol chemotypes (Lee et al., 2002). Deletion of *Tri13* in nivalenol producing *F. graminearum* generated deoxynivalenol producing mutants whereas heterologous expression of functional *Tri13* in deoxynivalenol chemotypes produced nivalenol producing mutants. When both a functional *Tri7* and *Tri13* were expressed in deoxynivalenol chemotypes, nivalenol and 4-acetylnivalenol were produced by mutants, which appeared to indicate that the *Tri7* gene product was responsible for acetylation of nivalenol (Lee et al., 2002). Understanding the genetic basis for chemotype determination in *F. graminearum* is important in light of the fact that there may be differences in toxicity between deoxynivalenol and nivalenol (Eudes et al., 2000). The ability to identify the predominant *F. graminearum* chemotype present in a contaminated field could potentially be of interest to cereal producers.

Functional analysis of another gene in the trichothecene cluster, *Tri5*, demonstrated that the encoded trichodiene synthase, which catalyzes the first reaction of the biosynthesis of trichothecenes, is a significant factor affecting *F. graminearum* virulence (Proctor et al., 1995). Insertional mutagenesis of *Tri5* generated non-trichothecene producing mutants that were observed to be less virulent on some varieties of wheat and rye, but did not have reduced virulence on maize. The effects of a lack of wild type *Tri5* expression appeared to be dependent on the host species and variety that was infected by the non-expressing mutant *F. graminearum* (Proctor et al., 1995). This host specificity effect was similarly noted in *Tri5*-disrupted mutant *F. sambucinum*, which exhibited wild type levels of virulence on infected potato, but had reduced virulence on parsnip roots (Desjardins et al., 1992; Hohn and Desjardins, 1992). Verification that reduced virulence was specifically the result of *Tri5* gene disruption and not due to an artifact of transformation such as an ectopic insertion was provided by the results of a complementation protocol in which a functional *Tri5* was reintroduced into the mutant *F. graminearum*. Complementation of the mutants resulted in the restoration of trichothecene production and the associated virulence to levels comparable to wild type *F. graminearum* (Proctor et al., 1997).

One option available to cereal growers to control diseases caused by *Fusarium* species is through fungicide use (Mullenborn et al., 2008). Fungicides like carbendazim (benzimidazole fungicide) have been used to control FHB in parts of Asia for a number of decades; however, resistance to carbendazim is relatively widespread (Yuan and Zhou, 2005; Zhang et al., 2009). This resistance may be the result of a mutation in one of three specific codons of the β_2-tubulin gene, with codon 167 being the predominant mutation site detected in carbendazim-resistant *F. graminearum* isolates collected from infected fields (Chen et al., 2008). Resistance to carbendazim does not appear to affect *F. graminearum* fitness nor does it influence FHB disease severity on wheat heads (Chen et al., 2007; Zhang et al., 2009). Determination of the effects of carbendazim resistance on trichothecene biosynthesis and, specifically, *Tri5* expression was achieved through the analysis of mutant *F. graminearum* generated through site directed mutagenesis of selected β_2-tubulin codons (Zhang et al., 2009). The strain that was used to introduce β_2-tubulin codon mutations was a wild type *F. graminearum* that was sensitive to carbendazim. Site directed mutations introduced into codon 167 resulted in carbendazim-resistant mutants that produced significantly greater amounts of trichothecenes as well as increased *Tri5* gene expression when compared to the wild type from which they were derived (Zhang et al., 2009). Additional gene replacement and site directed mutagenesis at specific β_2-tubulin sites (codons 50, 167, 198, or 200) introduced carbendazim-resistance to generated fungal mutants, which suggests that the amino acids encoded by the targeted codons may form part of the carbendazim binding site (Qiu et al., 2011). The efficiency in which trichothecenes are produced in *F. graminearum* may be higher in carbendazim-resistant strains. The effects of carbendazim resistance on trichothecene production may have implications on the selection of fungicides that are used for chemical control of FHB-causing *Fusarium* species and may also have relevance to the ongoing development of novel fungicides for effective fungal control.

CONCLUSION

The value of genomics-based research to characterize pathogenicity in agriculturally significant *Fusarium* species is apparent based on the many publications that result from research programs worldwide, some of which were described in this chapter. Ongoing initiatives to complete whole genome sequencing of phytopathogenic *Fusarium* species will likely benefit from new high-throughput, next-generation DNA pyrosequencing technology that is rapidly developing. Presently, large-scale DNA pyrosequencing efforts are primarily accomplished through the use of platforms developed by Illumina, Applied Biosystems (SOLID), and Roche/454 Life Sciences. South African researchers are currently attempting to use DNA pyrosequencing to determine the whole genome sequence of *F. circinatum* (Nayaka et al., 2011). In addition to whole genome sequencing, DNA pyrosequencing may also be relevant to studies involving SNP detection, species detection and identification, and genotyping (Marsh, 2007). While genomics has contributed to a more thorough understanding of numerous factors mediating pathogenicity, including secondary metabolism and fungicide resistance, successfully using the generated information to develop novel targeted disease management strategies is a considerable challenge for plant pathologists and will ultimately determine the true value of genomics-based research.

REFERENCES

Allen, T.D., Dawe, A.L., and Nuss, D.L. (2003). Use of cDNA microarrays to monitor transcriptional responses of the chestnut blight fungus *Cryphonectria parasitica* to infection by virulence-attenuating hypoviruses. Eukaryot. *Cell.* 2: 1253-1265.

Becher, R., Weihmann, F., Deising, H.B., and Wirsel, S.G.R. (2011). Development of a novel multiplex DNA microarray for *Fusarium graminearum* and analysis of azole fungicide responses. *BMC Genomics.* 12: 52.

Bhatnagar, D., Ehrlich, K.C., and Cleveland, T.E. (2003). Molecular genetic analysis and regulation of aflatoxin biosynthesis. *Appl. Microbiol. Biotechnol.* 61: 83-93.

Bingle, L.E.H., Simpson, T.J., and Lazarus, C.M. (1999) Ketosynthase domain probes identify two subclasses of fungal polyketide synthase genes. *Fungal Genet. Biol.* 26: 209-223.

Both, M., Csukai, M., Stumpf, M.P.H., and Spanu, P.D. (2005). Gene expression profiles of *Blumeria graminis* indicate dynamic changes to primary metabolism during development of an obligate biotrophic pathogen. *Plant Cell.* 17: 2107-2122.

Brown, D.W., Cheung, F., Proctor, R.H., Butchko, R.A.E., Zheng, L., Lee, Y., Utterback, T., Smith, S., Feldblyum, T., Glenn, A.E., Plattner, R.D., Kendra, D.F., Town, C.D., and Whitelaw, C.A. (2005). Comparative analysis of 87,000 expressed sequence tags from the fumonisin-producing fungus *Fusarium verticillioides. Fungal Genet. Biol.* 42: 848-861.

Brown, D.W., Butchko, R.A., Busman, M., and Proctor, R.H. (2007). The *Fusarium verticillioides FUM* gene cluster encodes a Zn(II)2Cys6 protein that affects *FUM* gene expression and fumonisin production. *Eukaryot. Cell.* 6: 1210-1218.

Brown, D.W., Butchko, R.A.E., and Proctor, R.H. (2006). *Fusarium* genomic resources: Tools to limit crop diseases and mycotoxin contamination. *Mycopathologia.* 162: 191-199.

Chen, C.J., Wang, J.X., Luo, Q.Q., Yuan, S.K., and Zhou, M.G. (2007). Characterization and fitness of carbendazim-resistant strains of *Fusarium graminearum* (wheat scab). *Pest Manage. Sci.* 62: 1201-1207.

Chen, C.J., Bi, C.W., Yu, J.J., Wang, J.X., Li, H.X., Luo, Q.Q., and Zhou, M.G. (2008). Carbendazim-resistance and its molecular mechanism in *Gibberella zeae*. *J. Plant Pathol.* 90 (Suppl. 2): 69.

Coleman, J.J., Rounsley, S.D., Rodriguez-Carres, M., Kuo, A., Wasmann, C.C., Grimwood, J., Schmutz, J., Taga, M., White, G.J., Zhou, S., et al. (2009). The genome of *Nectria haematococca*: contribution of supernumerary chromosomes to gene expansion. *PLoS Genet.* 5: e1000618.

Cuomo, C.A., Guldener, U., Xu, J.R., Trail, F., Turgeon, B.G., Di Pietro, A., Walton, J.D., Ma, L.J., Baker, S.E., Rep, M., et al. (2007). The *Fusarium graminearum* genome reveals a link between localized polymorphism and pathogen specialization. *Science.* 317: 1400-1402.

Desjardins, A.E., Hohn, T.M., and McCormick, S.P. (1992). Effect of gene disruption of trichodiene synthase on the virulence of *Gibberella pulicaris. Mol. Plant Microbe Interact.* 5: 214-222.

Eudes, F., Comeau, A., Rioux, S., and Collin, J. (2000). Phytotoxicite de huit mycotoxines associees a la fusariose de l'epi chez le ble. *Can. J. Plant Pathol.* 22: 286-292.

Gaffoor, I., Brown, D.W., Plattner, R., Proctor, R.H., Qi, W., and Trail, F. (2005). Functional analysis of the polyketide synthase genes in the filamentous fungus *Gibberella zeae* (anamorph *Fusarium graminearum*). *Eukaryot. Cell.* 4: 1926-1933.

Gale, L.R., Bryant, J.N., Calvo, S., Giese, H., Katan, T., O'Donnell, K., Suga, H., Taga, M., Usgaard, T.R., Ward, T.J., et al. (2005). Chromosome complement of the fungal plant pathogen *Fusarium graminearum* based on genetic and physical mapping and cytological observations. *Genetics.* 171: 985-1001.

Geiser D.M., Jimenz Gasco M.M., Kang S., Mkalowska I., Veeraraghavan N., Ward T.J., Zhang N., Kuldau G.A., and O'Donnell K. (2004). FUSARIUM-IDv.1.0: A DNA sequence database for identifying *Fusarium*. *European Journal of Plant Pathology* 110: 473-479.

Goswami, R.S., Xu, J.R., Trail, F., Hilburn, K., and Kistler, H.C. (2006). Genomic analysis of host-pathogen interaction between *Fusarium graminearum* and wheat during early stages of disease development. *Microbiology.* 152: 1877-1890.

Guldener, U., Mannhaupt, G., Munsterkotter, M., Haase, D., Oesterheld, M., Stumpflen, V., Mewes, H.W., and Adam, G. (2006). FGDB: a comprehensive fungal genome resource on the plant pathogen *Fusarium graminearum*. *Nucleic Acids Res.* 34: (Database issue): D456-458.

Guldener, U., Seong, K.Y., Boddu, J., Cho, S., Trail, F., Xu, J.R., Adam, G., Mewes, H.W., Muehlbauer, G.J., and Kistler, H.C. (2006). Development of a *Fusarium graminearum* Affymetrix GeneChip for profiling fungal gene expression in vitro and in planta. *Fungal Genet. Biol.* 43: 316-325.

Herrmann, M., Zocher, R., and Haese, A. (1996a). Effect of disruption of the enniatin synthetase gene on the virulence of *Fusarium avenaceum*. *Mol. Plant Microbe Interact.* 9: 226-232.

Herrmann, M., Zocher, R., and Haese, A. (1996b). Enniatin production by *Fusarium* strains and its effect on potato tuber tissue. *Appl. Environ. Microbiol.* 62: 393-398.

Hohn, T.M. and Desjardins, A.E. (1992). Isolation and gene disruption of the *tox5* gene encoding trichodiene synthase in *Gibberella pulicaris*. *Mol. Plant Microbe Interact.* 5: 249-256.

Hughes, T.R., Mao, M., Jones, A.R., Burchard, J., Marton, M.J., Shannon, K.W., Lefkowitz, S.M., Ziman, M., Schelter, J.M., Meyer, M.R., et al. (2001). Expression profiling using microarrays fabricated by an ink-jet oligonucleotide synthesizer. *Nat. Biotechnol.* 19: 342-347.

Hwang, L., Hocking-Murray, D., Bahrami, A.K., Andersson, M., Rine, J., and Sil, A. (2003). Identifying phase-specific genes in the fungal pathogen *Histoplasma capsulatum* using a genomic shotgun microarray. *Mol. Biol. Cell.* 14: 2314-2326.

Ichinoe, M., Kurata, H., Sugiura, Y., and Ueno, Y. (1983). Chemotaxonomy of *Gibberella zeae* with special reference to production of trichothecenes and zearalenone. *Appl. Environ. Microbiol.* 46: 1364-1369.

Jurgenson, J.E., Bowden, R.L., Zeller, K.A., Leslie, J.F., Alexander, N.J., and Plattner, R.D. (2002). A genetic map of *Gibberella zeae* (*Fusarium graminearum*). *Genetics.* 160: 1451-1460.

Kristensen, R., Gauthier, G., Berdal, K.G., Hamels, S., Remacle, J., and Holst-Jensen, A. (2007). DNA microarray to detect and identify trichothecene- and moniliformin-producing *Fusarium* species. *J. Appl. Microbiol.* 102: 1060-1070.

Kristensen, R., Torp, M., Kosiak, B., and Holst-Jensen, A. (2005). Phylogeny and toxigenic potential is correlated in *Fusarium* species as revealed by partial translation elongation factor I alpha gene sequences. *Mycol. Res.* 109: 173-186.

Kumar, L., Breakspear, A., Kistler, C., Ma, L.J., and Xie, X. (2010). Systematic discovery of regulatory motifs in *Fusarium graminearum* by comparing four *Fusarium* genomes. *BMC Genomics.* 11: 208.

Lee, T., Han, Y.K., Kim, K.H., Yun, S.H., and Lee, Y.W. (2002). *Tri13* and *Tri7* determine deoxynivalenol- and nivalenol-producing chemotypes of *Gibberella zeae*. *Appl. Environ. Microbiol.* 68: 2148-2154.

Lee, T., Oh, D.W., Kim, H.S., Lee, J., Kim, Y.H., Yun, S.H., and Lee, Y.W. (2001). Identification of deoxynivalenol- and nivalenol-producing chemotypes of *Gibberella zeae* using PCR. *Appl. Environ. Microbiol.* 67: 2966-2972.

Liao, S.M., Zhang, J., Jeffrey, D.A., Koleske, A.J., Thompson, C.M., Chao, D.M., Ciljoen, M., van Vuuren, H.J.J., and Young, R.A. (1995). A kinase-cyclin pair in the RNA polymerase II holoenzyme. *Nature.* 374: 193-196.

Lievens, B., Houterman, P.M., and Rep, M. (2009). Effector gene screening allows unambiguous identification of *Fusarium oxysporum* f.sp. *lycopersici* races and discrimination from other *formae speciales*. *FEMS Microbiol. Lett.* 300: 201-215.

Ma, L.J., van der Does, H.C., Borkovich, K.A., Coleman, J.J., Daboussi, M.J., Di Pietro, A., Dufresne, M., Freitag, M., Grabherr, M., Henrissat, B., et al. (2010). Comparative genomics reveals mobile pathogenicity chromosomes in *Fusarium*. *Nature.* 464: 367-373.

Marsh, S. (2007). Pyrosequencing applications. *Methods Mol. Biol.* 373: 15-24.

Mullenborn, C., Steiner, U., Ludwig, M., and Oerke, E.C. (2008). Effect of fungicides on the complex of *Fusarium* species and saprophytic fungi colonizing wheat kernels. *Eur. J. Plant Pathol.* 120: 157-166.

Mullins, E.D., Chen, X., Romaine, P., Raina, R., Geiser, D.M., and Kang, S. (2001). *Agrobacterium*-mediated transformation of *Fusarium oxysporum*: An efficient tool for insertional mutagenesis and gene transfer. *Phytopathology.* 91: 173-180.

Nayaka, S.C., Wulff, E.G., Udayashankar, A.C., Nandini, B.P., Niranjana, S.R., Mortensen, C.N., and Prakash, H.S. (2011). Prospects of molecular markers in *Fusarium* species diversity. *Appl. Microbiol. Biotechnol.* 90: 1625-1639.

Nicolaisen, M., Justesen, A.F., Thrane, U., Skouboe, P., and Holmstrom, K. (2005). An oligonucleotide microarray for the identification and differentiation of trichothecene producing and non-producing *Fusarium* species occurring on cereal grain. *J. Microbiol. Methods.* 62: 57-69.

O'Donnell, K., Cigelnik, E., and Nirenberg, H.I. (1998). Molecular systematic and phylogeography of the *Gibberella fujikuroi* species complex. *Mycologia.* 90: 465-493.

O'Donnell, K., Nirenberg, H.I., Aoki, T., and Cigelnik, E. (2000). A multigene phylogeny of the *Gibberella fujikuroi* species complex: detection of additional phylogenetically distinct species. *Mycoscience.* 41: 61-78

O'Donnell, K., Ward, T.J., Geiser, D.M., Kistler, H.C., and Aoki, T. (2004). Genealogical concordance between the mating type locus and seven other nuclear genes supports formal recognition of nine phylogenetically distinct species within the *Fusarium graminearum* clade. *Fungal Genet. Biol.* 41: 600-623.

Parry, D.W., Jenkinson, P., and McLeod, L. (1995). *Fusarium* ear blight (scab) in small grain cereals – a review. *Plant Pathol.* 44: 207-238.

Parry, D.W. and Nicholson, P. (1996). Development of a PCR assay to detect *Fusarium poae* in wheat. *Plant Pathol.* 45: 383-391.

Pieper, R., Kleinhauf, H., and Zocher, R. (1992). Enniatin synthetases from different fusaria exhibiting distinct amino acid specificities. *J. Antibiot.* 45: 1273-1277.

Pirttila, A.M., McIntyre, L.M., Payne, G.A., and Woloshuk, C.P. (2004). Expression profile analysis of wild-type and *fcc1* mutant strains of *Fusarium verticillioides* during fumonisin biosynthesis. *Fungal Genet. Biol.* 41: 647-656.

Proctor, R.H., Hohn, T.M., and McCormick, S.P. (1995). Reduced virulence of *Gibberella zeae* caused by disruption of a trichothecene toxin biosynthetic gene. *Mol. Plant Microbe Interact.* 8: 593-601.

Proctor, R.H., Hohn, T.M., and McCormick, S.P. (1997). Restoration of wild-type virulence to *Tri5* disruption mutants of *Gibberella zeae* via gene reversion and mutant complementation. *Microbiology.* 143: 2583-2591.

Qiu, J.B., Xu, J.Q., Yu, J.J., Bi, C.W., Chen, C.J., and Zhou, M.G. (2011). Localisation of the benzimidazole fungicide binding site of *Gibberella zeae* β_2-tubulin studied by site-directed mutagenesis. *Pest Manag. Sci.* 67: 191-198.

Relogio, A., Schwager, C., Richter, A., Ansorge, W., and Valcarcel, J. (2002). Optimization of oligonucleotide-based DNA microarrays. *Nucleic Acids Res.* 30: e51.

Rep, M. and Kistler, H.C. (2010). The genomic organization of plant pathogenicity in *Fusarium* species. *Curr. Opin. Plant Biol.* 13: 420-426.

Rodriguez-Carres, A., White, G., Tsuchiya, D., Taga, M., and VanEtten, H.D. (2008). The supernumerary chromosome of *Nectria haematococca* that carries pea-pathogenicity-related genes also carries a trait for pea rhizosphere competitiveness. *Appl. Environ. Microbiol.* 74: 3849-3856.

Schilling, A.G., Moller, E.M., and Geiger, H.H. (1996). Polymerase chain reaction-based assays for species-specific detection of *Fusarium culmorum*, *F. graminearum*, and *F. avenaceum*. *Phytopathology*. 86: 515-522.

Schnerr, H., Vogel, R.F., and Niessen, L., (2002). Correlation between DNA of trichothecene-producing *Fusarium* species and deoxynivalenol concentrations in wheat-samples. *Lett. Appl. Microbiol.* 35: 121-125.

Seifert, K.A. and Levesque, C.A. (2004). Phylogeny and molecular diagnosis of mycotoxigenic fungi. *Eur. J. Plant Pathol.* 110: 449-471.

Shim, W.B. and Woloshuk, C.P. (2001). Regulation of fumonisin B1 biosynthesis and conidiation in *Fusarium verticillioides* by a cyclin-like (C-type) gene, *FCC1*. *Appl. Environ. Mic*robiol. 67: 1607-1612.

Temporini, E.D. and VanEtten, H.D. (2004). An analysis of the phylogenetic distribution of the pea pathogenicity genes of *Nectria haematococca* MPVI supports the hypothesis of their origin by horizontal transfer and uncovers a potentially new pathogen of garden pea: *Neocosmospora boniensis*. *Curr. Genet.* 46: 29-36.

Torp, M. and Nirenberg, H.I. (2004). *Fusarium langsethiae* sp. nov. on cereals in Europe. *Int. J. Food Microbiol.* 95: 247-256.

Trail, F., Xu, J.R., San Miguel, P., Halgren, R.G., and Kistler, H.C. (2003). Analysis of expressed sequence tags from *Gibberella zeae* (anamorph *Fusarium graminearum*). *Fungal Genet. Biol.* 38: 187-197.

Turner, A.S., Lees, A.K., Rezanoor, H.N., and Nicholson, P. (1998). Refinement of PCR-detection of *Fusarium avenaceum* and evidence from DNA marker studies for phenetic relatedness to *Fusarium tricinctum*. *Plant Pathol.* 47: 278-288.

Waalwijk, C., de Koning, J.R.A., Baayen, R.P., and Gams, W. (1996). Discordant groupings of *Fusarium* spp. from sections of *Elegens*, *Liseola*, and *Dlaminia* based on ribosomal ITS1 and ITS2 sequences. *Mycologia.* 88: 361-368.

Waalwijk, C., van der Heide, R., de Vries, I., van der Lee, T., Schoen, C., Corainville, G.C., Hauser-Hahn, I., Kastelein, P., Kohl, J., Lonnet, P., Demarquet, T., and Kema, G.H.J. (2004). Quantitative detection of *Fusarium* species in wheat using TaqMan. *Eur. J. Plant Pathol.* 110: 481-494.

Wang, C.S., Hu, G., and St. Leger, R.J. (2005). Differential gene expression by *Metarhizium anisopliae* growing in root exudate and host (*Manduca sexta*) cuticle or hemolymph reveals mechanisms of physiological adaptation. *Fungal Genet. Biol.* 42: 704-718.

Wulff, E.G., Sorensen, J.S., Lubeck, M., Nielsen K.F., Thrane, U., and Torp, J. (2010). *Fusarium* spp. associated with rice Bakanae: ecology, genetic diversity, pathogenicity and toxigenicity. *Environ. Microbiol.* 12: 649-657.

Yuan, S.K. and Zhou, M.G. (2005). A major gene for resistance to carbendazim, in field isolates of *Gibberella zeae*. *Can. J. Plant Pathol.* 27: 58-63.

Zhang, Y.J., Yu, J.J., Zhang, Y.N., Zhang, X., Cheng, C.J., Wang, J.X., Hollomon, D.K., Fan, P.S., and Zhou, M.G. (2009). Effect of carbendazim resistance of trichothecene production and aggressiveness of *Fusarium graminearum*. *Mol. Plant Microbe Interact.* 22: 1143-1150.

In: Fusarium
Editors: Tony Ferarri Rios and Emilia Reyes Ortega
ISBN: 978-1-61942-539-2

Chapter 9

FUSARIUM HEAD BLIGHT AND DON CONTAMINATION MANAGEMENT IN SOFT AND DURUM WHEAT CULTIVATION*

Andrea Maiorano†, Massimo Blandino and Amedeo Reyneri
Department of Agronomy, Forest and Land Management,
University of Turin,via Leonardo da Vinci, Grugliasco (TO), Italy

ABSTRACT

Fusarium Head Blight (FHB), also known as scab, is a devastating disease of enormous economic importance throughout the world that attacks all classes of barley and wheat. Every year, all of the most important cereal producers in the world are affected by this disease. *Fusarium graminearum* and *Fusarium culmorum* are considered the most pathogenic and widespread agents of FHB in wheat. Both *F. graminearum* and *F. culmorum* produce deoxynivalenol (DON, also known as vomitoxin), a mycotoxin of the trichotecenes group, one of the most widespread mycotoxins in cereals, which can be synthesised in the field in small grain cereals and in maize. This mycotoxin has a great impact on the health of animals and humans, due to their cytotoxic activity and immunosuppressive effects. As a consequence, FHB and DON contamination are responsible for serious direct and indirect economic losses. Economic losses can be attributed to yeld loss due to fungal disease, reduced technological and nutritional quality of grain, reduced crop value due to mycotoxin contamination, reduced animal productivity, human health costs. Prevention, monitoring, sampling, chemical analysis, litigation, mitigation, and research costs also need to be taken into account. Moreover, FHB and DON limits, which have been established in many parts of the world for animal feeding and human consumption, have made the worldwide cereal market much more selective.

F. graminearum and F. culmorum inoculum is usually abundant and its production, dispersal and deposition are influenced by weather conditions. Warm, humid weather,

---

* A version of this chapter also appears in *Soybean and Wheat Crops: Growth, Fertilization, and Yield*, edited by Samuel Davies and George Evans, published by Nova Science Publishers, Inc. It was submitted for appropriate modifications in an effort to encourage wider dissemination of research.

† Corresponding author: Andrea Maiorano; Tel +39 011 670 8928, e-mail: andrea.maiorano@unito.it.

frequent rainfall and heavy dew favour spore germination. Temperature and water activity of the host are the fundamental factors that affect fungal growth inside grain. DON synthesis is mainly influenced by temperature and water ativity and its production seems to play a decisive role on the aggressiveness of *F. graminearum* and *F. culmorum*: it has been suggested that in the presence of DON, plant defence mechanisms are not triggered fast enough, thereby leading to an increased aggressiveness of the pathogen.

The accumulation of DON in the grain is intimately related to the development of the disease in the field. Prevention strategies through pre-harvest agronomic management can achieve the quality and safety standars required by the market and international regulations. Among the agricultural practices, FHB development and DON contamination in wheat grains are mainly affected by tillage, crop rotation, varieties and fungicide application. None of these strategies on their own are able to significantly reduce the impact of this disease. Nevertheless, careful planning of all the different management decisions canl lead to the designing of crop management systems able to reduce FHB and mycotoxin incidence and severity. In other words, it is necessary for producers to make use of the available wheat varieties that are known to be more resistant and to apply all of the good agricultural practices tuned to a determined cultivation area.

Useful tools for prevention are represented by FHB and DON predictive models: the use of a model to predict the outcome of a disease is desirable to enhance and trigger management opportunities with the aim of reaching high technological, nutritional and productivity quality and safety of the production.

Keywords: Fusarium Head Blight, deoxynivalenol, *F. graminearum* epidemiology, FHB economic impact, crop management systems, predictive models

1. INTRODUCTION

Fusarium Head Blight (FHB), also known as scab, is a devastating disease of enormous economic importance throughout the world that attacks all classes of barley and wheat. Every year, all of the most important cereal producers in the world are affected by this disease. As a consequence, FHB is responsible for serious direct and indirect economic losses.

FHB is caused by different species of the genus *Fusarium*, but two of them represent the most important species worldwide: *Fusarium graminearum* Schwabe (teleomorph *Gibberella zeae* (Schw.) Petch.) and *Fusarium culmorum* (W.G. Sm.) Sacc. (teleomorph unknown). Their temperature requirements determine their relative competitive strength and their consequent geographical predominance: *F. culmorum* tends to predominate in the cooler maritime regions of northwest Europe, where *Fusarium poae* (Peck) Wollenw., *Fusarium avenaceum* (Corda) Sacc. and *Microdochium nivale* (Fr.) Samuels & Hallet (syn.*Fusarium nivale*) also assume a greater importance (Birzele et al., 2002; Kosiak et al., 2003; Ioos et al., 2004). Other less frequently isolated species are *F. poae*, *F. cerealis* (syn. F. crookwellense), *F. equiseti* (syn. *F. scirpi*) (*G. intricans*), *F. sporotrichioides*, and *F. tricinctum*. Many other species may be sporadically encountered, including *F. acuminatum*, *F. subglutinans* (syn. *F. sacchari*), *F. solani*, F. *oxysporum*, *F. semitectum* (syn. *F. pallidoroseum*, *F. incarnatum*), *F. verticillioides* (syn.: F. *moniliforme*), and *F. proliferatum* (Bottalico and Perrone, 2002).

F. graminearum is considered the most pathogenic and widespread agent of FHB in wheat. Both *F. graminearum* and *F. culmorum* produce deoxynivalenol (DON, also known as vomitoxin), a mycotoxin of the trichotecenes group, one of the most widespread mycotoxins in cereals, which can be synthesised in the field in small grain cereals and in maize. Trichotecenes are associated with serious mycotoxicosis in humans and animals. This group of mycotoxins has a great impact on the health of animals and humans, due to their cytotoxic activity and immunosuppressive effects. Their main effects—related to their concentration in the commodity—are a reduced feed uptake, vomiting and immunosuppression (Rotter et al., 1996; Oswald et al., 2005). Therefore, FHB and DON limits, which have been established in many parts of the world for animal feeding and human consumption, have made the worldwide cereal market much more selective.

2. Occurrence and Economic Impact

2.1. Occurrence

FHB infection and DON contaminations have been reported by the most important wheat producers in the world, on all continents. The aetiological complexity of FHB is the main reason for the great diffusion of this disease and its geographical and seasonal variability. Table 1 illustrates a review of the natural occurrence of FHB and DON contamination of wheat grain in different part of the world in recent years (from 1984 to 2007). *F. graminearum* is the most prevalent species in most sites, except in England (Nicholson et al., 1993), Norway (Kosiak et al., 2003) and Latvia (Treikale et al., 2008), where the environmental conditions are more favourable to *F. avenaceum*, *F. poae* and *F. culmorum*.

Data taken in different periods in the Netherlands have shown an interesting change in the prevalence of the *Fusarium* spp. In a screening conducted in 2001 and 2002, Waalwijk et al. (2003) found *F. graminearum* to be the most prevalent species, a result that was in contrast with two previous surveys conducted in the 1980s by Daamen et al. (1991) and in the early 1990s by De Nijs et al. (1996), in which *F. culmorum* was reported to be the predominant agent for FHB. Waalwijk suggested some possible reasons for this change: i) the increase in maize production (*F. graminearum* is a major pathogen on maize and has the capacity to survive on maize stubble); ii) climatic changes might favour the propagation of *F. graminearum* over *F. culmorum*, as the former species has a higher temperature optimum; iii) the homothallic nature of *F. graminearum*, which allows the production of large masses of ascospores that can play a role in the epidemiology.

The data from the Netherlands show that FHB is a very dynamic disease, mainly due to its aetiological complexity. It is also possible that fungi may be spread from one country to another due to the increases in the global grain trade. However, as far as *Fusarium* fungi are concerned, this risk is likely to be minimal since these phytopathogens are field rather than storage organisms. On the other hand, there is now overwhelming evidence of global contamination of cereals and animal feeds with trichothecenes. Trade in wheat may, therefore, contribute to the worldwide dispersal of mycotoxins (Placinta et al., 1999).

Scab is prevalent in warm, humid regions where flowering coincides with rainy periods. In the United States, the incidence of this disease has been increasing over the last 10 years for a combination of reasons. Perhaps the most important reason concerns the increased area in which wheat is rotated with maize or other cereals. Other reasons are changes in the cropping system for soil protection purposes and changes in wheat cropping from traditional to more humid, non-traditional areas (McMullen et al., 1997; Gilchrist and Dubin, 2002).

Among the data reported in the table, the highest DON contaminations are reported in Michigan (20000 μg/kg in 2004), England (10626 μg/kg in 2003), Russia (8600 μg/kg, year not specified) and Argentina (8440 μg/kg, year not specified). The data from Africa (Kenya and Morocco) and from China show the lowest range of contamination, but these regions of the world are so large that more data would be needed to characterize their DON contamination in wheat.

2.2. Economic Impact

The economic costs of FHB and mycotoxins are very difficult to accurately determine, as so many factors are involved and all of the indirect effects are almost impossible to take into account. In general, economic losses in wheat due to FHB can be attributed to i) yield loss due to fungal disease and to ii) reduced technological and nutritional quality of wheat grain. Invasion of kernels by *Fusarium* spp. destroys the starch granules, storage proteins and cell wall, thus affecting grain quality. The embryo is usually not infected except in heavily invaded kernels. However, slightly infected kernels with apparently uninfected embryos exhibit reduced germination and vigour (Bechtel et al., 1985; Snijders, 1990). Economic losses due to DON contamination can be attributed to i) reduced crop value due to mycotoxin contamination, ii) losses in animal productivity due to mycotoxin-related contamination, iii) human health costs. Moreover, prevention, monitoring, sampling, chemical analysis, litigation, mitigation, and research costs also need to be taken into account.

Losses in quality lead to additional costs related to marketing, processing and exports. For instance, the milling industries purchase various classes of wheat for specific milling or baking purposes; therefore, high-quality grain is necessary for high prices, and grain that does not meet the minimum standards is docked to a lower grade and the producer receives a lower price (Windels, 2000). The differentiation of wheat grain quality has three main effects: 1) additional costs related to the separation of grain consignments into different lots; 2) the adaptation of all of the productive process in different production lines based on high/low quality products and their destination (industry, foods, feeds, etc.); and 3) price effects due to a reduction in high-quality grain supply.

In principle, reductions in wheat grain supply due to yield loss or to quality classification tend to increase future prices, but the beneficiaries of high price premiums for milling-quality grains are only the producers who have high-quality wheat, while a larger share of production may be discounted because of poor quality (Johnson et al., 1998).

Therefore, the economic impact of FHB and DON contamination is shared by all of the subjects in the productive chain up to the consumer: crop producers, animal producers, grain handlers, distributors, processors, consumers and society in general, because all of these costs contribute to elevate the final price (Charmley et al., 1994). Thus, the economic impact of FHB and DON is really difficult to quantify as so many factors should be taken into account, but some models have been proposed to estimate FHB and DON costs within limited contexts.

Wheat and barley losses caused by scab epidemics in the United States during the 1990s were estimated at close to 3 billion of US dollars (US$). American wheat farmers were estimated to have lost over 13.6 billion tons, evaluated at about 2.5 billion US$. Economic losses for wheat producers in Canada in the same period were estimated as 220 million US$

in Quebec and Ontario and 300 million US$ in Manitoba (Windels, 2000). In 1992, the average wheat yields in northeastern North Dakota and northwestern Minnesota were about 3.36 t/ha. In 1993, the yields lowered respectively to 1.68 t/ha and to 2.01 t/ha and from 1994 through 1998, the yields fluctuated but always remained between 1.75 and 2.29 (Windels, 2000).

In another simulation conducted for the same period (1991–1997), Johnson et al. (1998) measured the cumulative economic losses suffered by wheat producers in nine US states concerning three wheat classes (soft red winter, hard red spring, and durum), taking into account the loss of production (bushels) and changes in prices (US$/bushels). According to the analysis, during the 1991-1997 period, wheat producers in the affected regions suffered cumulative losses of 1.300 million US$ (61.8% hard red spring, 32.6% soft red winter, and 5.6% durum). Nganje et al. (2001) continued the analysis started by Johnson et al. for the 1998-2000 period, simulating the primary (production and prices) and secondary impact of losses incurred in other sectors of the economy which result from subsequent rounds of spending and re-spending within an economy. The direct combined effect of price discounts and yield reduction from FHB were estimated at 733 US$ million, while the combined direct and secondary losses were estimated at 2.300 US$ million. Income losses resulted in a reduction in farm numbers: about 2000 farms were lost in North Dakota during the 1992-1996 period, versus 500 during the previous four years. The economic cost of DON in wheat and corn was estimated in a simulation conducted by CAST (2003). In this simulation, the direct costs for yields that did not respect U.S. food and feed crop advisory limits, the direct costs for potential livestock losses, the costs of efforts to mitigate the contamination (dry storage, physical separation during milling, and testing) and human health hazard were taken into account. The simulation generated mean annual costs for DON of 637 million US$ in crop losses (wheat and corn) and 18 million US$ in feed losses. The mean simulated annual cost of livestock losses was about 2 million US$.

These simulation give an idea of the magnitude of the economic effects of FHB infection and DON contamination. They refer to contamination in some U.S. regions, but infections and contaminations have been reported in a wealth of literature from all over the world where wheat is cultivated (Table 1). Nevertheless, in order to reduce the impact of the exclusion of the food and feed industries for the DON contaminated wheat stocks, other industrial uses could be considered. Mesterházy (2004) indicated three possibilities that have greater chance: a) direct energy production, b) usage for gasoline production, c) alcohol production for industrial use or as fuel. In this way, the utilization of contaminated grain could be a feasible solution for the farmers who cannot sell their contaminated grain to the feed and food markets.

3. Ecology and Epidemiology of Fusarium Head Blight in Wheat and Synthesis of Deoxynivalenol

In order to understand how Fusarium head blight invades wheat grain and synthesizes DON, it is important to consider the *Fusarium* spp. life cycle on small grain cereals

Many papers have dealt with the epidemiology of Fusarium head blight, or more specifically of *F. graminearum*, which has always been considered the most important FHB pathogen.

Table 1. Natural occurrence of FHB and DON contamination of wheat grain in different part of the world in recent years (from 1984 to 2007)

Continent	State	Period	FHB, species and notes	DON contamination (µg/kg) and notes	Reference
Africa	Kenya	2004	*F. graminearum* the most virulent species. Prevalent species: *Fusarium poae, F. chlamydosporum* and *F. oxysporum*	Range: 110–290 µg/kg	Muthomi et al. (2008)
	Morocco	2001–2005		17 samples: 7 samples contaminated. Mean of positives= 65.9 µg/kg Max=128 µg/kg	Hajjaji et al. (2006)
	Tunisia	Not specified Recent years	Presence of *F. culmorum* and *F. pseudograminearum*	All isolated strains were DON producers	Kammoun et al. (2008)
America (North)	Canada	1988–1991	Prevalent species: *F. graminearum*. Others: *F. culmorum, F. acuminatum, F. avenaceum, F. equiseti, F. poae, F. sporotrichioides*		Wong et al. (1992)
	Kansas	1993	29 out of 30 samples infected by *F. graminearum*	26 out of 116 samples contaminated	Trigostockli et al. (1995)
	Michigan	2004	FHB survey: average number of infected heads=77.6%; range=12%-100%	Range: 1000–20,000 µg/kg Average: 8300 µg/kg	Hart et al. (2004)
	Minnesota	1984–1986	Prevalent species: *F. graminearum, F. poae* Others: *F. equiseti, F. acuminatum, F. sporotrichiodes, F. semitectum*		Wilcoxson et al. (1988)

Continent	State	Period	FHB, species and notes	DON contamination (µg/kg) and notes	Reference
America (South)	Argentina	2 years (years not specified)	All samples positive to *F. graminearum*.	Range: 0–8440 µg/kg	Lori et al. (2003)
	Argentina	1993	*F. graminearum* prevalent species.	Range: 300–4500 µg/kg	Dalcero et al. (1997)
	Brasil	2005		National (94% contaminated, mean 332 µg/kg) and imported (Paraguay and Argentina, 88% contaminated, mean 92 µg/kg) stored wheat samples.	Calori-Dominguesi et al. (2007)
	Uruguay	2001–2002	Prevalent species: *F. graminearum* Others: *F. avenaceum, F. culmorum, F. poae, F. equiseti*		Pereira et al. (2004)
Asia	China	1989		30 samples (47% positive) Range: 7–309 µg/kg	Luo et al. (1990)
	Japan		Prevalent species: *F. graminearum, M. nivale* Others: *F. avenaceum, F. tricinctum*		Parry et al. (1995)
	Nepal	1997	Prevalent species: *F. graminearum* and *F. equiseti* Others: *G. fujikuroi MP-D, F. oxysporum, F. avenaceum F. semitectum, F. torulosum, F. acuminatum, F. chlamydosporum*.		Desjardins et al. (2000)
	Russia	1989–2002		Range: 0–8600 µg/kg (2166 samples)	Tutelyan (2004)

Table 1. (Continued)

Continent	State	Period	FHB, species and notes	DON contamination (μg/kg) and notes	Reference
	Russia	Not Specified Recent years	Widespread and dominant species: *F. poae*, *F. sporotrichioides*, *F. avenaceum*. Widespread but infrequent species: *F. tricinctum*, *F. equiseti*, *F. semitectum*. Prevalent species distribution by region: *F. graminearum* (Central and Northwestern regions), *F. acuminatum* (Central Siberia, Far East), *F. culmorum* (North-western, Central, Ural), *F. cerealis* (South, Far East), *F. sambucinum* and *G. fujikuroi* complex (South, Far East, Central)		Gagkaeva and Gavrilova (2008)
Europe	Bulgaria	1995	140 samples *Fusarium* spp and *Alterrnaria* spp	Mean=180 μg/kg Max=1800 μg/kg	Vrabcheva et al. (1996)
	Croatia	2001–2003	Prevalent species: *F. graminearum* Other: *F. culmorum*		Krstanovic et al. (2005)
	England	1991	Prevalent species: *F. avenaceum, F. culmorum*. Others: *F. lateritirum, F. poae*		Nicholson et al. (1993)
	France	2000–2002	Prevalent species: *F.graminearum, F. avenaceum* and *F. poae*. Others: *M. nivale, F. culmorum*. 749 samples		Ioos et al. (2004)

Continent	State	Period	FHB, species and notes	DON contamination (µg/kg) and notes	Reference
	Germany	2000–2001		Mean=24 µg/kg, Max=309 µg/kg (41 samples, Incidence 95%)	Schollenberger et al.(2006)
	Germany	1989–1993		Mean for each year from 167 to 735 µg/kg	Muller et al.(2001)
	Hungary	1991–1998		367 samples Range:70-1560 µg/kg	Rafai et al. (2000)
	Italy	2004–2006	Prevalent species: *F. graminearum* and *F. culmorum*.	Range: 25-2660 µg/kg	Maiorano et al. (2008)
	Italy	2000		Durum wheat: max=1000 µg/kg Soft wheat: max=330 µg/kg	Logrieco et al. (2003)
	Latvia	2006-2007	Prevalent species: *F. poae*, *F. culmorum*, *F. gibbosum*, *F. avenaceum* var. *herbarum* Other species: *F. oxysporum* var. *orthoceras*, *F. sambucinum*, *F. sporotrichioides*, *F. moniliforme*, *F. semitectum*	2006: 64 samples, 4.7% contaminated, range 28-36 µg/kg 2007: 81 samples, 27.2 contaminated, range 1.0-615.0 µg/kg	Treikale et al.(2008)
	Lithuania	1998		84 samples. Low contamination	Keblys et al.(2001)
	Norway	1994–1996	695 samples (wheat, barley and oats) Prevalent species: *F. avenaceum, F. poae, F. tricinctum* and *F. culmorum* Others: *F. graminearum, F. sporotrichioides*		Kosiak et al.(2003)
	Norway	1988–1996		About 5000 samples (wheat, barley, oats). 61.5% of wheat samples positive	Langseth and Elen (1997)

Table 1. (Continued)

Continent	State	Period	FHB, species and notes	DON contamination (µg/kg) and notes	Reference
	Poland	1994–1995	17% of kernels with visible symptoms of scab (*F. graminearum*)	Mean=9628 µg/kg	Perkowski et al.(1997)
	Slovakia	2004-2006		139 samples, 9.3% exceeding EU food limit (1250 µg/kg)	Slikova et al.(2008)
	Spain	2003	*F. graminearum, F. culmorum, F. equiseti*		Jurado et al. (2004)
	The Czech Republic	1999–2001		Almost all samples contaminated. Only two samples exceed the EU food limit (1250 µg/kg)	Hajslova et al. (2007)
	The Netherlands	1974–1986	Prevalent species: *M. nivale*, *F. culmorum*. Other species: *F. avenaceum*, *F. graminearum*.		Daamen et al. (1991)
	The Netherlands	1991 and 1993	Prevalent species in 1991: *Fusarium culmorum*, *Fusarium avenaceum* Prevalent species in 1993: *Fusarium poae*, *Fusarium culmorum*, *Fusarium crookwellense*		De Nijs et al. (1996)
	The Netherlands	2001–2002	Prevalent species: *F. graminearum*, in contrast with Daamen et al., (1991) and De Nijs et al. (1996)		Waalwijk et al. (2003)
	The United Kingdom	2001–2005		954 samples 2001: mean=90 µg/kg, Max=5175 µg/kg 2002:mean=116 µg/kg, Max=3065 µg/kg	Edwards (2004b)

Continent	State	Period	FHB, species and notes	DON contamination (µg/kg) and notes	Reference
				2003:mean=218 µg/kg, Max=10626 µg/kg	
Oceania	New Zealand	1999–2000	Prevalent species: *F. graminearum* Other species: *F. avenaceum*, *F. culmorum*, *F. poae* FHB incidence (%): 0.0–2.4 in 1999, 0.0–3.5 in 2000	Range: 0.00-260 µg/kg in 1999, 60-3820 µg/kg in 2000	Cromey et al.(2002)
	Australia (New South Wales)	2000	Outbreak of Fusarium head blight caused by higher rainfall than average occurred during anthesis in November. *F. graminearum*, *F. culmorum*, *F. cerealis* (= *F. crookwellense*), *F. pseudograminearum* and *F. avenaceum*.		Tan et al. (2004)

Atanasoff, in 1920, first described the development of FHB, and in particular of *Gibberella saubinetii* (=*Gibberella zeae*). The first attempt to describe the complete life cycle of *F. graminearum* on cereals was, however, made by Sutton (1982) who described the disease cycle of the fungus in relation to the pathosystem formed by maize, wheat and the fungus itself in which the fungus survives, spreads, develops and invades the cereal tissues, producing mycotoxin in grain. The study by Sutton, was followed by others (Snijders, 1990; Parry et al., 1995; Gilchrist and Dubin, 2002; Logrieco et al., 2003; Gilbert and Fernando, 2004; Xu and Berrie, 2005; Wagacha and Muthomi, 2007) which described the epidemiology of head blight Fusaria, and other which updated the description of the pathosystem with new knowledge about (i) the influence of the environmental and climatic conditions on the fungi development (Doohan et al., 2003; Xu, 2003), (ii) the competitive relationships between different *Fusarium* spp (Xu et al., 2007), (iii) the inoculum dispersal dynamics (Horberg, 2002; Rossi et al., 2002a; Paul et al., 2004; Maldonado-Ramirez et al., 2005), and so on. One of the most recent descriptions of the epidemiology of FHB was given by Osborne and Stein in 2007.

A brief description of the epidemiology of FHB in wheat is given in this section, through a review of the main literature.

3.1. Inoculum Sources and Inoculum Structures

The principal *Fusarium* spp. sources of inoculum are crop debris like maize stalks, grain, ear pieces and cob, and the stubble and straw of barley, wheat, and other cereals. Soil itself and contaminated seeds can be also sources of inoculum, but their role is marginal compared to the former sources. The fungi are present and survive in colonized crop residues as mycelium, and may develop saprophytically on residues on the soil surface for 2 or 3 years (Sutton, 1982; Osborne and Stein, 2007). Moreover, *Fusarium* spp can survive parasitically and saprohytically on leaves throughout the wheat growing season, which means that these leaves may contribute additional inoculum to the FHB development (Ali and Francl, 2001). In some locations, the leaves have also shown large macroconidial concentrations, suggesting that the fungus can also grow epiphytically, resulting in a higher inoculum level within the canopy (Gilbert and Fernando, 2004). In addition, noncereal residues such as soybeans, sunflower, pasture, and many gramineous weed species have been reported to be important sources of inoculum for *Fusarium* spp. (Jenkinson and Parry, 1994; Inch and Gilbert, 2003; Pereyra and Dill-Macky, 2008).

In general, the inoculum structures are spores (ascospores, macroconidia, chlamidospores) and fungal mycelia (Sutton, 1982), but macroconidia and ascospores have been reported to be the most important types of inoculum and both of them are abundantly produced on the infected residues (Khonga and Sutton, 1988; Fernando et al., 2000). *F. graminearum* has an additional epidemiological advantage because it regularly and abundantly forms perithecia of its sexual stage (*Gibberella zeae*), resulting in the production of ascospores on the colonized residues (Xu and Berrie, 2005).

3.2. Inoculum Production, Dispersal and Deposition

The production of conidia and ascospores is critically influenced by temperature, the water activity (a_w) of the host, rainfall, relative humidity and light (UV light required). Sutton (1982) reported 29°C and 25-28°C as the optimum temperature for perithecia and ascospore production, respectively. Moisture is required for formation of both macroconidia and ascospores: Xu (2003) reported that when the soil moisture content is below 30%, ascospore production is not possible. When it is greater than 80%, ascospore production is at its maximum. Macroconidia production is also dependent on the temperature o a great extend: Rossi et al. (2002b) reported that they can be produced in the 5-35°C range while the optimum temperature for *M. nivale* is 26°C, 28°C for *F. avenaceum* and 32°C for *F. graminearum* and *F. culmorum*. Macroconidia are formed in mucilage in sporodochia and rain and wetting are necessary for their liberation.

The dispersal of both macroconidia and ascospores is associated with rainfall events and high relative humidity (Paulitz, 1996; Horberg, 2002). In a study conducted in Italy, very few macroconidia were sampled in the air before rainfall, but their number progressively increased with the beginning of a rain event. In the presence of high humidity, conidia continued to be sampled at high densities for some hours after rain had ceased and they usually reached their peak under these conditions. Finally, the density of the airborne conidia rapidly decreased when the relative humidity dropped (Rossi et al., 2002a).

Turgor pressure represents the driving force behind ascospore discharge. It has been suggested that the discharge of ascospores results from the buildup of turgor pressure generated by ion fluxes and the accumulation of mannitole, which is the main simple sugar component in the fluid (the epiplasm) with which the ascospores are discharged from asci from within the ascus (Trail et al., 2002). Ascospores are discharged to distances from <1 mm to nearly 10 mm with an average between 4.6 mm to 3.9 mm. This distance is sufficient for the ascospores to surpass the laminar boundary layer of air over which they become airborne and can encounter turbulent air currents which help them ascend into the atmosphere. This mainly happens during daylight hours, when the laminar boundary layer is considered to be only millimetres thick, while during the night, it can extend to nearly 10 m above ground level. Therefore, ascospores of *G. zeae* discharged during the daylight hours have the greatest probability of becoming airborne in turbulent air currents (Schmale et al., 2005). Once ascospores become airborne, they are subjected to air turbulence: during the day, energy from the sun heats the earth's surface and there is an upward transfer of heat toward the cooler atmosphere. Ascospores can be transported in the atmosphere to significant vertical and horizontal distances over the surface of the earth by turbulent mixing masses of air. At night, the surface of the earth cools rapidly in comparison to the atmosphere and there is a downward transfer of heat, which means that at night, as turbulent mixing slows down, the ascospores may settle out of the atmosphere over the surface of the earth (Maldonado-Ramirez et al., 2005). This could explain the apparent contradiction with the interpretation given by Paulitz (1996) concerning the results of his experiments in which higher atmospheric concentrations of *G. zeae* were reported during the night hours: the higher concentration of *G. zeae* at night might be the effect of spore deposition and not of spore release as explained by the author. On the contrary, the low number of spores detected during the daytime might be due to the presence of mixed or turbulent layers. Thus, spore deposition may be separated from spore release in both time and space (Schmale et al., 2006).

3.3. Germination, Head Infection, Growth

Warm, humid weather, frequent rainfall and heavy dew favour spore germination. Pinkish-red mycelium and conidia develop abundantly in infected spikelets and the infection can spread to adjacent spikelets or through the entire head. Infected kernels become shrivelled and discoloured with a white, pink, or light-brown scaly appearance, as a result of mycelial outgrowths from the pericarp (Logrieco et al., 2003).

Temperature, relative humidity and water activity of the host are the fundamental factors that affect fungal germination and growth. In general, *F. graminearum* mycelium grows well over a wide range of temperatures, up to 30°C, and is associated with the warmer regions of the world, whereas *F. poae* is found more frequently in temperate climates. Most of the species can be found in much of the geographical area affected by FHB, but individual species usually dominate a specific region and *F. graminearum* dominates in a majority of regions (Osborne and Stein, 2007).

Rossi et al. (2001), carrying out *in vitro* experiments, studied the effect of relative humidity (RH) and temperature on the infection by *F. graminearum*, *F. culmorum*, *F. avenaceum* and *M. nivale*. They reported that when spikes were incubated at different RH the

Table 2. Minimum (Tmin), optimum (Topt), and maximum (Tmax) temperatures for the germination of four different fungal species that cause Fusarium head blight in wheat

Species	Tmin	Topt	Tmax
F. avenaceum	14	20–28	35
F. culmorum	16	20–26	33
F. graminearum	5–10	24–29	35
F. poae		20–25	
M. nivale	10	15–20	28

(Rossi *et al.*, 2001; Brennan *et al.*, 2003; Hope *et al.*, 2005; Ramirez *et al.*, 2006).

frequency of the infected glumes was very high when wetted at 100% RH and very low at all the other RH tested (95%, 85%, 75% and 65%) for all the species except for *F. culmorum* whose frequency only started to significantly increase at 75% RH. As far as temperature is concerned, they found a minimum temperature between 10.0 and 16.5°C, an optimum one between 18 and 29°C and a maximum one between 28.5 and 35.5°C (Table 4). Brennan et al. (2003) reported lower optimum temperatures for the same fungal species (differences from 2 to 8°C in the case of *F. nivale*). Ramirez et al. (2006) reported similar results concerning the optimum and maximum temperature for *F. graminearum* but a lower minimum temperature (5°C). These difference could be related to the different growth substrate that was used. Ramirez et al. (2006) also studied the effect of a_w and reported an optimum level for *F. graminearum* growth at the highest tested level: 0.995 a_w. No growth was observed at 0.900 a_w.

Once the spores have germinated, the propagation of the hyphae is more favoured in the flower parts than in the other organs. The fungus can develop both inter- and intracellularly and cause severe damages. In 2001, Bushnell provided a summary on of the way that head blight Fusaria can infect the spikelets. The stomates can be entered by *F. graminearum* (Kang and Buchenauer, 2000a; Pritsch et al., 2000). The glume, palea, lemma and the awn can all be colonized by *Fusarium* mycelia. Hyphae can enter into the floret mouth opening where it can colonize caught and retained anthers or the apical brush of the caryopsis. Another potential pathway of entry is the crevice between the palea and lemma, especially near the floret base. Whether entering from the mouth or crevice, the fungus can develop abundantly on and within interior tissues of the lemma and palea. Surface tissues of the ovary and lodicules are especially susceptible. However, the mode of penetration of *F. graminearum* into these interior tissues has not been determined yet. *F. graminearum* can grow within tissues between cells instead of entering them, establishing a biotrophic relationship with the host tissues. How and where the intercellular fungus penetrates the cells for its subsequent growth within the cells is still not known. Once established within the floret, the fungus colonizes and follows vascular tissues in the floret stalk through the rachilla or rachis into other florets (Kang and Buchenauer, 2000a; Pritsch et al., 2000; Bushnell, 2001; Wagacha and Muthomi, 2007).

Stress-related genes are known to be activated in both resistant and susceptible cultivars in response to *F. graminearum* invasion. The genes activation is observed in both the invaded and in the adjacent uncolonized spikelet, which demonstrates that this response is expressed systemically. However, the molecular and physiological responses of heads to invasion by

Fusarium head blight have not been fully investigated (Pritsch et al., 2000; Pritsch et al., 2001; Kong et al., 2005).

On the other hand, Fusarium head blight may produce cell wall degrading enzymes such as cellulases, xylanases and pectinases during infection and colonization of wheat spike tissues (Kang and Buchenauer, 2000b).

3.4. Deoxynivalenol Biosynthesis and its Role in the Pathogenesis

Deoxynivalenol (DON, vomitoxin) is a type B trichothecene, an epoxy-sesquiter-penoid (Figure 1). DON is probably the most frequently encountered mycotoxin in the FHB of wheat throughout the world (Parry et al., 1995; FAO, 2003).

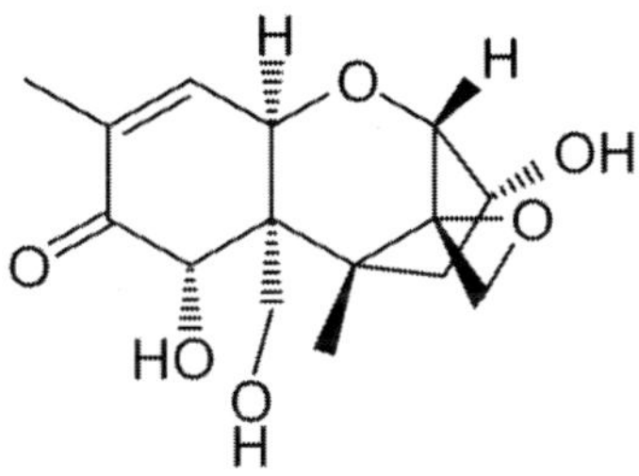

Figure 1. Chemical structure of deoxynivalenol.

Its occurrence is associated above all with *F. graminearum* and *F. culmorum*. DON is associated with serious mycotoxicosis in humans and animals. It has cytotoxic activity (protein synthesis inhibition, effects on DNA and RNA synthesis, mitochondrial function inhibition, effects on cellular membranes and on correct cellular division and apoptosis) and an immunosuppressor effect which reduce the resistance to microbial infections (Rotter et al., 1996; Minervini et al., 2004).

According to the results of in vitro experiments with *F. graminearum*, DON is synthesized in a range of temperatures between 5°C and over 30°C, with an optimum production between 25°C and 30°C, depending on the a_w. The optimum a_w for DON production has been reported to be near 1 (0.995) and the minimum requirement is between 0.900-0.930 (Hope et al., 2005; Ramirez et al., 2006). In field conditions, high FHB incidence and DON contamination are not always directly correlated and contrasting cases have been reported (Hermann et al., 1998; Höhn et al., 2002; Menniti et al., 2003; Miedaner et al., 2003), suggesting that other complex factors such as year, site, cultivar and agronomic technique are very important in DON contamination. However, DON contamination has been reported to depend to a greater extend on rainfalls around heading and anthesis and on temperatures between 10°C and 32°C around the same period (Hooker et al., 2002; Koch et al., 2006).

DON seems to play a decisive role on the aggressiveness of *F. graminearum* and *F. culmorum*. DON is a strong protein synthesis inhibitor and this may cause the inhibition of defence response genes in susceptible host-cultivars, leading to a rapid increase of the disease (Mesterhazy, 2002). Moreover, DON can induce the complete loss of chloroplast pigments at a sub-lethal concentration (Rotter et al., 1996). It has been suggested that in the presence of

DON, plant defence mechanisms are not triggered fast enough, thereby leading to an increased aggressiveness of the pathogen (Wagacha and Muthomi, 2007).

4. Crop Management Systems to Reduce the Occurrence of FHB and DON Contamination in Wheat

The accumulation of DON in the grain is intimately related to the development of the disease in the field (McMullen et al., 1997). Therefore, the current strategy to control DON in cereals is to manage FHB. Most practices aimed at FHB prevention are essentially crop management practices whose goal is to reduce infection or fungal growth by toxigenic fungi (Parry et al., 1995). Among the agricultural practices, FHB development and/or DON contamination in wheat grains are mainly affected by tillage, crop rotation, varieties and fungicide application (Pirgozliev et al., 2003). None of these strategies on their own are able to significantly reduce the impact of this disease (Edwards, 2004a).

Although information is available on the basic effect of single agricultural practices on *Fusarium* infection and DON contamination in wheat, only a few studies have been made to quantify the relative importance of each of these factors compared to the others that are involved and to verify their interactions and combined effects.

Preliminary information on the relative effect of management options on FHB incidence and DON contamination could be obtained through a simplified approach to calculate the severity of the effect of the single factors, as proposed by Koch et al. (2006). In this study, the severity of the effect of the single agricultural practices was calculated as follows: the mean DON value of the treatment with the highest DON concentration is divided by the mean value of the lowest treatment.

The data obtained in the experiments conducted by Koch et al. (2006) are reported and compared in table 5 with other information obtained from the literature and with information from similar trials conducted in 2005–2007 in 4 sites in northern and central Italy. In all these studies, the effect of at least two agricultural practices on DON contamination were compared, in natural inoculated conditions.

Based on these data, the following ranking order can be obtained: susceptibility of wheat cultivar (1.6–5.6) = the preceding crop (2.0–4.3) > soil tillage (1.4–6.5) ≥ fungicide application at anthesis of wheat (1.7–2.7). These data confirm that wheat crop quality management, aiming at a low DON concentration, must start in the field and first focus on agronomic factors that influence FHB infection. Above all, conditions such as preceding host crops, especially maize, which can leave infected residues in the field and the cultivation of a susceptible cultivar contribute to increase *Fusarium* infections of wheat crops. These data confirm the results obtained in France in 7 years of trials by researchers from Arvalis (Barrier-Guillot et al., 2006; Gourdain, 2008), who obtained a grid with 7 levels of risk for DON contamination in wheat grains based on 3 combined risk factors: previous crop, tillage and varietal sensibility to FHB.

4.1. Crop Residue and Proceeding Crops

It is generally accepted that if a wheat crop follows an alternative host for *Fusarium* pathogens it is at a greater risk to FHB and DON contamination of grain. All species of

Fusarium are able to survive through the winter as mycelia in non-decomposed residues from the previous crop, with the subsequent production of ascospores in the perithecia located on the surface of the pathogen-infested residues (Osborne and Stein, 2007). The primary reservoir of inoculum is debris from the previous crop (Bai and Shaner, 1994) and pathogens survive longer on residues that do not degrade easily, such as stem nodes (Sutton, 1982). Thus, FHB disease and DON contamination are more severe if the preceding crops are maize or sorghum, rather than wheat or barley and even less contamination is observed following other crops (Dill-Macky and Jones, 2000; Krebs et al., 2000; Champeil et al., 2004b; Maiorano et al., 2008).

This may be accounted for in two ways: first, it is related to the density of the residues left by the preceding crop, second, the nature of the preceding crop can affect the composition of the pathogen complex throughout the following year.

As clearly reported in the literature, limited soil tillage or no-till increase the frequency of head blight, whereas deep tillage, such us ploughing, decreases it (Teich, 1989; McMullen et al., 1997; Miller et al., 1998; Krebs et al., 2000; Maiorano et al., 2008), because it reduces the density of the residues on the surface of the soil and helps to decrease the production of inoculum (Teich and Hamilton, 1985) and the quantity of spores available for dispersal (Bateman et al., 1998).

The general effect of tillage practices that enhance *Fusarium* infection is highly variable: in experiments conducted in Italy, direct sowing after maize increased the DON concentration in wheat grains by a factor of 6.5 on average in all conditions. This effect was much smaller when direct sowing was applied after wheat or other crops (Teich and Hamilton, 1985; Dill-Macky and Jones, 2000). In general, the data presented clearly show that the tillage system is less important than the influence of the preceding crop and the susceptibility of the wheat cultivar. Furthermore, the relative importance of tillage practices for a low DON concentration in wheat grain is closely related to the level of FHB pressure: under highly infectious conditions, for example with grain maize as the previous crop, a highly susceptible cultivar and abundant rainfall during anthesis, minimum tillage or no-tillage could result in a considerably increased DON concentration.

4.2. Cultivar Susceptibility

Among the other factors examined in the several studies, the cultivation of a susceptible cultivar showed a clearly higher DON contamination compared to a moderately resistant one. With the exception of the experiment conducted in North Italy in 2002-2003 (Blandino et al., 2006), in all the other works, the use of a more susceptible variety increased the content of this mycotoxin in wheat grain by a factor of 4.6. These results, in each situation, reported a comparison between a moderately resistant and a susceptible variety. It is assumed that if wheat cultivars with an intermediate *Fusarium* rating instead of a moderately resistant and susceptible or highly susceptible rating are tested, a considerable decrease of the varietal effect could be expected. Furthermore, all the wheat cultivars used in these experiments were commercial and commonly present on the farm fields. This means that the choice of the less susceptible cultivars available on the markets is a key tool that both the farmer and end user could apply to reduce the risk of high DON content.

Table 3. Severity of the effect of several agricultural practices on deoxynivalenol (DON) contamination in winter wheat grain on average of the other experimental factors included and with no-inoculated conditions

Country and year	Effect	Value in			Reference
		Numerator	Denominator	Severity of effect	
USA (Minnesota); 1995-1997	Preceding crop	Maize	Soybean	2.0	Dill-Macky and Jones, 2000
	Tillage	Direct sowing	Ploughing (30 cm)	1.4	
Canada (Ontario); 1996-1999	Preceding crop	Maize	Soybean	2.9	Schaafsma et al., 2001
	Tillage(*)	Minimum tillage	Ploughing (30 cm)	1.6	
	Cultivar	Highly susceptible	Moderately resistant	3.7	
Germany; 2001-2002 (Exp. 1)	Preceding crop	Winter wheat	Sugar beet	4.3	Koch et al., 2006
	Tillage(*)	Direct sowing	Ploughing (30 cm)	2.7	
	Cultivar	Highly susceptible	Moderately resistant	4.3	
Germany; 2001-2002 (Exp. 2)	Tillage (after wheat)	Direct drilling	Ploughing (30 cm)	3.4	Koch et al., 2006
	Cultivar	Highly susceptible	Moderately resistant	5.6	
	Fungicide application	Without	With	2.1	
Italy; 2002-2003	Cultivar	Highly susceptible	Moderately resistant	1.6	Blandino et al., 2006
	Fungicide application	Without	With	1.7	
Italy; 2006-2007	Tillage (after maize)	Direct sowing	Ploughing (30 cm)	6.5	Blandino, unpublished
	Cultivar	Highly susceptible	Moderately resistant	4.8	
	Fungicide application	Without	With	2.7	

* Data obtained from commercial wheet farm fields with different previous crops.

Table 4. Effect of combined agricultural practices on grain yield, FHB incidence and severity and DON contamination

Year	Treatment		Soil tillage[V]	Cultivar[W]	Fungicide treatment[X]	FHB incidence (%)[Y]	FHB severity (%)[Z]	Yield ($t\ ha^{-1}$)	DON ($\mu g\ kg^{-1}$)
2006	HR	high risk	Direct sowing	HS	untreated	39.3 a	10.2 a	6.7 c	1594 a
	R	risk	Direct sowing	HS	prochloraz at heading	18.9 c	4.9 b	7.0 bc	598 b
	MR	medium risk	Ploughing	HS	untreated	26.8 b	4.8 b	6.8 c	161 c
	RI	right	Ploughing	MR	untreated	14.3 c	1.4 c	7.3 ab	44 c
	CA	careful	Ploughing	MR	prochloraz at heading	6.1 d	0.2 d	7.5 a	13 c
2007	HR	high risk	Direct sowing	HS	untreated	68.3 a	12.3 a	4.7 c	531 a
	R	risk	Direct sowing	HS	prochloraz at heading	47.8 b	7.5 b	5.4 b	221 b
	MR	medium risk	Ploughing	HS	untreated	65.9 a	8.1 b	5.8 b	156 b
	RI	right	Ploughing	MR	untreated	49.2 b	4.3 c	6.5 a	38 c
	CA	careful	Ploughing	MR	prochloraz at heading	34.5 c	2.0 d	6.6 a	13 c

V: The previous crop was maize in each year and site.

W: HS: highly susceptible; MR: moderately resistant.

X: triazole fungicide, A.I. prochloraz (Sportak 45 EW®, BASF) was applied once at 0.59 kg (A.I.) at heading.

Y: FHB incidence was calculated as the percentage of ears with symptoms of disease at the soft dough stages (GS 85).

Z: FHB severity was calculated as the percentage of kernels per ear with symptoms of disease at the soft dough stages (GS 85).

Values, in the same column, followed by the same letter are not significantly different, according to the Student-Newman-Keuls test. The data reported in the table are the means of 4 sites in northern and central Italy.

At present, no durable, fully FHB-resistant wheat cultivars exist, therefore control relies on the use of cultivars with partial resistance (Mesterhazy, 2002), but in recent studies by Mesterhazy et al. (2005) and Tóth et al. (2008), the wheat varieties that are more resistant to FHB were shown to reduce DON production to almost zero.

The development of FHB resistant varieties is one of the main goals of numerous breeding programmes across the world (Yuen and Schoneweis, 2007).

Five types of varietal resistance are currently know: they affect the penetration of the fungus into the plant (type I) (Schroeder and Christensen, 1963), the infection kinetics, with resistance to colonisation within the ear (type II) (Schroeder and Christensen, 1963), the degradation of the mycotoxins (type III) (Miller and Young, 1985; Miller et al., 1985), the consequences of infection, with tolerance to high DON concentrations (type IV) (Wang and Miller, 1988) and the resistance to grain infection, with differences in yield despite similar levels of attack (type V) (Mesterházy, 1995).

Type I and II resistances have been associated with certain morphological characteristics of wheat cultivars. Hilton et al., (1999) observed a significant negative relationship between plant height and resistance to FHB. The earliest variety seems to accumulate more deoxynivalenol that the late-flowering varieties (Jenny et al., 2000), probably due to a greater coincidence of the phase of maximum susceptibility in plants with the period most favourable for spore dispersion.

Such observations have been supported by genetic mapping of the quantitative trait loci (QTL), where the QTLs of the plant morphological traits, such as plant height and heading date, coincide with the QTLs of lower FHB severity and DON concentration (Zhu et al., 1999; Ma et al., 2000).

Several research groups have identified molecular markers linked to QTLs associated with various aspects of resistance (Buerstmayr et al., 2002; Buerstmayr et al., 2003). The most common sources of genetic resistance are derived from the Chinese variety (Zhang et al., 2004; Yang et al., 2005; Yu et al., 2008).

As far as the most recent employment of genetic engineering is concerned, Anand et al. (2003) produced transgenic wheat lines, expressing chitinase and ß-1,3-glucanase genes, that provided moderate resistance to FHB in greenhouse trials. In addition, transgenic wheat expressing the *F. sporotrichioides* Tri101 gene, which encodes a protein that detoxifies the trichothecene mycotoxins produced by *F. graminearum*, conferred partial protection against the spread of FHB in infected spikes (Okubara et al., 2002).

Combining all the desired agronomic characters, plus resistance to other important disease and insects, with a high degree of resistance to FHB is still a major challenge. The successes in enhancing resistance to FHB and DON contamination in cereals using conventional breeding, molecular markers and through transgenic approaches have been thoroughly discussed in several reviews (Dahleen et al., 2001; Kolb et al., 2001; Ruckenbauer et al., 2001; Hollins et al., 2003; Miedaner et al., 2003; Bai and Shaner, 2004; Champeil et al., 2004a; Snijders, 2004).

4.3. Fungicide Application

Adapting the crop sequence, incorporating the previous crop residues and using a resistant cultivar are precautionary measures to control FHB infection that must be taken at the beginning of wheat cultivation. During crop growth, a fungicide application may be applied at anthesis to diminish DON formation (Pirgozliev et al., 2003).

Although several studies have demonstrated that good levels of control can be achieved with fungicides in *in vitro* experiments (Matthies et al., 1999; Ramirez et al., 2004; Mullenborn et al., 2008) or in trials in which wheat is artificially inoculated (Boyacoglu et al., 1992; Mielke and Weinert, 1996; Matthies and Buchenauer, 2000; Mesterhazy et al., 2003; Haidukowski et al., 2005), the situation under natural infection condition is less clear. In the field, there is conflicting evidence as to the ability of fungicides to reduce FHB and to reduce DON contamination, due to the complex interaction between fungicide, mycotoxin production, other fungal species and weather conditions (Milus and Parsons, 1994; D'Mello et al., 1998; Simpson et al., 2001).

The outcome of the use of fungicides seems to depend on the fungal species that is present and the effect that the particular fungicide has on these species. Fungicides containing triazoles (the most important are bromuconazole, epoxiconazole, metconazole, propioconazole, tebuconazole and tetraconazole), imidazoles (Prochloraz) or triazolinthiones (prothioconazole) active ingredients which inhibit the biosynthesis of ergosterol, have proved to be the most active molecules against FHB infection and DON contamination (Jennings et al., 2000; Edwards et al., 2001; Siranidou and Buchenauer, 2001; Menniti et al., 2003; Ioos et al., 2005).

Strobilurins have instead shown poor efficacy against FHB caused by toxigenic *Fusarium* spp. (Pirgozliev et al., 2003) and often both in vitro and in field studies have revealed an increase in DON concentration following an application of these fungicides compared to unsprayed controls (D'Mello et al., 2001; Simpson et al., 2001).

Magan et al. (2002) suggested that the fungicide action could influence the different fungal species by changing the ratio between the toxigenic and non toxigenic fungi. Different authors (Bertelsen et al., 1999; Jennings et al., 2000; Simpson et al., 2001) observed that the application of azoxystrobin, unlike triazoles, has little effect on the reduction of *F. culmorum*, *F. avenaceum* and *F. graminearum*, while it guarantees a significant reduction in *M. nivale*, which is not able to synthesise DON (Edwards, 2004a). As suggested by Pirgozliev et al. (2003), the increase in DON that was observed is therefore probably due to an increase in the infection of the *Fusarium* species, following the reduction in the presence of *M. nivale*. To assure lower mycotoxin contamination in wheat grain, the application of strobilurin fungicides is only recommended in a mixture with azoles (Chala et al., 2003; Pirgozliev et al., 2003). The effect on *Fusarium* also depends on the timing and frequency of applications (Homdork et al., 2000). The best control of FHB pathogens using fungicide has been observed following applications at wheat anthesis, whereas their efficacy declined rapidly with time after or before this growth stage (Wiersma and Motteberg, 2005).

In the studies reported in Table 5, the fungicide application controlled FHB infection and DON contamination in the regions more prone to the disease and in the wheat cultivars with low levels of resistance. Thus, fungicide may be useful when weather or other cultural conditions are particularly favourable for disease development and FHB disease pressure is high.

4.4. Effect of other Agricultural Practices

Mineral fertilization, weed control, sowing date and planting density seem to play a minor role on FHB infection and DON contamination in wheat compared to the previously reported agricultural practices (Champeil et al., 2004b).

Martin et al. (1991), Lemmens et al. (2004) and Heier et al. (2005) reported that FHB and DON contamination may be directly correlated to an increase in mineral N-input. However, the effect of mineral nutrition on head blight infection is still unclear: other works have reported that the level of the disease is similar with or without a nitrogen fertilizer application (Teich and Hamilton, 1985; Fauzi and Paulitz, 1994; Schaafsma et al., 1998) or, on the contrary, have shown that nitrogen application can limit the disease (Teich and Nelson, 1984). Teich (1987) observed a higher FHB infection in Ontario, Canada, with the application of ammonium nitrate than when urea was applied. In other studies the application of different types of nitrogen fertilizer did not lead to any significantly different results (Lemmens et al., 2004). Potassium, phosphorus and pH did not appear to affect the DON content to any great extent (Teich and Hamilton, 1985).

The isolation of *Fusarium* species, pathogenic to winter wheat from common grasses (Cassini and 1970; Lager and Wallenhammer, 2003) and broad-leaved weeds (Jenkinson and Parry, 1994), suggests that an effective weed control can reduce the inoculum. Moreover, weeds can also favour *Fusarium* infection by modifying the microclimate and increasing heat and humidity in the canopy.

Sowing date could have an indirect effect on the FHB infection. It in fact influences the moment in which the flower appears, a moment which can sometimes coincide or not with the *Fusarium* spore release (Lacey et al., 1999). Caron (1993) reported that a late sowing date favoured the development of head blight, whereas an early sowing date favoured the development of foot rot.

Severe infections of head blight have been reported with high planting densities (Mesterházy, 1995), which could increase the humidity of the canopy (Deadman and Cooke, 1997), favouring spore germination. On the other hand, Champeil et al. (2004b) suggested that an elevated density of the canopy could be a greater obstacle to spore dispersal.

Nevertheless, at the moment, the magnitude of the effects that these agricultural practises have on FHB disease development, if indeed there is one, remain unclear and each one of them needs to be evaluated alone and in combination with the other crop techniques. Although only a few studies have been conducted on the effects of agricultural practices, it is possible to say they are probably very dependent on the climatic and soil conditions of each region. Thus, it is difficult to extrapolate the results obtained for these preventive measures from each wheat growing area and to generalize them for application in other regions.

In the future, biological control could be used to offer an additional strategy and as part of an integrated management of FHB. In vitro assays, trials in glasshouses and under field conditions have shown that some bacteria included in the genera *Bacillus* and *Pseudomonas*, some yeasts belonging to the genera *Rhodotorula*, *Sporobolomyces* and *Cryptococcus* and some fungi of the genera *Trichoderma*, are able to reduce *F. graminearum* growth (Fernandez, 1992; Schisler et al., 2000; Schisler et al., 2002; da Luz et al., 2003; Khan et al., 2004). These potential biocontrol agents could be applied at wheat anthesis in order to prevent *Fusarium* infection, when conditions for the disease exist. In a recent work under greenhouse conditions, Palazzinia et al. (2007) observed a significant reduction in DON contamination in wheat grain inoculated with bacterial strains. Unfortunately, under field conditions, the biological control effects have been shown to be too variable (McMullen et al., 2002).

4.5. Effect of Crop Management Systems

In order to optimise the control of FHB and DON contamination of grain, it could be interesting to evaluate the effect of different crop management systems.

At present, there is a lack of information about the effects of entire cropping systems on FHB.

Full-factorial experiments that take into account tillage (tilled vs not tilled), wheat cultivar (susceptible vs moderately resistant) and fungicide application (triazoles at heading vs not treated) were conducted in naturally inoculated conditions in 4 sites in Italy during the 2006-2007 period.

The results were discussed taking into account some of the possible combinations of the tested agricultural practices, representing different crop management systems classified for the expected risk in FHB infection and DON contamination (Table 6), according to the following scheme:

1) HR: high risk agricultural practices, characterized by direct sowing of an untreated highly susceptible cultivar;
2) R: risk agricultural practices, with direct sowing of a highly susceptible cultivar, treated with fungicide at heading;
3) MR: medium-risk agricultural practices, with ploughing and cultivation of a highly susceptible, untreated cultivar;
4) RI: right agricultural practices, characterized by ploughing, a moderately resistant cultivar and no fungicide application;
5) CA: careful agricultural practices, in which a moderately resistant cultivar was sown after ploughing and was treated with fungicide at heading.

The crop management systems showed significant differences as far as yield, incidence and the severity of FHB, and DON contamination are concerned.

As expected, FHB severity decreased going from the higher risk crop management system to the more careful ones. Compared to the HR treatment, the reduction in the severity of FHB was 45, 43, 76 and 91% for R, MR, RI and CA, respectively. On the other hand, grain yield generally increased with the application of crop management systems that are less favourable to disease development.

The crop management system R reduced DON contamination by 60% compared to HR, while a higher reduction was observed for both the MR and RI protocols. Compared to RI, a further, not significant reduction, was observed with the fungicide application at heading (CA).

If the data of DON contamination presented by Koch et al. (2006) are extrapolated, and the agricultural practices in the same previously reported cropping systems are compared, similar results can be obtained. Compared to the HR treatment, the head blight severity reduction was 71, 80, 96 and 96% for R, MR, RI and CA, respectively.

These results underline that in years and environments with low FHB pressure the main factors that contribute to heavy *Fusarium* infections of wheat crops are the presence of residues of *Fusarium*-host crops on the soil surface and the cultivation of susceptible cultivars. Moreover, no further significant reductions were observed with fungicide applications at heading on a resistant wheat cultivar sown after ploughing.

Obst et al. (2000), in a four-year study in Germany, determined five FHB risk factors: (i) maize as previous crop, (ii) minimum tillage after maize, (iii) use of a moderate or highly susceptible wheat variety and (iv) application of strobilurin products. Again in this experiment, the combination of these risk factors showed synergic effects: a single risk factor showed a threefold relative risk of DON contamination, while four combined risk factors showed a fifty-six fold relative risk of DON contamination.

The data available at the moment concerning the comparison of different cropping systems underline that crop management strategies used to minimize DON contamination risk in wheat grains must first consider the effects of the preceding crop and cultivar. Soil tillage and fungicide application at anthesis could lead to a significant reduction in DON contamination in years and environments with higher FHB disease pressure. For this reason, these preventive measures should be taken into consideration in all the wheat growing areas most prone to FHB infection and DON contamination.

In the next few years the effects of all crop techniques on all mycotoxins must be combined to provide guidelines on Good Agricultural Practices (GAP), in order to minimise mycotoxin contamination of cereal products.

5. Predictive Models of Fusarium Head Blight and Deoxynivalenol Contamination in Wheat

A model is a simplified representation of a system that is a limited part of the reality which contains interrelated elements. The behaviour of the system and the relations between the elements are simulated by mathematical and statistical equations and/or logical operations (France and Thornley, 1984; de Wit, 1993; Jones and Luyten, 1998). The great development and diffusion of information technologies and computer programming, above all during the 1980s, have led to the possibility of elaborating a large amount of data and developing very complex models (Barrett and Nearing, 1998). Nevertheless, the level of complexity and detail needed for a specific model depends on the objectives and questions being asked of the model and on the amount of information (data) and time that is available for model building and testing. In agriculture, models are mainly used for three purposes: i) by researchers to examine scientific hypothesis or to study and better comprehend biological/-epidemiological/agricultural systems; ii) by governmental agencies as policy analysis tools, to assist in best management solutions for problem or susceptible areas; iii) by agronomic technical services or producers as part of computerized decision support systems to evaluate optimum management practices. The use of a model to predict the outcome of a disease is desirable to enhance and trigger management opportunities with the aim of reaching high technological and/or nutritional and/or productivity quality and/or safety of the production (Boote *et al.*, 1996; Schaafsma and Hooker, 2007; Maiorano, 2008; Maiorano *et al.*, 2009). Kranz and Royle (1978) classified the epidemiological models according to their main objective in i) descriptive (models that provide generalise experimental results), ii) predictive (models that allow prediction of epidemics), and iii) conceptual (models that allow the identification of the problem by distinguishing cause from effect). The descriptive models are easy to comprehend, often require fewer inputs, and often are easier to use and apply, but they have to be calibrated to each new site and year. On the other hand, the conceptual models are better able to model genotype X environment interaction, but their complexity makes them more difficult to understand, to use and to apply, and they also require more input information (Prandini et al., 2008).

The relevant economic impact (section 1) of FHB for the main cereal producing countries in the world has stimulated many researchers to develop epidemiological models in order to simulate FHB infection and DON contamination. Most of these models have been included in computerized decision support systems mainly used to help the operator in the filed to decide weather or not applying and the correct timing of application of fungicides against FHB.

A review of the models that have been developed for FHB and for DON synthesis is presented in this section. A similar review has recently been presented by Prandini et al. (2008).

5.1. Argentina

Moschini and Fortugno (1996) developed nine empirical equations, in the region of Pergamino, Argentina, associating mean head blight incidence data with temperature and moisture variables. The models were obtained with linear regression techniques by fitting a 12-year dataset (1978-1990). The two equations showing the best results were validated with data from Pergamino for the years 1991, 1992 and 1993 and subsequently also in the northern Argentinean Pampas Region (from 1993 to 1995) by Moschini et al. (2001). The meteorological variables (temperature, humidity, rainfall) and the time segment of observation (beginning eight days prior to the heading date and finishing after the accumulation of 530 degree days) where chosen for their agreement with the fungal environmental requirements. The authors also incorporated a factor for cultivar susceptibility in the model for its validation in the Pampas Region in 1993-1995, thus improving the fit between the observed and predicted head blight incidence.

Both models developed in Pergamino satisfactorily predicted head blight incidence in both Pergamino (Moschini and Fortugno, 1996) and in the Pampas region (Moschini et al., 2001) showing that these models could be successfully used to predict FHB incidence in those regions.

5.2. Argentina, Brazil, Uruguay

Fernandes et al. (2004) used a linked process based modelling approach to explain FHB epidemics that developed at three sites in South America: Pergamino (Argentina), La Estanzuela (Uruguay), and Passo Fundo (Brazil). In order to model FHB in wheat, they used a wheat development model (Cropsim-Ceres 2002 included in DSSAT 4.0) and an FHB model which calculate three parameters, i) proportion of susceptible tissue, ii) infection rate, iii) spore cloud density. The final risk is calculated from the summation of the partial indices.

The rates and rules in the model are influenced by weather variables such as daily mean temperature, daily mean relative humidity, daily solar radiation, and consecutive rainy days.

In Passo Fundo, Brazil, the model evaluation with disease data from 5 years of epidemics, showed that the risk estimated by the model explained over 95% of variation in disease field epidemics. Details on development of the model development still have to be published.

5.3. Belgium

Detrixhe et al. (2003) developed an agrometeorological model that is able to estimate the risk of FHB infection at a regional scale (1km x 1km grid) in winter wheat. Moisture on plant surfaces has a considerable influence on fungal pathogen development. Therefore, estimation of leaf wetness and duration is an important agrometeorological parameter that needs to be taken into account in the development of FHB in wheat. This model is based on a leaf wetness duration calculation module (Oger et al., 2002), which estimates the duration of surface wetness from data provided by a weather station network (for standard meteorological data) and a weather radar which cover the whole Belgian territory (for precipitation data). Weather data (temperature, relative humidity, wind speed, short-wave and long-wave radiation) are collected over a period that starts 8 days before wheat flowering and ends 7 days after and these data are interpolated to a grid size of 1 km x 1 km, which cover the entire Belgian territory. A first evaluation of the model was carried out in 2002 with the collection of 43 wheat samples and their analyses for *Fusarium* spp. incidence. The first evaluations highlighted some instrumental limitations concerning the risk assessment in regions located at more than 120 km from the weather radar and the necessity to include FHB parameters like crop rotation, cultivar characteristics and field disease history in the model. Nevertheless, the use of spatial interpolation of meteorological data and of the estimation of leaf wetness makes this model very interesting and original. Further calibration/validation tests are in progress to optimise the model. The institute (Agricultural Research Center, Gemloux, Belgium) is also still working on the question of leaf wheat duration (Oger, 2008, *personal communication*).

5.4. Canada

DONcast is a prediction model that has been used successfully in Ontario, Canada, as a forecasting tool mainly to help with fungicide spray decisions at heading and for grain marketing decisions (Hooker et al., 2002). Growers and crop advisors in Ontario have used it since 2000 and it continues to evolve with the inclusion of environmental and agronomic diversities. It was developed from data collected from over 750 farms across Ontario, since 1996, using multiple regression techniques which allowed the most important climatic variables associated with the increase or decreases of DON to be indentified. The model is based on three equations, related to three different critical periods around wheat heading: i) from 4 to 7 days before heading, ii) from 3 to 6 days after heading, and iii) from 7 to 10 days after heading. The first critical period (i) corresponds to inoculum production: in this period, rainfall amounts of>5 mm per day trigger and increase the DON potential while daily minimum air temperatures of less than 10°C limit the DON potential. Similarly, weather variables in critical periods ii and iii correspond to infection during flowering and fungal growth. Here, the number of rainy days and days with relative humidity over 75% at 11.00 h increase the DON potential; daily maximum temperatures over 32°C and average temperatures of less than 12°C instead limit the DON potential. Daily weather data are converted to binary values using a set of criteria for each variable within each critical period. The binary values are summed within each weather variable and critical period and the complex summations are plugged into empirical equations to forecast the concentration of DON in wheat grain at harvest. Validation analysis has shown a greater accuracy than 80% in determining whether a fungicide application is able to reduce DON. In 2004, a web-based interactive model was developed for industry in Ontario. This model allows input of field-specific weather and agronomic variables to be used as input for more accurate predictions. The model is available on the website www.ontarioweathernetwork.com/DONcast.cfm.

Since 2000, DONcast has been validated and calibrated not only in Canada, but also in other regions including the United States, Uruguay and France (Hooker and Schaafsma, 2004; Schaafsma and Hooker, 2006; Schaafsma et al., 2006; Schaafsma and Hooker, 2007).

5.5. The Czech Republic

Data about DON contamination in wheat grain, weather conditions during the growing season and cultivation practices from two field experiments conducted in 2002-2003 were used for the development of a neural network model designed for DON contamination prediction in The Czech Republic (Klem et al., 2007). Using the data from field experiments, Klem et al. trained neural networks to predict DON content, on the basis of weather data, as a continuous input variable and preceding crop as a categorical input variable. The neural network works with five input variables: the categorial variable (preceding crop) and continuous variables (average April temperature, sum of April precipitation, average temperature five days prior to anthesis, sum of precipitation five days prior to anthesis).

5.6. Italy

Rossi et al. (2003) developed a dynamic simulation model for the risk of FHB in wheat. The model calculates a daily infection risk based on sporulation, spore dispersal and infection of the host tissue of the four main species that cause the disease (*G. zeae*, *F. culmorum*, *Gibberella. avenacea*, *Monographella nivalis*). Spore yield and dispersal are calculated as functions of temperature, rainfall and relative humidity, while the main factors affecting the infection rate are temperature, wetness and the host growth stage. The model also calculates the risk for mycotoxin production by *G. zeae* and *F. culmorum* in the infected head tissue. The model was validated with data from different wheat growing areas in northern Italy from 2002 to 2004 using different soft and durum wheat cultivars. The validation showed a general agreement between the model simulations and actual FHB epidemics and DON contamination under different epidemiological conditions.

The interesting aspect of this model is that it was developed using the System Analysis principles (Leffelaar, 1993) and unlike the others, this is the first known attempt to build a mechanistic model of FHB development and DON contamination in wheat.

5.7. Switzerland

Musa et al. (2007) developed the internet-based decision support system FusaProg. The model included in FusaProg takes into account the effects of cropping factors, previous crops, soil and straw management, as well as the susceptibility of the wheat varieties to FHB. These factors are used as driving variables and are combined with the prevailing weather conditions and growth stage in order to predict the deoxynivalenol content of a specific wheat plot before harvest. Hourly measured and forecasted data from 60 stations of a Swiss weather service and private stations are transferred to a server where the data are automatically analyzed for periods conducive for *F. graminearum* infection, according to weather rules based on three parameters: daily mean temperature, sum of precipitation and daily mean of the relative humidity. Weather conditions over the last three days as well as the weather forecast are taken into account for the DON risk assessment. The FusaProg model assesses the risk of DON contamination in three steps: i) a specific field is classified according to the previous crop, the straw management, and the seedbed tillage; ii) the susceptibility of the variety and the growth stage of the host is evaluated; iii) weather data are taken into account. In 2006, the systems was evaluated by Swiss cantonal plant protection officers and in 2007, the system was made available to Swiss wheat producers on the internet site http://www.fusaprog.ch/.

5.8. The United States

In the United States, the Fusarium Head Blight Risk Assessment Tool was developed by a cooperative research project at five universities. Using historical records of the weather and observations of disease severity in field plots, logistic regression models were developed to predict the probability of the mean disease severity exceeding 10%. Several empirical models were developed that used weather data for seven days prior to flowering, seven or 10 days starting from flowering, or both pre- and post-flowering time windows. The currently-used models, which were 80% accurate in predicting the data used in model development, utilize only pre-flowering environmental data. The implicit assumption underlying the models is that scab epidemics are determined (at least in part) by inoculum availability at flowering and that the weather immediately preceding flowering determines the magnitude of the available inoculum. In 2004, Penn State and Ohio State Universities deployed the predictive system in a Web-based format for evaluation purposes in 23 states. Separate models were used for winter wheat with a low level of corn residue, winter wheat with a high level of corn residue, and spring wheat production systems. Scab risk maps were produced at a 20-km resolution using temperature and relative humidity data. The model is available on the Internet site http://www.wheatscab.psu.edu/riskTool.html

6. Conclusions and Future Outlook on the Prevention of FHB and DON Contamination in Wheat

In these sections it has been illustrated that Fusarium head blight and *Fusarium* toxin contamination of soft and durum wheat depends on many factors. This means that a unique solution to this problem cannot exist.

Thus, the producers' objective must be to find technical strategies that can minimize the risk of DON contamination as much as possible in wheat grain, flour, semolina and in foodstuffs, taking into account that a total absence of the toxin is almost impossible in the field.

In order to reduce the risk of contamination, it is necessary for producers to adopt a strategy that integrates and modulates all of the preventive measures presented in section 4. In other words, it is necessary for producers to make use of the available wheat varieties that are known to be more resistant and to apply all of the good agricultural practices tuned to a determined cultivation area.

Even though this kind of strategy could appear to be easy to apply, the use of this type of approach is only just a beginning. In fact, until now, soft and durum wheat production has been dominated by strategies aimed at maximising the productivity and the technological quality for the transformation industry. These aspects will doubtlessly continue to guide the decisions of the producers and of the whole food and feed production chain, but now they will also have to pay attention to the safety aspects, which are considered a fundamental pre-requisite in cereal production. Thus, all of the agricultural practices aimed at the optimization of productivity and technological quality should now be inserted into production strategies aimed at the reduction of the presence of contaminants like DON, not the other way around.

In this context, the higher the probability of running into high DON contamination, the tighter the agricultural practice constraint necessary to reduce the risk of FHB and consequently the more the other cultivation possibilities will need to be reduced to improve other technical aspects.

In an environment with a high DON contamination risk, the agronomic constraint to prevent this toxin could become so tight that the production destined to specific products (such as baby food) becomes economically inconvenient, or that the premium prices necessary to compensate the losses or the higher costs for more expensive agricultural practices result to be no longer sustainable for the chain production. In this perspective, the expansion of durum wheat cultivation towards cooler areas—as observed, for example, in Italy in recent years—could be more hampered by the uncertainty about DON contamination than productivity or technological quality difficulties (e.g., protein content, flour indexes) that this cultivation could encounter in new environments.

In the future, should stricter limits be established for the most widespread mycotoxins—e.g., DON—wheat cultivation will have to abandon some cultivation areas unless more effective solutions to contrast the toxigenic fungi are found. To this end, it would be necessary to have i) more resistant wheat varieties with the same productivity and quality characteristics, ii) more effective and innovative fungicides, and iii) competitor microorganisms against the toxigenic ones, or those that can destroy the toxic molecules. Nevertheless, at the moment none of these solutions appear to be solely able to solve the problem, if the capability of the fungal species to develop resistances or adaptive strategies is also taken into account.

Thus, we come back to the initial affirmation: FHB development and DON synthesis depend on many factors and, consequently, the elimination of the contamination is not a realistic objective. Only an integrated strategy is realistic and will be able to answer the need for reducing the presence of this contaminant as much as possible.

References

Ali, S.,Francl, L., 2001. *Progression of Fusarium species on wheat leaves from seedling to adult stages in North Dakota.* In: S.M. Canty, J. Lewis, L. Siler and R.W. Ward (Editors), 2001 National Fusarium Head Blight Forum. Michigan State University, Erlanger, KY, pp. 99.

Anand, A., Zhou, T., Trick, H.N., Gill, B.S., Bockus, W.W.,Muthukrishnan, S., 2003. Greenhouse and field testing of transgenic wheat plants stably expressing genes for thaumatin-like protein, chitinase and glucanase against Fusarium graminearum. *Journal of Experimental Botany,* 54(384): 1101-1111.

Atanasoff, D., 1920. Fusarium blight (scab) of wheat and other cereals. *J. Agric. Res,* 20(1): 1-32.

Bai, G.H.,Shaner, G., 1994. Scab of Wheat - Prospects for Control. *Plant Disease,* 78(8): 760-766.

Bai, G.H.,Shaner, G., 2004. Management and resistance in wheat and barley to Fusarium head blight. *Annual Review of Phytopathology,* 42: 135-161.

Barrett, J.R.,Nearing, M.A., 1998. Humanization of decision support using information from simulation. In: R.M. Peart and R.B. Curry (Editors), *Agricultural systems modeling and simulation.* Marcel Dekker, Inc., Gainesville, Florida (USA), pp. 1-18.

Barrier-Guillot, B., Delambre, M., Morel, A., Maumene, C., Gouet, H., Grosjean, F.,Leuillet, M., 2006. Identification of agronomic factors that influence the level of deoxynivalenol (DON) in wheat grown in France. In: H. Njapau, S. Trujillo, H.P. van Egmond and D.L. Park (Editors), *Mycotoxins and phycotoxins. Advances in determination, toxicology and exposure management.* Wageningen Academic Publishers, The Netherlands, pp. 239-247.

Bateman, G.L., Murray, G., Gutteridge, R.J.,Coskun, H., 1998. Effects of method of straw disposal and depth of cultivation on populations of Fusarium spp in soil and on brown foot rot in continuous winter wheat. *Annals of Applied Biology,* 132(1): 35-47.

Bechtel, D.B., Kaleidau, L.A., Gaines, R.L.,Seitz, L.M., 1985. The Effects of Fusarium Graminearum Infection on Wheat Kernels. *Cereal Chemistry,* 62: 191-197.

Bertelsen, J.R., de-Neergaard, E.,Smedegaaed-Petersen, V., 1999. Reason for improved yield when using azoxystrobin in winter wheat, 16th Danish Plant Protection Conference. *Crop Protection in Organic Farming, Pests and Diseases,* Tjele, Denmark.

Birzele, B., Meier, A., Hindorf, H., Krämer, J.,Dehne, H.W., 2002. Epidemiology of Fusarium Infection and Deoxynivalenol Content in Winter Wheat in the Rhineland, Germany. *European Journal of Plant Pathology,* 108(7): 667-673.

Blandino, M., Minelli, L.,Reyneri, A., 2006. Strategies for the chemical control of Fusarium head blight: Effect on yield, alveographic parameters and deoxynivalenol contamination in winter wheat grain. *European Journal of Agronomy,* 25(3): 193-201.

Boote, K.J., Jones, J.W.,Pickering, N.B., 1996. Potential Uses and Limitations of Crop Models. *Agron J,* 88(5): 704-716.

Bottalico, A.,Perrone, G., 2002. Toxigenic Fusarium species and Mycotoxins Associated with Head Blight in Small-Grain Cereals in Europe. *European Journal of Plant Pathology,* 108(7): 611-624.

Boyacoglu, D., Hettiarachy, N.S.,Stack, R.W., 1992. Effect of three systemic fungicides on deoxynivalenol (vomitoxin) production by *Fusarium graminearum* in wheat. Canadian Journal of Plant Science, 72: 93-101.

Brennan, J.M., Fagan, B., van Maanen, A., Cooke, B.M.,Doohan, F.M., 2003. Studies on in vitro growth and pathogenicity of European Fusarium fungi. *European Journal of Plant Pathology,* 109(6): 577-587.

Buerstmayr, H., Lemmens, M., Hartl, L., Doldi, L., Steiner, B., Stierschneider, M.,Ruckenbauer, P., 2002. Molecular mapping of QTLs for Fusarium head blight resistance in spring wheat. I. Resistance to fungal spread (Type II resistance). *TAG Theoretical and Applied Genetics,* 104(1): 84-91.

Buerstmayr, H., Steiner, B., Hartl, L., Griesser, M., Angerer, N., Lengauer, D., Miedaner, T., Schneider, B.,Lemmens, M., 2003. Molecular mapping of QTLs for Fusarium head blight resistance in spring wheat. II. Resistance to fungal penetration and spread. *Theoretical and Applied Genetics,* 107(3): 503-508.

Bushnell, W.R., 2001. What is known about infection pathways in Fusarium Head Blight? In: S.M. Canty, J. Lewis, L. Siler and R.W. Ward (Editors), 2001 *National Fusarium Head Blight Forum.* Michigan State University, Erlanger, KY, pp. 105.

Calori-Dominguesi, M.A., de Almeida, R.R., Tomiwaka, M.M., Gallo, C.R., da Gloria, E.M.,Dias, C.T.S., 2007. Occurrence of deoxynivalenol in national and imported wheat used in Brazil. *Ciencia E Tecnologia De Alimentos,* 27(1): 181-185.

Caron, D., 1993. Les Fusarioses. In: ITCF (Editor), Maladies des blés et orges, pp. 30-39.

Cassini, R.,, E., Dijon, pp. 271-279., 1970. Facteurs favorables ou défavorables au development des fusarioses et septorioses du blé, *Proceedings of the Meeting of sections Cereals and Physiology*, Eucarpia, Dijon, pp. 271-279.

CAST, 2003. Mycotoxins: risk in plant, animal, and human systems, Task Force Report 139. *Council for Agricultural Science and Technology,* Ames, IA, USA, 199 pp.

Chala, A., Weinert, J.,Wolf, G.A., 2003. An integrated approach to the evaluation of the efficacy of fungicides against Fusarium culmorum, the cause of head blight of wheat. *Journal of Phytopathology,* 151(11-12): 673-678.

Champeil, A., Dore, T.,Fourbet, J.F., 2004a. Fusarium head blight: epidemiological origin of the effects of cultural practices on head blight attacks and the production of mycotoxins by Fusarium in wheat grains. *Plant Science,* 166(6): 1389-1415.

Champeil, A., Fourbet, J.F., Dore, T.,Rossignol, L., 2004b. Influence of cropping system on Fusarium head blight and mycotoxin levels in winter wheat. *Crop Protection,* 23(6): 531-537.

Charmley, L.L., Rosenburg, A.,Trenholm, H.L., 1994. Factors responsible for economic losses due to *Fusarium* mycotoxin contamination of grains, foods, and feedstuffs. In: J.D.

Miller and H.L. Trenholm (Editors), Mycotoxins in grain: Compounds other than aflatoxin. *American Association of Cereal Chemists,* St. Paul, MN.

Cromey, M.G., Shorter, S.C., Lauren, D.R.,Sinclair, K.I., 2002. Cultivar and crop management influences on fusarium head blight and mycotoxins in spring wheat (*Triticum aestivum*) in New Zealand. *New Zealand Journal of Crop and Horticultural Science,* 30: 235-247.

D'Mello, J.P.F., Macdonald, A.M.C.,Rinna, R., 2001. Effects of azoxystrobin on mycotoxin production in a carbendazim-resistant strain of Fusarium sporotrichioides. *Phytoparasitica,* 29(5): 431-440.

D'Mello, J.P.F., Macdonald, A.M.C., D., P.,Dijksma, W.T.P., 1998. 3-Acetyl deoxynivalenol and esterase production in a fungicide-insensitive strain of *Fusarium culmorum. Mycotoxin Research,* 14: 8-9.

da Luz, W.C., Stockwell, C.A.,Bergstrom, G.C., 2003. Biological control of Fusarium graminearum. In: K.J.L.a.W.R. Bushnell (Editor), *Fusarium Head Blight of Wheat and Barley. Amer. Phytopathol.* Soc. Press, St. Paul, MN, pp. 381–394.

Daamen, R.A., Langerak, C.J.,Stol, W., 1991. Surveys of cereal diseases and pests in the Netherlands. 3. Monographella nivalis and Fusarium spp. in winter wheat fields and seed lots. *European Journal of Plant Pathology,* 97(2): 105-114.

Dahleen, L.S., Okubara, P.A.,Blechl, A.E., 2001. Transgenic approaches to combat fusarium head blight in wheat and barley. *Crop Science,* 41(3): 628-637.

Dalcero, A., Torres, A., Etcheverry, M., Chulze, S.,Varsavsky, E., 1997. Occurrence of deoxynivalenol and Fusarium graminearum in Argentinian wheat. *Food Additives and Contaminants,* 14(1): 11-14.

de Nijs, M., Soentoro, P., Asch, E.D.-v., Kamphuis, H., Rombouts, F.M., Notermans, S.,H.W, 1996. Fungal Infection and Presence of Deoxynivalenol and Zearalenone in Cereals Grown in The Netherlands. *Journal of Food Protection,* 59: 772-777.

de Wit, C.T., 1993. Philosophy and terminology. In: P.A. Leffelaar (Editor), *On systems analysis and simulation of ecological processes.* Kluwer Academic Publishers, Wageningen, pp. 3-4.

Deadman, M.L.,Cooke, B.M., 1997. Wheat disease and cereal-clover bi-crops. *Manage, Focus* 2: 15-19.

Desjardins, A.E., Manandhar, G., Plattner, R.D., Maragos, C.M., Shrestha, K.,McCormick, S.P., 2000. Occurrence of Fusarium species and mycotoxins in nepalese maize and wheat and the effect of traditional processing methods on mycotoxin levels. *Journal of Agricultural and Food Chemistry,* 48(4): 1377-1383.

Detrixhe, P., Chandelier, A.,Cavelier, M., 2003. Development of an agrometeorological model integrating leaf wetness duration estimation to assess the risk of head blight infection in wheat. *Aspects of Applied Biology* 68: 199-204.

Dill-Macky, R.,Jones, R.K., 2000. The effect of previous crop residues and tillage on Fusarium head blight of wheat. *Plant Disease,* 84(1): 71-76.

Doohan, F.M., Brennan, J.,Cooke, B.M., 2003. Influence of climatic factors on Fusarium species pathogenic to cereals. *European Journal of Plant Pathology,* 109(7): 755-768.

Edwards, S.G., 2004a. Influence of agricultural practices on fusarium infection of cereals and subsequent contamination of grain by trichothecene mycotoxins. *Toxicology Letters,* 153(1): 29-35.

Edwards, S.G., 2004b. Investigation of Fusarium mycotoxins in UK wheat production. In: S.M. Canty, T. Boring, K. Versdahl, J. Warwell and R.W. Ward (Editors), *2nd International Symposium on Fusarium Head Blight.* Michigan State University, Orlando, FL, USA, pp. 398-400.

Edwards, S.G., Pirgozliev, S.R., Hare, M.C.,Jenkinson, P., 2001. Quantification of Trichothecene-Producing Fusarium Species in Harvested Grain by Competitive PCR To Determine Efficacies of Fungicides against Fusarium Head Blight of Winter Wheat. *Appl. Environ. Microbiol.,* 67(4): 1575-1580.

FAO, 2003. Manual on the application of the HACCP system in mycotoxin prevention and control. *FAO Food and Nutrition Papers,* 73. FAO, Rome, 122 pp.

Fauzi, M.T.,Paulitz, T.C., 1994. The Effect of Plant-Growth Regulators and Nitrogen on Fusarium Head Blight of the Spring Wheat Cultivar Max. *Plant Disease,* 78(3): 289-292.

Fernandes, J.M., Cunha, G.R., Del Ponte, E., Pavan, W., Pires, J.L., Baethgen, W., Gimenez, A., Magrin, G.,Travasso, M.I., 2004. Modeling fusarium head blight in wheat under climate change using linked process-based models. In: S.M. Canty, T. Boring, K. Versdahl, J. Wardwell and R. Ward (Editors), *2nd International Symposium on Fusarium Head Blight.* Michigan State University, Orlando, FL, USA, pp. 441-444.

Fernandez, M.R., 1992. The Effect of Trichoderma-Harzianum on Fungal Pathogens Infesting Wheat and Black Oat Straw. *Soil Biology & Biochemistry,* 24(10): 1031-1034.

Fernando, W.G.D., Miller, J.D., Seaman, W.L., Seifert, K.A.,Paulitz, T.C., 2000. Daily and seasonal dynamics of airborne spores of *Fusarium graminearum* and other *Fusarium* sepcies sampled over wheat plots. *Canadian Journal of Botany,* 78: 497-505.

France, J.,Thornley, J.H.M., 1984. Role of mathematical models in agriculture and agricultural research. In: J. France and J.H.M. Thornley (Editors), *Mathematical models in agriculture.* Butterworths, Hurley, Maidenhead, UK, pp. 1-14.

Gagkaeva, T.Y.,Gavrilova, O.P., 2008. Current distribution of *Fusarium* fungi on small grain cereals in Russia. *Journal of Plant Pathology - Abstracts of presentations of the X International Fusarium and Fusarium Genomics Workshop* 2008, Alghero, Sardinia (Italy), August 30-September 2, 2008, 90(3S): S3.47-S3.48.

Gilbert, J.,Fernando, W.G.D., 2004. Epidemiology and biological control of Gibberella zeae Fusarium graminearum. *Canadian Journal of Plant Pathology-Revue Canadienne De Phytopathologie,* 26(4): 464-472.

Gilchrist, L.,Dubin, H.J., 2002. *Fusarium Head Blight.* In: B.C. Curtis, S. Rajaram and H. Gómez Macpherson (Editors), Bread Wheat. Improvement and Production.

Gourdain, E., 2008. Maîtriser le risque sur les cultures de blés: quels outils pour quelles utilisations? In: I.d.v. ARVALIS (Editor), Maîtriser le risque fusariotoxines. *ARVALIS,* Paris, Salons de l'Aveyron, pp. 27-40.

Haidukowski, M., Pascale, M., Perrone, G., Pancaldi, D., Campagna, C.,Visconti, A., 2005. Effect of fungicides on the development of Fusarium head blight, yield and deoxynivalenol accumulation in wheat inoculated under field conditions with Fusarium graminearum and Fusarium culmorum. *Journal of the Science of Food and Agriculture,* 85(2): 191-198.

Hajjaji, A., El Otmani, M., Bouya, D., Bouseta, A., Mathieu, F., Collin, S.,Lebrihi, A., 2006. Occurrence of mycotoxins (ochratoxin A, deoxynivalenol) and toxigenic fungi in Moroccan wheat grains: impact of ecological factors on the growth and ochratoxin A production. *Molecular Nutrition & Food Research,* 50(6): 494-499.

Hajslova, J., Lancova, K., Sehnalova, M., Krplova, A., Zachariasova, M., Moravcova, H., Nedelnik, J., Markova, J.,Ehrenbergerova, J., 2007. Occurrence of trichothecene mycotoxins in cereals harvested in the Czech Republic. *Czech Journal of Food Sciences,* 25(6): 339-350.

Hart, P., Munn, B., Ward, R.,Siler, L., 2004. Fusarium head blight in Michigan in 2004. In: S.M. Canty, T. Boring, K. Versdahl, J. Warwell and R.W. Ward (Editors), *2nd International Symposium on Fusarium Head Blight.* Michigan State University, Orlando, FL, USA, pp. 453-457.

Heier, T., Jain, S.K., Kogel, K.H.,Pons-Kuhnemann, J., 2005. Influence of N-fertilization and fungicide strategies on Fusarium head blight severity and mycotoxin content in winter wheat. *Journal of Phytopathology,* 153(9): 551-557.

Hermann, W., Kuebler, E.,Aufhammer, W., 1998. Aehrenbefall mit fusarien und toxingehalt im korngut bei verschiedenen wintergetreidearten. *Pflanzenbauwiss. Ger. J. Agron.,* 6(1-8).

Hilton, A.J., Jenkinson, P., Hollins, T.W.,Parry, D.W., 1999. Relationship between cultivar height and severity of Fusarium ear blight in wheat. *Plant Pathology,* 48(2): 202-208.

Höhn, D., Rosenberger, A., Kaul, H.-P., Kübler, E.,Aufhammer, W., 2002. Infection with Fusarium head blight (F. culmorum) and deoxynivalenol accumulation of winter wheat as affected by varied maturation. *Pflanzenbauwissenschaften,* 6(1): 1-8.

Hollins, T.W., Ruckenbauer, P.,De Jong, H., 2003. Progress towards wheat varieties with resistance to fusarium head blight. *Food Control,* 14(4): 239-244.

Homdork, S., Fehrmann, H.,Beck, R., 2000. Effects of field application of tebuconazole on yield, yield components and the mycotoxin content of Fusarium-infected wheat grain. *Journal of Phytopathology,* 148(1): 1-6.

Hooker, D.C.,Schaafsma, A.W., 2004. The DONcast model: predicting deoxynivalenol (DON) in wheat. In: S.M. Canty, T. Boring, J. Wardwell and R. Ward (Editors), Proceedings of the Second International Symposium on Fusarium Head Blight. Michigan State University, Orlando, FL, USA, pp. 458.

Hooker, D.C., Schaafsma, A.W.,Tamburic-Ilincic, L., 2002. Using weather variables pre- and post-heading to predict deoxynivalenol content in winter wheat. Plant Disease, 86(6): 611-619.

Hope, R., Aldred, D.,Magan, N., 2005. Comparison of environmental profiles for growth and deoxynivalenol production by *Fusarium culmorum* and *F. graminearum* on wheat grain. *Letters in Applied Microbiology,* 40(4): 295-300.

Horberg, H.M., 2002. Patterns of splash dispersed conidia of Fusarium poae and Fusarium culmorum. *European Journal of Plant Pathology,* 108(1): 73-80.

Inch, S.,Gilbert, J., 2003. The incidence of Fusarium species recovered from inflorescences of wild grasses in southern Manitoba. *Canadian Journal of Plant Pathology-Revue Canadienne De Phytopathologie,* 25(4): 379-383.

Ioos, R., Belhadj, A.,Menez, M., 2004. Occurrence and distribution of Microdochium nivale and Fusarium species isolated from barley, durum and soft wheat grains in France from 2000 to 2002. *Mycopathologia,* 158(3): 351-362.

Ioos, R., Belhadj, A., Menez, M.,Faure, A., 2005. The effects of fungicides on Fusarium spp. and Microdochium nivale and their associated trichothecene mycotoxins in French naturally-infected cereal grains. *Crop Protection,* 24(10): 894-902.

Jenkinson, P.,Parry, D.W., 1994. Isolation of fusarium species from common broad-leaved weeds and their pathogenicity to winter wheat. *Mycological Research,* 98: 776-780.

Jennings, P., Turner, J.A.,Nicholson, P., 2000. Overview of Fusarium ear blight in the UK—effect of fungicide treatment on disease control and mycotoxin production, *BCPC Crop Protection Conference—Pest and Disease.* Farnham: British Crop Protection Council, pp. 707–712.

Jenny, E., Hecker, A., Kessler, P., Külling, C.,H.-R., F., 2000. Getreidefusariosen: Sortenresistenz und Toxingehalte. *Agrarforschung*, 7(270-273).

Johnson, D.D., Flaskerud, G.K., Taylor, R.D.,Satyanarayana, V., 1998. *Economic impacts of fusarium head blight in wheat,* North Dakota State University, Department of Agribusiness and Applied Economics, Fargo.

Jones, J.W.,Luyten, J.C., 1998. Simulation of biological processes. In: R.M. Peart and R.B. Curry (Editors), *Agricultural systems modeling and simulation.* Marcel Dekker, Inc., Gainesville, Florida (USA), pp. 19-62.

Jurado, M., Vàzquez, C., Lòpez-Errasquin, E., Patiño, B.,Gonzàlez-Jaén, M.T., 2004. Analysis of the occurrence of *Fusarium* species in spanish cereals by PCR assay. In: S.M. Canty, T. Boring, K. Versdahl, J. Warwell and R.W. Ward (Editors), *2nd International Symposium on Fusarium Head Blight.* Michigan State University, Orlando, FL, USA, pp. 460-464.

Kammoun, G.L., Barreau, C., Richard-Forget, F., Gargouri, S.,Hajlaoui, M.R., 2008. Ocurrence of trichothecene mycotoxins in Tunisian *Fusarium* species isolated from wheat. *Journal of Plant Pathology - Abstracts of presentations of the X International Fusarium and Fusarium Genomics Workshop 2008,* Alghero, Sardinia (Italy), August 30-September 2, 2008, 90(3S): S3.54.

Kang, Z.,Buchenauer, H., 2000a. Cytology and ultrastructure of the infection of wheat spikes by Fusarium culmorum. *Mycological Research*, 104(9): 1083-1093.

Kang, Z.,Buchenauer, H., 2000b. Ultrastructural and Cytochemical Studies on Cellulose, Xylan and Pectin Degradation in Wheat Spikes Infected by <i>Fusarium culmorum</i>. *Journal of Phytopathology,* 148(5): 263-275.

Keblys, M., Flaoyen, A.,Langseth, W., 2001. The occurrence of type A and B trichothecenes in Lithuanian cereals. *Acta Agriculturae Scandinavica Section B-Soil and Plant Science,* 50(3-4): 155-160.

Khan, N.I., Schisler, D.A., Boehm, M.J., Lipps, P.E.,Slininger, P.J., 2004. Field testing of antagonists of Fusarium head blight incited by Gibberella zeae. *Biological Control,* 29(2): 245-255.

Khonga, E.B.,Sutton, J.C., 1988. Inoculum production and survival of Gibberella zeae in maize and wheat residues. *Can. J. Plant Pathol.(*10): 232-239.

Klem, K., Vanova, M., Hajslova, J., Lancova, K.,Sehnalova, M., 2007. A neural network model for prediction of deoxynivalenol content in wheat grain based on weather data and preceding crop. *Plant Soil and Environment,* 53(10): 421-429.

Koch, H.-J., Pringas, C.,Maerlaender, B., 2006. Evaluation of environmental and management effects on Fusarium head blight infection and deoxynivalenol concentration in the grain of winter wheat. *European Journal of Agronomy,* 24(4): 357-366.

Kolb, F.L., Bai, G.H., Muehlbauer, G.J., Anderson, J.A., Smith, K.P.,Fedak, G., 2001. Host plant resistance genes for fusarium head blight: Mapping and manipulation with molecular markers. *Crop Science,* 41(3): 611-619.

Kong, L., Anderson, J.M.,Ohm, H.W., 2005. Induction of wheat defense and stress-related genes in response to *Fusarium graminearum*. *Genome,* 48: 29-40.

Kosiak, B., Torp, M., Skjerve, E.,Thrane, U., 2003. The Prevalence and Distribution of *Fusarium* species in Norwegian Cereals: a Survey. Acta Agriculturae Scandinavica, *Section B - Plant Soil Science,* 53(4): 168 - 176.

Kranz, J.,Royle, D.J., 1978. Perspective in mathematical modelling of plant disease epidemics. In: P.R. Scott and A. Bainbridge (Editors), *Plant Disease Epidemiology.* Blackwell Scientific Pubblications, Oxford, pp. 111-120.

Krebs, H., Streit, B.,Forrer, H.-R., 2000. Effect of tillage and preceding crops on Fusarium infection and deoxynivalenol content of wheat. In: T. Alfoldi (Editor), 13th International IFOAM Scientific Conference, *The World Grows Organic* IOS, Amsterdam, Basel, Convention Center, pp. 28-31.

Krstanovic, V., Klapec, T., Velic, N.,Milakovic, Z., 2005. Contamination of malt barley and wheat by Fusarium graminearum and Fusarium culmorum from the crop years 2001-2003 in eastern Croatia. *Microbiological Research,* 160(4): 353-359.

Lacey, J., Bateman, G.L.,Mirocha, C.J., 1999. Effects of infection time and moisture on development of ear blight and deoxynivalenol production by Fusarium spp. in wheat. *Annals of Applied Biology,* 134(3): 277-283.

Lager, J.,Wallenhammer, A.C., 2003. Crop loss from soil-borne pathogens in white clover stands assessed by chemical treatments. Z. Pflanzenk. *Pflanzen,* 110: 120-128.

Langseth, W.,Elen, O., 1997. The occurrence of deoxynivalenol in Norwegian cereals - Differences between years and districts, 1988-1996. *Acta Agriculturae Scandinavica Section B-Soil and Plant Science,* 47(3): 176-184.

Leffelaar, P.A., 1993. *On systems analysis ans simulation of ecological processes.* Kluwer Academic Publlisher, Dordrecht, The Netherlands, 294 pp.

Lemmens, M., Haim, K., Lew, H.,Ruckenbauer, P., 2004. The effect of nitrogen fertilization on Fusarium head blight development and deoxynivalenol contamination in wheat. *Journal of Phytopathology,* 152(1): 1-8.

Logrieco, A., Bottalico, A., Mule, G., Moretti, A.,Perrone, G., 2003. Epidemiology of toxigenic fungi and their associated mycotoxins for some Mediterranean crops. *European Journal of Plant Pathology,* 109(7): 645-667.

Lori, G.A., Sisterna, M.N., Haidukowski, M.,Rizzo, I., 2003. Fusarium graminearum and deoxynivalenol contamination in the durum wheat area of Argentina. *Microbiological Research,* 158(1): 29-35.

Luo, Y., Yoshizawa, T.,Katayama, T., 1990. Comparative study on the natural occurrence of Fusarium mycotoxins (trichothecenes and zearalenone) in corn and wheat from high- and low-risk areas for human esophageal cancer in China. *Applied and Environmental Microbiology,* 56(12): 3723-3726.

Ma, Z.P., Steffenson, B.J., Prom, L.K.,Lapitan, N.L.V., 2000. Mapping of quantitative trait loci for Fusarium head blight resistance in barley. *Phytopathology,* 90(10): 1079-1088.

Magan, N., Hope, R., Colleate, A.,Baxter, E.S., 2002. Relationship between growth and mycotoxin production by Fusarium species, biocides and environment. *European Journal of Plant Pathology,* 108(7): 685-690.

Maiorano, A., 2008. Control of fusarium-toxins in cereal grain: effect of the agronomic practices and development of decision support systems. *PhD Thesis.,* University of Turin, Turin (Italy), 103 pp.

Maiorano, A., Blandino, M., Reyneri, A.,Vanara, F., 2008. Effects of maize residues on the Fusarium spp. infection and deoxynivalenol (DON) contamination of wheat grain. *Crop Protection,* 27(2): 182-188.

Maiorano, A., Reyneri, A., Sacco, D., Magni, A.,Ramponi, C., 2009. A dynamic risk assessment model (FUMAgrain) of fumonisin synthesis by Fusarium verticillioides in maize grain in Italy. *Crop Protection,* 28(3): 243-256.

Maldonado-Ramirez, S.L., Schmale Iii, D.G., Shields, E.J.,Bergstrom, G.C., 2005. The relative abundance of viable spores of Gibberella zeae in the planetary boundary layer suggests the role of long-distance transport in regional epidemics of Fusarium head blight. *Agricultural and Forest Meteorology,* 132(1-2): 20-27.

Martin, R.A., Macleod, J.A.,Caldwell, C., 1991. Influences of Production Inputs on Incidence of Infection by Fusarium Species on Cereal Seed. *Plant Disease,* 75(8): 784-788.

Matthies, A.,Buchenauer, H., 2000. Effect of tebuconazole (Folicur R) and prochloraz (Sportak R) treatments on Fusarium head scab development, yield and deoxynivalenol (DON) content in grains of wheat following artificial inoculation with *Fusarium culmorum. Z. Pflanz. Pflanzenschutz.,* 1: 33-52.

Matthies, A., Walker, F.,Buchenauer, H., 1999. Interference of selected fungicides, plant growth retardants as well as piperonyl butoxide and 1-aminobenzotriazole in trichothecene production of Fusarium graminearum (strain 4528) in vitro. *Zeitschrift Fur Pflanzenkrankheiten Und Pflanzenschutz-Journal of Plant Diseases and Protection,* 106(2): 198-212.

McMullen, M., Jones, R.,Gallenberg, D., 1997. Scab of Wheat and Barley: A Re-emerging Disease of Devastating Impact. *Plant Disease,* 81(12): 1340-1348.

McMullen, M., Lukach, J., McKay, K.,Schatz, B., 2002. Fungicide and biological agent effects on Fusarium head blight across two wheat classes. *J. Appl. Genet.,* 43A: 223-226.

Menniti, A.M., Pancaldi, D., Maccaferri, M.,Casalini, L., 2003. Effect of Fungicides on Fusarium Head Blight and Deoxynivalenol Content in Durum Wheat Grain. *European Journal of Plant Pathology,* 109(2): 109-115.

Mesterhazy, A., 2002. Role of deoxynivalenol in aggressiveness of Fusarium graminearum and F-culmorum and in resistance to Fusarium head blight. *European Journal of Plant Pathology,* 108(7): 675-684.

Mesterházy, A., 1995. Types and components of resistance to *Fusarium* head blight of wheat. *Plant Breeding*, 114(5): 377-386.

Mesterházy, A., 2004. Possible ways to utilize mycotoxin contaminated grain. In: S.M. Canty, T. Boring, K. Versdahl, J. Warwell and R.W. Ward (Editors), *2nd International Symposium on Fusarium Head Blight.* Michigan State University, Orlando, FL, USA, pp. 413-414.

Mesterhazy, A., Bartok, T., Kaszonyi, G., Varga, M., Toth, B.,Varga, J., 2005. Common resistance to different Fusarium spp. causing Fusarium head blight in wheat. *European Journal of Plant Pathology,* 112(3): 267-281.

Mesterhazy, A., Bartok, T.,Lamper, C., 2003. Influence of wheat cultivar, species of Fusarium, and isolate aggressiveness on the efficacy of fungicides for control of Fusarium head blight. *Plant Disease,* 87(9): 1107-1115.

Miedaner, T., Schneider, B.,Geiger, H.H., 2003. Deoxynivalenol (DON) content and Fusarium head blight resistance in segregating populations of winter rye and winter wheat. *Crop Science,* 43(2): 519-526.

Mielke, H.,Weinert, J., 1996. Investigations on the effect of various fungicides on the pathogen of partial head blight (*Fusarium culmorum* (W. G. Sm.) Sacc.). *Nachr bl Dtsch Pflanzenschutzd,* 48: 93-95.

Miller, J.D., Culley, J., Fraser, K., Hubbard, S., Meloche, F., Ouellet, T., Seaman, W.L., Seifert, K.A., Turkington, K.,Voldeng, H., 1998. Effect of tillage practice on fusarium head blight of wheat. *Canadian Journal of Plant Pathology-Revue Canadienne De Phytopathologie,* 20(1): 95-103.

Miller, J.D.,Young, J.C., 1985. Deoxynivalenol in an experimental *F. graminearum* infection of wheat. *Canadian Journal of Plant Pathology,* 7: 132-135.

Miller, J.D., Young, J.C.,Sampson, D.R., 1985. Deoxynivalenol and Fusarium head blight resistance in spring cereals. *Journal of Phytopathology,* 113: 359-367.

Milus, E.A.,Parsons, C.E., 1994. Evaluation of foliar fngicide for controlling Fusarium Head Blight of wheat. *Plant Disease*, 78: 697-699.

Minervini, F., Fornelli, F.,Flynn, K.M., 2004. Toxicity and apoptosis induced by the mycotoxins nivalenol, deoxynivalenol and fumonisin B-1 in a human erythroleukemia cell line. *Toxicology in Vitro, 1*8(1): 21-28.

Moschini, R.C.,Fortugno, C., 1996. Predicting wheat head blight incidence using models based on meteorological factors in Pergamino, Argentina. *European Journal of Plant Pathology,* 102(3): 211-218.

Moschini, R.C., Pioli, R., Carmona, M.,Sacchi, O., 2001. Empirical Predictions of Wheat Head Blight in the Northern Argentinean Pampas Region. *Crop Sci,* 41(5): 1541-1545.

Mullenborn, C., Steiner, U., Ludwig, M.,Oerke, E.C., 2008. Effect of fungicides on the complex of Fusarium species and saprophytic fungi colonizing wheat kernels. *European Journal of Plant Pathology,* 120(2): 157-166.

Muller, H.M., Reimann, J., Schumacher, U.,Schwadorf, K., 2001. Further survey of the occurrence of Fusarium toxins in wheat grown in southwest Germany. *Archives of Animal Nutrition-Archiv Fur Tierernahrung,* 54(2): 173-182.

Musa, T., Hecker, A., Vogelgsang, S.,Forrer, H.R., 2007. Forecasting of Fusarium head blight and deoxynivalenol content in winter wheat with FusaProg. *EPPO Bulletin,* 37(2): 283-289.

Muthomi, J.W., Ndung'u, J.K., Gathumbi, J.K., Mutitu, E.W.,Wagacha, J.M., 2008. The occurrence of Fusarium species and mycotoxins in Kenyan wheat. *Crop Protection,* 27(8): 1215-1219.

Nganje, W.E., Johnson, D.D., Wilson, W.W., Leistritz, F.L., Bangsund, D.A.,Tiapo, N.M., 2001. *Economic impacts of Fusarium Head Blight in wheat and barley: 1998-2000,* North Dakota State University, Department of Agribusiness and Applied Economics, Agribusiness & Applied Economics Report, Fargo, ND.

Nicholson, P., Jenkinson, P., Rezanoor, H.N.,Parry, D.W., 1993. Restriction fragment length polymorphism analysis of variation in Fusarium species causing ear blight of cereals. *Plant Pathology,* 42(6): 905-914.

Obst, A., Lepschy, J., Beck, R., Bauer, G.,Bechtel, A., 2000. The risk of toxins by Fusarium graminearum in wheat — interactions between weather and agronomic factors. *Mycotoxin Research,* 16(0): 16-20.

Oger, R., Buffet, D.,Kestemont, M.H., 2002. *Estimation of leaf wetness duration using standard weather station data and radar images to assess the risk of wheat contamination by mycotoxins at regional scale in Belgium.*

Okubara, Okubara, P., Blechl, Blechl, A., McCormick, McCormick, S., Alexander, Alexander, N., Dill, M., Dill-Macky, R., Hohn,Hohn, T., 2002. Engineering deoxynivalenol metabolism in wheat through the expression of a fungal trichothecene acetyltransferase gene. *TAG Theoretical and Applied Genetics,* 106(1): 74-83.

Osborne, L.E.,Stein, J.M., 2007. Epidemiology of Fusarium head blight on small-grain cereals. *International Journal of Food Microbiology,* 119(1-2): 103-108.

Oswald, I.P., Marin, D.E., Bouhet, S., Pinton, P., Taranu, I.,Accensi, F., 2005. Immunotoxicological risk of mycotoxins for domestic animals. *Food Additives and Contaminants,* 22(4): 354-360.

Palazzinia, J.M., Ramirez, M.L., Torres, A.M.,Chulze, S.N., 2007. Potential biocontrol agents for Fusarium head blight and deoxynivalenol production in wheat. *Crop Protection,* 26(11): 1702-1710.

Parry, D.W., Jenkinson, P.,Mcleod, L., 1995. Fusarium Ear Blight (Scab) in Small-Grain Cereals - a Review. *Plant Pathology,* 44(2): 207-238.

Paul, P.A., El-Allaf, S.M., Lipps, P.E.,Madden, L.V., 2004. Rain Splash Dispersal of Gibberella zeae Within Wheat Canopies in Ohio. *Phytopathology,* 94(12): 1342-1349.

Paulitz, T.C., 1996. Diurnal release of ascospores by Gibberella zeae in inoculated wheat plots. *Plant Disease,* 80(6): 674-678.

Pereyra, S.A.,Dill-Macky, R., 2008. Colonization of the Residues of Diverse Plant Species by Gibberella zeae and Their Contribution to Fusarium Head Blight Inoculum. *Plant Disease,* 92(5): 800-807.

Pereyra, S.A., Dill-Macky, R.,Castro, M., 2004. *Fusarium* species present in wheat and barley grains in Uruguay. In: S.M. Canty, T. Boring, K. Versdahl, J. Warwell and R.W. Ward (Editors), *2nd International Symposium on Fusarium Head Blight.* Michigan State University, Orlando, FL, USA, pp. 488.

Perkowski, J., Stachowiak, J., Kiecana, I., Golinski, P.,Chelkowski, J., 1997. Natural occurrence of Fusarium mycotoxins in Polish cereals. *Cereal Research Communications,* 25(3): 379-380.

Pirgozliev, S.R., Edwards, S.G., Hare, M.C.,Jenkinson, P., 2003. Strategies for the control of Fusarium head blight in cereals. *European Journal of Plant Pathology,* 109(7): 731-742.

Placinta, C.M., D'Mello, J.P.F.,Macdonald, A.M.C., 1999. A review of worldwide contamination of cereal grains and animal feed with Fusarium mycotoxins. *Animal Feed Science and Technology,* 78(1-2): 21-37.

Prandini, A., Sigolo, S., Filippi, L., Battilani, P.,Piva, G., 2008. Review of predictive models for Fusarium head blight and related mycotoxin contamination in wheat. *Food and Chemical Toxicology,* In Press, Corrected Proof.

Pritsch, C., Muehlbauer, G.J., Bushnell, W.R., Somers, D.A.,Vance, C.P., 2000. Fungal Development and Induction of Defense Response Genes During Early Infection of Wheat Spikes by Fusarium graminearum. *Molecular Plant-Microbe Interactions,* 13(2): 159-169.

Pritsch, C., Vance, C.P., Bushnell, W.R., Somers, D.A., Hohn, T.M.,Muehlbauer, G.J., 2001. Systemic expression of defense response genes in wheat spikes as a response to Fusarium graminearum infection. *Physiological and Molecular Plant Pathology,* 58(1): 1-12.

Rafai, P., Bata, r., Jakab, L.,Vanyi, A., 2000. Evaluation of mycotoxin-contaminated cereals for their use in animal feeds in Hungary. *Food Additives & Contaminants: Part A,* 17(9): 799 - 808.

Ramirez, M.L., Chulze, S.,Magan, N., 2004. Impact of environmental factors and fungicides on growth and deoxinivalenol production by Fusarium graminearum isolates from Argentinian wheat. *Crop Protection,* 23(2): 117-125.

Ramirez, M.L., Chulze, S.,Magan, N., 2006. Temperature and water activity effects on growth and temporal deoxynivalenol production by two Argentinean strains of Fusarium graminearum on irradiated wheat grain. *International Journal of Food Microbiology,* 106(3): 291-296.

Rossi, V., Giosuè, S., Pattori, E., Spanna, F.,Del Vecchio, A., 2003. A model estimating the risk of *Fusarium* head blight on wheat. *EPPO/OEPP Bulletin,* 33(3): 421-425.

Rossi, V., Languasco, L., Pattori, E.,Giosue, S., 2002a. Dynamics of airborne Fusarium macroconidia in wheat fields naturally affected by head blight. *Journal of Plant Pathology,* 84(1): 53-64.

Rossi, V., Pattori, E., Ravanetti, A.,Giosue, S., 2002b. Effect of constant and fluctuating temperature regimes on sporulation of four fungi causing head blight of wheat. *Journal of Plant Pathology,* 84(2): 95-105.

Rossi, V., Ravanetti, A., Pattori, E.,Giosuè, S., 2001. Influence of temperature and humidity on the infection of wheat spikes by some fungi causing fusarium head blight. *Journal of Plant Pathology,* 83(3): 189-198.

Rotter, B.A., Prelusky, D.B.,Pestka, J.J., 1996. Toxicology of deoxynivalenol (vomitoxin). *Journal of Toxicology and Environmental Health,* 48(1): 1-34.

Ruckenbauer, P., Buerstmayr, H.,Lemmens, M., 2001. Present strategies in resistance breeding against scab (Fusarium spp.). *Euphytica,* 119(1-2): 121-127.

Schaafsma, A.W.,Hooker, D.C., 2006. Applications in forecasting deoxynivalenol in wheat using DONcast. In: D. Barug, D. Bhatnagar, H.P. van Egmond, J.W. van der Kamp, W.A. van Osenbruggen and A. Visconti (Editors), *The mycotoxin factbook. Food & feed topics. Wageningen Academic Publishers,* The Netherlands, pp. 211-222.

Schaafsma, A.W.,Hooker, D.C., 2007. Climatic models to predict occurrence of Fusarium toxins in wheat and maize. *International Journal of Food Microbiology,* 119(1-2): 116-125.

Schaafsma, A.W., Hooker, D.C., Pineiro, M., Diaz de Ackermann, M., Pereyra, S.,Castaño, J.P., 2006. Pre-harvest forecasting of deoxynivalenol for regulatory action in wheat grain in Uruguay using readily available weather inputs. In: H. Njapau, S. Trujillo, H.P. van Egmond and D.L. Park (Editors), *Mycotoxins and Phycotoxins. Advances in determination, toxicology and exposure management.* Wageningen Academic Publishers, The Netherlands, pp. 227-238.

Schaafsma, A.W., Tamburic-Illincic, L.,Miller, J.D., 1998. The effect of agronomic practice on the accumulation of deoxynivalenol (DON) in winter wheat fields in Ontario, 1996-1997. In: M.S. University (Editor), *National Fusarium Head Blight Forum,* USA, pp. 7-9.

Schisler, D.A., Khan, N.I.,Boehm, M.J., 2002. Biological control of Fusarium head blight of wheat and deoxynivalenol levels in grain via use of microbial antagonists. *Mycotoxins and Food Safety,* 504: 53-69.

Schisler, D.A., Slininger, P.J., Hanson, L.E.,Loria, R., 2000. Potato cultivar, pathogen isolate and antagonist cultivation medium influence the efficacy and ranking off bacterial antagonists of Fusarium dry rot. *Biocontrol Science and Technology,* 10(3): 267-279.

Schmale, D.G., Arntsen, Q.A.,Bergstrom, G.C., 2005. The forcible discharge distance of ascospores of Gibberelia zeae. *Canadian Journal of Plant Pathology-Revue Canadienne De Phytopathologie,* 27(3): 376-382.

Schmale, D.G., Shields, E.J.,Bergstrom, G.C., 2006. Night-time spore deposition of the fusarium head blight pathogen, Gibberella zeae, in rotational wheat fields. *Canadian Journal of Plant Pathology-Revue Canadienne De Phytopathologie,* 28(1): 100-108.

Schollenberger, M., Muller, H.M., Rufle, M., Suchy, S., Plank, S.,Drochner, W., 2006. Natural occurrence of 16 Fusarium toxins in grains and feedstuffs of plant origin from germany. *Mycopathologia,* 161(1): 43-52.

Schroeder, H.W.,Christensen, J.J., 1963. Factors aff ecting resistance of wheat to scab caused by *Gibberella zeae. Phytopathology,* 53: 831-838.

Simpson, D.R., Weston, G.E., Turner, J.A., Jennings, P.,Nicholson, P., 2001. Differential control of head blight pathogens of wheat by fungicides and consequences for mycotoxin contamination of grain. *European Journal of Plant Pathology,* 107(4): 421-431.

Siranidou, E.,Buchenauer, H., 2001. Chemical control of Fusarium head blight on wheat. *Zeitschrift Fur Pflanzenkrankheiten Und Pflanzenschutz-Journal of Plant Diseases and Protection,* 108(3): 231-243.

Slikova, S., Sudyova, V.,Gregova, E., 2008. Deoxynivalenol in wheat from the growing areas of Slovakia. *Cereal Research Communications,* 36(2): 279-287.

Snijders, C.H.A., 1990. Fusarium head blight and mycotoxin contamination of wheat, a review. *Netherland Journal of Plant Pathology,* 96: 187-198.

Snijders, C.H.A., 2004. Resistance in wheat to Fusarium infection and trichothecene formation. *Toxicology Letters, 1*53(1): 37-46.

Sutton, J.C., 1982. Epidemiology of wheat head blight and maize ear rot caused by *Fusarium graminearum. Canadian Journal of Plant Pathology,* 4: 195-209.

Tan, M.K., Simpfendorfer, S., Backhouse, D.,Murray, G.M., 2004. Occurrence of Fusarium head blight (FHB) in southern NSW in 2000: identification of causal fungi and determination of putative chemotype of Fusarium graminearum isolates by PCR. *Australasian Plant Pathology,* 33(3): 385-392.

Teich, A.H., 1987. Less wheat scab with urea than with ammonium nitrate fertilisers. *Cereal Research Communications,* 15: 35-38.

Teich, A.H., 1989. Epidemiology of wheat (Triticum aestivum L.) scab caused by *Fusarium* spp. In: J. Chelkowsky (Editor), Fusarium: mycotoxins, taxonomy and pathogenicity. Elsevier, Amsterdam, pp. 269-282.

Teich, A.H.,Hamilton, J.R., 1985. Effect of Cultural Practices, Soil Phosphorus, Potassium, and pH on the Incidence of Fusarium Head Blight and Deoxynivalenol Levels in Wheat. *Appl. Environ. Microbiol.,* 49(6): 1429-1431.

Teich, A.H.,Nelson, K., 1984. Survey of Fusarium head blight and possible effects of cultural practices in wheat fields in Lambton County in 1983. *Can. Plant Dis. Surv.,* 64(1): 11-13.

Toth, B., Kaszonyi, G., Bartok, T., Varga, J.,Mesterhazy, A., 2008. Common resistance of wheat to members of the Fusarium graminearum species complex and F-culmorum. *Plant Breeding,* 127(1): 1-8.

Trail, F., Xu, H., Loranger, R.,Gadoury, D., 2002. Physiological and environmental aspects of ascospore discharge in Gibberella zeae (anamorph Fusarium graminearum). *Mycologia,* 94(2): 181-189.

Treikale, O., Priekule, I., Pugacheva, J.,Lazareva, L., 2008. Fusarium species in winter wheat in Latvia: distribution and toxigenic effect. *Journal of Plant Pathology - Abstracts of presentations of the X International Fusarium and Fusarium Genomics Workshop* 2008, Alghero, Sardinia (Italy), August 30-September 2, 2008, 90(3S): S3.59.

Trigo-Stockli, D.M., Curran, S.P.,Pedersen, J.R., 1995. Distribution and occurrence of mycotoxins in 1993 kansas wheat. *Cereal Chemistry,* 72(5): 470-474.

Tutelyan, V.A., 2004. Deoxynivalenol in cereals in Russia. *Toxicology Letters,* 153(1): 173-179.

Vrabcheva, T., Gessler, R., Usleber, E.,Martlbauer, E., 1996. First survey on the natural occurrence of Fusarium mycotoxins in Bulgarian wheat. *Mycopathologia,* 136(1): 47-52.

Waalwijk, C., Kastelein, P., de Vries, I., Kerényi, Z., van der Lee, T., Hesselink, T., Köhl, J.,Kema, G., 2003. Major Changes in Fusarium spp. in Wheat in the Netherlands. *European Journal of Plant Pathology,* 109(7): 743-754.

Wagacha, J.M.,Muthomi, J.W., 2007. Fusarium culmorum: Infection process, mechanisms of mycotoxin production and their role in pathogenesis in wheat. *Crop Protection,* 26(7): 877-885.

Wang, Y.Z.,Miller, J.D., 1988. Effects of *Fusarium graminearum* Metabolites on Wheat Tissue in Relation to Fusarium Head Blight Resistance. *Journal of Phytopathology,* 122(2): 118-125.

Wiersma, J.J.,Motteberg, C.D., 2005. Evaluation of five fungicide application timings for control of leaf-spot diseases and fusarium head blight in hard red spring wheat. *The Canadian Journal of Plant Pathology,* 27: 25-37.

Wilcoxson, R.D., Kommedahl, T., Ozmon, E.A.,Windels, C.E., 1988. Occurrence of Fusarium Species in Scabby Wheat from Minnesota and their Pathogenicity to Wheat. *Phytopathology,* 78: 586-589.

Windels, C.E., 2000. Economic and Social Impacts of Fusarium Head Blight: Changing Farms and Rural Communities in the Northern Great Plains. *Phytopathology,* 90(1): 17-21.

Wong, L.S.L., Tekauz, A., Leisle, D., Abramson, D.,McKenzie, R.I.H., 1992. Prevalence, distribution, and importance of fusarium head blight in wheat in Manitoba. *Canadian Journal of Plant Pathology-Revue Canadienne De Phytopathologie,* 14(3): 233-238.

Xu, X.M., 2003. Effects of environmental conditions on the development of Fusarium ear blight. *European Journal of Plant Pathology,* 109(7): 683-689.

Xu, X.M.,Berrie, A.M., 2005. Epidemiology of mycotoxigenic fungi associated with Fusarium ear blight and apple blue mould: A review. *Food Additives & Contaminants:* Part A, 22(4): 290 - 301.

Xu, X.M., Nicholson, P.,Ritieni, A., 2007. Effects of fungal interactions among Fusarium head blight pathogens on disease development and mycotoxin accumulation. *International Journal of Food Microbiology,* 119(1-2): 67-71.

Yang, J., Bai, G.,Shaner, G., 2005. Novel quantitative trait loci (QTL) for Fusarium head blight resistance in wheat cultivar Chokwang. *TAG Theoretical and Applied Genetics,* 111(8): 1571-1579.

Yu, J.B., Bai, G.H., Zhou, W.C., Dong, Y.H.,Kolb, F.L., 2008. Quantitative trait loci for Fusarium head blight resistance in a recombinant inbred population of Wangshuibai/Wheaton. *Phytopathology,* 98(1): 87-94.

Yuen, G.Y.,Schoneweis, S.D., 2007. Strategies for managing Fusarium head blight and deoxynivalenol accumulation in wheat. *International Journal of Food Microbiology,* 119(1-2): 126-130.

Zhang, X., Zhou, M., Ren, L., Bai, G., Ma, H., Scholten, O.E., Guo, P.,Lu, W., 2004. Molecular characterization of Fusarium head blight resistance from wheat variety Wangshuibai. *Euphytica,* 139(1): 59-64.

Zhu, H., Gilchrist, L., Hayes, P., Kleinhofs, A., Kudrna, D., Liu, Z., Prom, L., Steffenson, B., Toojinda, T.,Vivar, H., 1999. Does function follow form? Principal QTLs for Fusarium head blight (FHB) resistance are coincident with QTLs for inflorescence traits and plant height in a doubled-haploid population of barley. *Theoretical and Applied Genetics,* 99(7-8): 1221-1232.

In: Fusarium
Editors: Tony Ferarri Rios and Emilia Reyes Ortega
ISBN: 978-1-61942-539-2

Chapter 10

ANTAGONISTIC EFFECT OF MICROBIALLY-TREATED MIXTURE OF AGRO-INDUSTRIAL WASTES AND INORGANIC INSOLUBLE PHOSPHATE TO *FUSARIUM* WILT DISEASE*

N. Vassilev[1,†], M. Fenice[2], E. Jurado[1], A. Reyes[1], I. Nikolaeva[3], and M. Vassileva[1]

[1]Department of Chemical Engineering, Faculty of Sciences, University of Granada, Granada, Spain
[2]Department of Agrobiology and Agrochemistry, University of Tuscia, Italy
[3]Department of Public Technology, Malardalen University, Vasteras, Sweden

ABSTRACT

A microcosm studies was carried out to determine the effect of *Aspergillus niger*-treated mixture of two agro-industrial waste (AIW) materials, dry olive wastes and sugar beet press mud, on tomato (*Lycopersicon esculentum*) plants grown in a soil inoculated with *Fusarium oxysporum* f. sp. *lycopersici* (*Fol*). These waste materials were selected as they constitute a major environmental problem especially for Mediterranean countries. Agro-wastes were treated in conditions of solid-state fermentation in the presence of Morocco apatite (RP) and further applied at a rate of 50 g/kg soil. Soil-plant systems were additionally inoculated or not with the arbuscular mycorrhizal (AM) fungus *Glomus intraradices*. Plant growth and nutrition, symbiotic developments and soil enzymatic activities were stimulated in *Fol*-free soil supplemented with treated agro-wastes and significantly greater in treatments where AM fungus was introduced compared with the non-amended control. The introduction of *Fol* into the soil-plant system adversely influenced all studied parameters. AM fungus alone reduced the effect of the plant pathogen on tomato plants and significantly decreased (by $2.3x10^3$) the number of *Fol*-

* A version of this chapter also appears in *Progress in Environmental Microbiology,* edited by Myung-Bo Kim, published by Nova Science Publishers, Inc. It was submitted for appropriate modifications in an effort to encourage wider dissemination of research.

† Corresponding Author: nbvass@yahoo.com.

CFU compared with the non-mycorrhizal control grown in non-amended soil. Higher levels of pathogen control were achieved in *A. niger*/AIW/RP-amended mycorrhized plant-soil system. In separate, *in vitro* studies, *A. niger* demonstrated a strong suppressive effect on *Fol* with a 5-fold reduction of the pathogen colony diameter compared to the control. Production of siderophore-like metabolites was also detected in chrome azurol S plate assays with *A. niger*.

Keywords: *Aspergillus niger*; Sugar beet wastes; Dry olive wastes; Rock phosphate; Solid-state fermentation product; Biocontrol; Soil-plant system; *Glomus intraradices*; Fungal pathogen

INTRODUCTION

The direct application of insoluble rock phosphates (RP) is increasing globally because of the rapid expansion in the biological-organic agriculture and the need of inexpensive phosphate. Accordingly, the efficacy of free and immobilized cells of phosphate solubilizing microorganisms (PSM) applied in RP solubilization in fermentation systems or as biofertilizers in establishing plant cover and enhancing crop yields is well documented in several review papers (Kucey et al., 1989; Vassilev et al., 2001; Vassilev and Vassileva, 2003). Particularly fungal microorganisms were widely accepted as excellent phosphate solubilizers (Whitelaw, 2000). Their application is considered as an environmentally-mild alternative in substituting highly polluting chemical RP processing (Goldstein, 2000) and/or in preventing frequent applications of soluble forms of inorganic phosphate fertilizers which often results in P leaching to the groundwater thus causing eutrophication of natural water reservoirs (Del Campillo et al., 1999).

A number of *in vitro* and *in vivo* studies have shown the ability of PSM to release metabolites such as organic acids which through their hydroxyl and carboxyl groups chelate the cations bound to phosphate, the latter being converted to soluble forms (Sagoe et al., 1998). In any case, metabolizable C compounds must be applied to the microbes to ensure their growth, organic acid production, and, simultaneously, RP solubilization. In general, the interaction of minerals such as phosphate rocks with organic matter and microorganisms has an enormous impact on biodiversity, global climate, biological productivity, human nutrition and the toxicity of metal pollutants (Huang, 2002).

During the last decade, we have developed a biotechnological scheme for RP solubilization by fungal microorganisms grown on agro-industrial wastes (Vassilev et al., 1995). Introduction of the resulting fermentation product, containing mineralized organic matter, soluble phosphate and fungal mycelium, into soil-plant systems in greenhouse and field conditions resulted in a significant plant growth enhancement, higher level of mycorrhization and increased soil enzyme activities (Vassilev et al., 1996; 1998; Vassileva et al., 1999; Cereti et al., 2004; Medina et al., 2004; Vassilev et al., 2006). The same approach was successfully proved in revegetation and bioremediation strategies (Caravaca et al., 2004; Medina et al., 2005). Bearing in mind that micronutrients found in natural phosphates such as zinc and cooper, are known to suppress pathogens (Duffy and Defago, 1999) and the effect of mineralized organic matter and PSM on phytopathogenic fungi (Ros et al., 2005), the further step in our investigations was to test the potential biocontrol properties of the above mentioned fermentation product.

As most of the soil-borne pathogens are fungi, the two main methods for disease control currently available in crop production are repeated applications of fungicides and the use of cultivars resistant or tolerant to the pathogens. However, the overuse of chemicals to combat plant diseases has caused soil pollution and had harmful effect on human health. On the other hand, cultivars resistant to pathogens have been developed but their use is limited, especially in fruit and vegetable crops. During the last years, a large number of fungal microorganisms have been reported as antagonists of soil-borne fungal pathogens (Kiss, 2003).

The aim of this work was to study the effect of an acid-producing strain of *Aspergullus niger* grown on agro-industrial wastes (AIW) in the presence of insoluble inorganic phosphate (rock phosphate, RP) and further introduced into a soil-plant-phytopathogen fungus system.

MATERIALS AND METHODS

Fermentation Stage

Fermentation process details and analytical methods have been published in previous articles (Vassilev et al., 1995; 1996; 2006; Vassileva & Vassilev 1999). However, we will briefly mention the most important methodological points of the fermentation experiment.

The strain of *Aspergillus niger* NB2 used in this study had previously been selected as producing citric acid on complex substrates including lignocellulosic materials (Vassilev et al., 1995). It was maintained on potato-dextrose agar slants at 4° C. For inoculum preparation, *A. niger* was grown on a slant at 30° C for 7 days and spores were scraped in sterile distilled water enriched with 0.1 ml Tween 80.

Sugar beet wastes (sugar beet press mud) were obtained from the local fabric for sugar production (Azucarera de Jaen, Spain). Dry olive wastes were kindly provided by "COLGRA", Spain. Portions of 15 g of the solid wastes (dry olive wastes:sugar beet wastes, 1:0.5, w/w), ground to pass a 2-mm-pore screen, were placed in 250-ml Erlenmeyer flasks and mixed with medium strength Czapek-Dox mineral salt solution at a ratio of 1:1 (solid particles:liquid phase, w/v). All flasks were sterilized by autoclaving at 120° C for 30 min and inoculated with spore suspension of *A. niger* (1.2 x 10^7 spores/flask). Morocco low-grade rock phosphate (12.8 % P, 1 mm mesh) was sterlized separately and added at a rate of 3 g/l (0.15 g per flask) prior spore inoculation. Fermentation experiments were carried out at 30° C for 15 days. Characteristics of the resulting product, which was further used in soil-plant experiments, was pH, 3.8; titratable acidity, 62 mmol/l (83% of which was determined as citric acid); electrical conductivity (1:10), 1023 μS/cm; total P, 0.83%; total N, 0.7%; water soluble C, 1210 μg/g.

Soil-Plant Experiment

The treatments (five replications each) used in this experiment were as follows: (i) (C) control, soil without amendments; (ii) (+AM), control + *Glomus intraradices*; (iii) (+AM+*A.*

niger/AIW/RP), control amended with dry olive/sugar beet wastes treated with *A. niger* in the presence of rock phosphate + *G. intraradices*. The fermentation products obtained at the fermentation stage were mixed at a rate of 5 % with a steam-sterilized (100° C/1h, 3 consecutive days) soil-sand mixture (1:1; w/w) and left for equilibration for 3 weeks at room temperature. The soil used for the experiments was the top 0-20 cm of Granada province (Spain) soil, with pH 8.1 in a 1:1 soil-water ratio, organic carbon 0.46%, 2.1 mg/kg N, 1.7 mg/kg P ($NaHCO_3$-extractable P). The soil texture was 358 g/kg sand, 436 g/kg silt, 205 g/kg clay and 12 g/kg organic matter. When necessary, soil was inoculated with *Fusarium oxysporum* f. sp. *lycopersici*. The inoculum of the pathogen, maintained on potato-dextrose agar slants, was prepared after 7-day cultivation in flasks containing 100 ml of sterile Czapek-Dox liquid medium as described by De Cal et al. (2000). Each pot received 50 ml of a microconidial suspension to reach a final concentration of 4.5 x 10^5 microconidia of the pathogen per g of soil.

Surface sterilized (70% ethanol, 2 min; 0.5% sodium hypochlorite, 2 min; sterile distilled water, rinsed 4-5 times) seeds of tomato (*Lycopersicon esculentum*) were sown in sterilized soil. Two weeks after germination the seedlings were transplanted in pots (three seedlings per pot) containing 300 g of amended or non-amended soil according to the treatments i-iii. The seedlings were inoculated (treatments ii and iii) with *in vitro* produced spores of *G. intraradices* Schenck & Smith as described by Vimard et al. (1999).

Experiments were performed under controlled conditions in a climate room at 21 to 28° C, 16h light, 50% relative humidity. Water loss from field capacity was replaced daily by top watering.

Analyses

Plants were harvested after 7 weeks and the shoot and root fresh and dried weights were recorded. P contents were determined by the dry ashing digestion method of Jones et al. (1991) followed by the molybdo-vanado determination as described by Lachica et al. (1973). The nitrogen concentration was assayed by using the Kjeldahl digestion method following a modification of Jones et al. (1991).

Roots were carefully washed and the percentage of root length colonized by *G. intraradices* was determined by the gridline intersect method (Giovannetti and Mosse, 1980) after staining with 0.05 % Trypan blue in lactophenol (v/v) (Phillips and Hayman, 1970).

Rhizosphere soil samples (closely associated with the plant roots) were collected, stored at 4° C, and subsequently assayed for acid phosphatase and β-glucosidase activity using as substrates *p*-nitrophenyl disodium phosphate (PNPP, 0.115 M) and *p*-nitro-phenyl-□-D-glucopyranoside (PNG, 0.05M), respectively (Naseby and Lynch, 1997; Masciandaro et al., 1994). The amount of *p*-nitrophenol formed was determined spectrophotometrically at 398 nm (Tabatabai and Bremner, 1969).

Severity of symptoms was determined after 10, 20, 30 and 49 days using the following index: A, plant without symptoms; B, lower leaves yellow; C, lower leaves dead and some upper leaves wilted; D, lower leaves dead and upper leaves wilted; E, dead plant. Rhizospheric population of *F. oxysporum* f. sp. *lycopersici* was determined using a dilution plate method. Ten grams of air-dried rhizosphere soil were shaken for 1h in a 250-ml Erlenmeyer flask containing 100 ml of 0.2% agar-water. One ml of the suspension from each sample (100-fold dilution) was spread on Petri dishes containing selective medium. The plates were further incubated for 7 days and the number of fungal colony-forming units per gram soil was determined by colony counting.

In vitro Agar Plate Bioassays

A modified chrome azurol S (CAS)-agar plate assay was carried out according to Milagres et al. (1999) to test the ability of *A. niger* to produce siderophores. *A. niger* was also assayed for inhibition of *F. oxysporum* on potato-dextrose agar. 10-mm disks of fresh culture of the pathogen fungus grown for 7 days at 28° C were cut out and placed in the center of a 9-cm Petri plate with PDA. Spores of *A. niger* were further inoculated on either side of the *F. oxysporum* disk at the corner of the plate. Fungal growth for individual (*F. oxysporum*, control) and dual (*F. oxysporum+A. niger*) fungal cultures was observed daily during 7 days at 28° C. All in vitro experiments were carried out in triplicate.

New Duncan's multiple range tests was used where appropriate to test for significant difference.

RESULTS

Microscopic observation of plant roots showed that only AM-inoculated plants were root colonized (Table 1). The percentage of AM root length colonization was 16 % higher in pathogen-non-infected treatments amended with the fermentation product, compared with treatments where only *G. intraradices* was introduced into soil. In this latter case, the fungal pathogen significantly reduced AM colonization while no significant difference was observed in treatments amended with microbially-treated agro-wastes and RP compared with pathogen-non-infected mycorrhizal plants.

Table 1. Mycorrhizal colonization of tomato plants grown in amended or non-amended soil infected or not with *Fusarium oxysporum*

Treatment	AM colonization (%)	
	- *F. oxysporum*	+ *F. oxysporum*
+AM	43*b*	31*b*
+AM+*A. niger*/AIW/RP	59*a*	53*a*

Column values followed by the same superscript letters are not significantly different ($P<0.05$) using Duncan's multiple range test.

The second goal of this set of experiments was to assess growth and nutrient (nitrogen, N and phosphate, P) uptake of tomato plants under different conditions of cultivation (Table 2 and 3). After 7-weeks culture, fresh and dry weights of shoots and roots of tomato plants non-infected with *F. oxysporum* were significantly higher in treatments where the arbuscular mycorrhizal fungus was introduced compared to the non-mycorrhizal control. However, the greatest increase in plant weight was registered in treatments amended with the fermentation mixture containing *A. niger*, treated AIW, and partially solubilized RP. Similarly, mineral N and P contents in tomato shoots and roots were significantly greater in treatments where *G.*

intraradices was introduced into the soil-plant system compared with the control. These parameters were further enhanced in the treatments amended with microbially-treated AIW and RP. It should be noted, however, that particularly the contents of P, expressed as a percentage of tomato shoot dry weight, showed no significant differences between mycorrhizal plants and mycorrhizal plants +*A. niger*/AIW/RP which was possibly due to a dilution effect.

Table 2. Fresh weight (FW), dry weight (DW), nitrogen (N) and phosphate (P) content of tomato shoots in soil amended or non-amended with *Aspergillus niger* (*A.n.*) treated agro-industrial wastes (AIW) in the presence of rock phosphate (RP)

Parameter	FW (g)	DW (g)	N (mg)	N* (%)	P (mg)	P* (%)
- *Fusarium oxysporum*						
Control	4.75^{bc}	0.44^{c}	22.1^{d}	5.0	0.9^{bc}	0.21
+AM	6.91^{b}	0.7^{b}	32.9^{c}	4.7	1.2^{b}	0.17
+AM+ *A.n.*/AIW/RP	14.5^{a}	1.53^{a}	65.8^{a}	4.3	2.8^{a}	0.18
+ *Fusarium oxysporum*						
Control	3.1^{c}	0.28^{d}	8.71^{e}	3.1	0.61^{d}	0.21
+AM	5.26^{b}	0.51^{bc}	20.4^{d}	4.0	0.97^{bc}	0.19
+AM+ *A.n.*/AIW/RP	13.4^{a}	1.31^{a}	49.8^{b}	3.8	2.34^{a}	0.18

* % of DW.

Column values followed by the same superscript letters are not significantly different ($P<0.05$) using Duncan's multiple range test.

Table 3. Fresh weight (FW), dry weight (DW), N and P content of tomato roots in soil amended or non-amended with microbially-treated AIW in the presence of rock phosphate

Parameter	FW (g)	DW (g)	N (mg)	N* (%)	P (mg)	P* (%)
- *Fusarium oxysporum*						
Control	1.34^{c}	0.11^{c}	2.53^{c}	2.3	0.2^{c}	0.18
+AM	1.71^{b}	0.15^{b}	3.0^{bc}	1.99	0.28^{b}	0.19
+AM+ *A.n.*/AIW/RP	2.53^{a}	0.22^{a}	4.16^{a}	1.73	0.59^{a}	0.24
+ *Fusarium oxysporum*						
Control	0.93^{d}	0.1^{c}	2.0^{d}	2.0	0.16^{c}	0.16
+AM	1.36^{c}	0.13^{bc}	2.5^{c}	1.95	0.23^{bc}	0.18
+AM+AIW/RP/*A.n.*	2.1^{ab}	0.19^{a}	3.4^{b}	1.78	0.41^{a}	0.21

* % of DW.

Column values followed by the same superscript letters are not significantly different ($P<0.05$) using Duncan's multiple range test.

The introduction of *F. oxysporum* into the soil-plant system adversely affected the plant growth and development. This effect was more pronounced in the control treatment when the fresh plant biomass of 3.1 g was 35% lower compared with 4.75 g obtained in the control non-infected with *F. oxysporum*. *G. intraradices* significantly reduced the effect of *F. oxysporum* and fresh biomass (5.26 g) of mycorrhizal plants was 69.6 % higher than that recorded in the pathogen-infected control treatments. The highest plant growth and nutrient uptake among all pathogen-infected treatments were measured in soil-plant systems amended with *A. niger*/AIW/RP and inoculated with *G. intraradices*. In this treatment, differences between levels of plant growth, nutrient (particularly P) uptake and root mycorrhization of plants grown in the presence and absence of the pathogen were found to be insignificant. It should be noted, that the introduction of fermentation mixture improved soil enzyme activities involved in the P (acid phosphatase) and C (β-glucosidase) cycles compared with the same biochemical parameters in the control and +AM treatments (Table 4).

Table 4. Phosphatase and β-glucosidase activity in rhizosphere of tomato plants grown in amended or non-amended soil infected with *Fusarium oxysporum*

Treatment	Phosphatase (μg PNF per g per h)	β-glucosidase (μg PNF per g per h)
Control	23.65*c*	36.5*c*
+AM	63.84*b*	59.89*b*
+AM+*A.n.*/AIW/RP	117.32*a*	102.54*a*

Values within a column followed by same letter are not significantly different (P<0.05) as determined by Duncan's multiple range test.

Under these conditions, it was found that the presence of *G. intraradices* significantly reduced the effect of *F. oxysporum* on tomato plants accompanied by a significant decrease of 2.3×10^3 in the number of CFU compared with the control treatment (Table 5). However, the AM symbiosis only partly limited the negative effect of the pathogen. Further introduction of *A. niger*/AIW/RP into soil demonstrated higher capacity to control the pathogen effect and only 6×10^2 CFU were found per gram soil at the end of the experiment.

Table 5. Observations of disease severity and number of *Fusarium oxysporum* colony-forming units (CFU) in the rhizosphere of tomato plants grown in amended or non-amended soil inoculated or not with *Glomus intraradices* (AM)

Treatment	Days/severity of symptoms				*F. oxysporum*
+*F. oxysporum*	10	20	30	45	(log CFU/g soil)
Control	A	C	D	E	3.83
+AM	A	B	B	C	3.64
+AM+*A.n.*/AIW/RP	A	A	A/B	B	2.77

In a separate *in vitro* experiment, *A. niger* showed a strong suppressive effect on *F. oxysporum* with a 5-fold reduction of the pathogen colony diameter (14±0.6 mm) compared with the control (72±2.1mm) after a 7-day cultivation. In addition, *A. niger* produced a rapid (after 2 days of growth) color-change reaction from blue to purplish-red in chrome azurol S-agar plate assay which evidenced siderophore-like substance production.

DISCUSSION

It is now well established that biological control of fungal pathogens offers a potential alternative to chemical fungicides. A fungal biocontrol agent can prevent or suppress the pathogen by mechanisms involving antibiosis, parasitism, and competition for space and/or nutrition although the antagonistic effect can be based on more than one mode of action (Kiss, 2003).

In this work, the role of natural microflora in the suppressiveness of *F. oxysporum* was concisely avoided in order to determine the effect of *G. intraradices* alone and in combination with the fermentation product containing *A. niger*/AIW/partially solubilized RP.

Arbuscular mycorrhizal fungi are accepted as an important part of biocontrol fungi (Johansson et al., 2004) as they provoke increased resistance to certain wilt and root rot pathogens including *Fusarium* (Filion et al., 1999; Dar et al., 1997) due to an increased nutritional status of the host plant and/or releasing unspecified substances. These statements were confirmed in our study as *G. intraradices* not only significantly improved the growth and nutrient (N and P) uptake of tomato plants but in the presence of *F. oxysporum*, although at lower root mycorrhizal colonization (in comparison with the treatment free of pathogen), simultaneously decreased the pathogen population density (CFU/g rhizosphere soil) as compared with the non-mycorrhizal control. Improved pathogen performance is not uncommon in P deficient soil (as used in this study), where the addition of AM fungi produces vigorous plants better able to support pathogens (Borowitcz, 2001). In addition, poorer pathogen performance may be due to competition between pathogens and AM fungi for plant metabolites or infection sites (Traquair, 1995). Although fungal pathogens reduced root colonization by AM fungi, the latter were shown to provoke protection through increased hydrolytic enzyme activity including that directly involved in the regulation of the symbioses (Lambais and Mehdy, 1995; Pozo et al., 1998).

These effects of *G. intraradices* were further enhanced in soil-plant systems amended with *A. niger*/AIW/RP fermentation product. Availability of P was enhanced by the arbuscular mycorrhizal fungus which may be attributed to the overall increase in surface area and phosphatase activity of extraradical hyphae. The addition of microbially-treated organic matter to soil has a positive effect on soil fertility and increases its physical, chemical and biochemical characteristics as reported in previous studies using fermentation product with similar characteristics (Vassilev et al., 1996; Medina et al., 2004). Several studies assessed ecologically sustainable options for controlling fungal pathogens by combined protective actions of organic soil amendments and biocontrol microorganisms (Cotxarrera et al., 2002; Bardin et al., 2004).

In this study, results of *in vitro* experiments with *A. niger* showed a significant suppressive effect on *F. oxysporum* although the exact mechanisms of pathogen inhibition were not investigated. However, the biocontrol properties of *A. niger* could be attributed to siderophore production demonstrated in plate assays. In general, elaboration of ferrichrome-type siderophores by strains of genus *Aspergillus* has been observed which gives a competitive advantage (particularly for iron) over other microorganisms (Renshaw et al., 2002). Moreover, the presence of soluble phosphate obtained as a result of RP *A. niger*-mediated solubilization, most likely stimulated the production of siderophores as reported by Gram (1996). In addition, *A. niger* is known as a producer of a wide range of hydrolytic enzymes as confirmed in soil conditions by this study. This fungal microorganism can be considered as a biocontrol agent as recently suggested by other authors who demonstrated the biocontrol properties of P-solubilizing bacteria and fungi (Khan and Khan, 2002). Therefore, it was not surprising that *A. niger*/AIW/RP amendment reduced the capacity of *F. oxysporum* to harm the tomato plants by significantly reducing fungal-pathogen growth.

Conclusion

A solid-state fermentation process seems to be the preferred mass production method for obtaining large quantities of biocontrol agents (Larena et al., 2004). Direct applications of the resulting fermented products in soil results in a rapid mortality of plant pathogens (Adams et al., 2002). In our study, the fermented material additionally contains partially solubilized soluble P which is important bearing in mind the soil characteristics in the Mediterranean region. As agriculture moves toward more biological and environmentally mild methods, the combined effect of dual application of *A. niger*/AIW/RP and arbuscular mycorrhizal fungus *G. intraradices* on suppressing *F. oxysporum* should be further studied in natural conditions to critically assess a possible use in a large scale.

Acknowledgments

This work was supported by Project MEC CTM2005-06955, EC Project "Waste to Products", and R&C Programme (CICYT-MEC).

References

Adams, T.T., Eitman, M.A. & Hanel, B.M. (2002). Solid state fermentation of broiler litter for production of biocontrol agents. *Bioresource Technology, 82*, 33-41.

Bardin, S.D., Huang, H.C. & Moyer, J.R. (2004). Control of *Pythium* damping-off of sugar beet by seed treatment with crop straw powders and a biocontrol agent. *Biological Control, 29*, 453-460.

Borowitcz, V. (2001). Do arbuscular mycorrhizal fungi alter plant-pathogen relations? *Ecology, 82*, 3057-3068.

Caravaca, F., Alguacil, M.M., Vassileva, M., Diaz, G. & Roldan, A. (2004). AM fungi inoculation and addition of microbially treated dry olive cake-enhanced afforestation of a desertified Mediterranean site. *Land Degradation and Development, 15*, 153-161.

Cereti, C.F., Rossini, F., Federici, F., Quarantino, D., Vassilev, N. & Fenice, M. (2004). Reuse of microbially treated olive mill wastewater as fertilizer for wheat (*Triticum durum* Desf.). *Bioresource Technology, 91*, 135-140.

Cotxarrera, L., Trillas-Gay, M.I., Steinberg, C. & Alabouvette, C. (2002). Use of sewage sludge compost and *Trichoderma asperellum* isolates to suppress Fusarium wilt of tomato. *Soil Biology and Biochemistry, 34*, 467-476.

Dar, Gh. Hassan, Zargar, M.Y. & Beigh, G.M. (1997). Biocontrol of *Fusarium* root rot in the common bean (*Phaseolus vulgaris* L.) by using symbiotic *Glomus mosseae* and *Rhizobium leguminosarum*. *Microbial Ecology, 34*, 74-80.

De Cal, A., Garcia-Lopez, R., Melgarejo, P. (2000). Induced resistance by *Penicillium oxalicum* against *Fusarium oxysporum* f. sp. *lycopersici*: Histological studies of infected and induced tomato stems. *Phytopatology, 90*, 260-268.

Del Campillo, S.E., Van der Zee, S.E.A.T.M. & Torrent, J. (1999). Modelling long-term phosphorus leaching and changes in phosphorus fertility in excessively fertilized acid sandy soils. *European Journal of Soil Science, 50*, 391-399.

Duffy, B.K.; Defago, G. Trace mineral amendments in Agriculture for optimizing the biocontrol activity of plant-associated bacteria. In: P Berthelin, M Huang, JM Bollag, F Andreux editors. *Effect of mineral-organic-microorganism interactions on soil and freshwater environments*. New York: Kluwer Academic/Plenum Publishers, 1999, 295-304.

Filion, M., St-Arnaud, M. & Fortin, J.A. (1999). Direct interaction between the arbuscular mycorrhizal fungus *Glomus intraradices* and different rhizosphere microorganisms. *New Phytologist, 141*, 525-533.

Giovannetti, M. & Mosse, B. (1980). An evaluation of techniques for measuring vesicular-arbuscular mycorrhizal infection in roots. *New Phytologist, 84*, 489-500.

Goldstein, A.H. (2000). Bioprocessing of rock phosphate ore: essential technical considerations for the development of a successful commercial technology. IFA Technical Conference, New Orleance, pp. 1-21, http://goldsteinlab.alfred.edu/publications.html.

Gram, L. (1996). The influence of substrate on siderophore production by fish spoilage bacteria. *Journal of Microbiological Methods, 25*, 199-205.

Huang, P.M. Foreseeable impacts of soil mineral-organic component-microorganism interactions on society ecosystem health. In: A Violante, PM Huang, J-M Bollag and L Gianfreda, editors. *Soil Mineral-Organic Matter-Microorganism Interactions and Ecosystem Health*. Boston: Elsevier, 2002, 163-176.

Johansson, J.F., Paul, L.R. & Finlay, R.D. (2004). Microbial interactions in the mycorrhizosphere and their significance for sustainable agriculture. *FEMS Microbiology Ecology, 48*, 1-13.

Jones, J.B., Wolf, B. & Mills, H.A. Plant analysis handbook. Athens: Micro-Macro Publishing, 1991, 195-203.

Khan, M.R. & Khan, S.M. (2002). Effects of root-dip treatment with certain phosphate solubilizing microorganisms on the fusarium wilt of tomato. *Bioresource Technology, 85*, 213-215.

Kiss, L. (2003). A review of fungal antagonists of powdery mildews and their potential as biocontrol agents. *Pest Management Science, 59*, 475-483.

Kucey, R.M.N., Jansen, H.H. & Legett, M.E. (1989). Microbially mediated increases in plant-available phosphorus. *Advances in Agronomy, 42*, 198-228.

Lachica, M,, Aguilar, A. & Yanez, J. (1973). Metodos analiticos en la Estacion Experimental del Zaidin. *Anales en Edafología y Agrobiologia, 32*, 1033-1047.

Lambais, M.R. & Mehdy, M.C. (1995). Differential expression of defense-related genes in arbuscular mycorrhiza. *Canadian Journal of Botany, 73*, 533- 540.

Larena, I., De Cal, A. & Melgarejo, P. (2004). Solid substrate production of *Epicoccus nigrum* conidia for biological control of brown rot on stone fruits. *International Journal of Food Microbiology, 94*, 161-167.

Masciandaro, G., Ceccanti, B. & Garcia, C. (1994). Anaerobic digestion of straw and piggery wastewater: II. Optimization of the process. *Agrochimica, 3*, 195-203.

Medina, A., Vassilev, N., Alguacil, M., Roldan, A. & Azcon, R. (2004). Increased plant growth, nutrient uptake, and soil enzymatic activities in a desertified Mediterranean soil amended with treated residues and inoculated with native mycorrhizal fungi and a plant growth-promoting yeast. *Soil Science, 169*, 260-270.

Medina, A., Vassilev, N., Barea, J.M., Azcon, R. (2005). Application of *Aspergillus niger*-treated agrowaste residue and *Glomus mossae* for improving growth and nutrition of *Trifolium repens* in a Cd-contaminated soil. *Journal of Biotechnology, 116,* 369-378.

Milagres, A.M.F., Manchuca, A. & Napoleao, D. (1999). Detection of siderophore production from several fungi and bacteria by a modification of chrome azurol S (CAS) agar plate assay. *Journal of Microbiological Methods, 37*, 1-6.

Naseby, D.C. & Lynch, J.M. (1997). Rhizosphere soil enzymes as indicators of perturbation caused by a genetically modified strain of *Pseudomonas fluorescens* on wheat seeds. *Soil Biology and Biochemistry, 29*, 1353-1362.

Phillips, J.M. & Hayman, D.S. (1970). Improved procedure for clearing roots and staining parasitic and vesicular arbuscular mycorrhizal fungi for rapid assessment of infection. *Transactions of the British Mycological Socety, 55*, 159-161.

Pozo, M.J., Azcon-Aguilar, C. & Dumas-Gaudot, E. (1998). Chitosanase and chitinase activities of tomato roots during interactions with arbuscular mycorrhizal fungi or *Phytophthora parasitica*. *Journal of Experimental Botany, 49*, 1729-1739.

Renshaw, J., Robson, G., Trinci, P.J., Wiebe, M.G., Livens, F.R., Collison, D. & Taylor, R.J. (2002). Fungal siderophores: structures, functions and applications. *Mycological Research, 106*, 1123-1142.

Ros, M., Hernandez, M.T., Garcia, C., Bernal, A. & Pascual, J.A. (2005). Biopesticide effect of green compost against fusarium wilt on melon plants. *Journal of Applied Microbiology, 98*, 845-854.

Sagoe, C.I., Ando, T., Kouno, K. & Nagaoka, T. (1998). Relative importance of protons and solution calcium concentration in phosphate rock dissolution by organic acids. *Soil Science and Plant Nutrition, 44*, 617-625.

Tabatabai, M.A. & Bremner, J.M. (1969). Use of *p*-nitrophenol phosphate in assay of soil phiosphatase activity. *Soil Biology and Biochemistry, 1*, 301-307.

Traquair, J.A. (1995). Fungal biocontrol of root diseases: endomycorrhizal suppression of cylindrocarpon root rot. *Canadian Journal of Botany, 73*, S89-S95.

Vassileva, M. & Vassilev, N. (1999). Rock phosphate solubilization by *Aspergillus niger* on olive cake-based medium and its further application in soil-plant system. *World Jounal of Microbiology and Biotechnology, 14*, 281-284.

Vassilev, N. & Vassileva, M. (2003). Biotechnological solubilization of rock phosphate on media containing agro-industrial wastes. *Applied Microbiology and Biotechnology, 61*, 435-440.

Vassilev, N., Baca, M., Vassileva, M. & Franco, I. (1995). Rock phosphate solubilization by *Aspergillus niger* grown on sugar beet waste medium. *Applied Microbiology and Biotechnology, 44*, 546-549.

Vassilev, N., Franco, I. & Vassileva, M. (1996). Improved plant growth with rock phosphate solubilized by *Aspergillus niger* grown on sugar beet wastes. *Bioresource Technology, 55*, 237-241.

Vassilev, N., Medina, A. & Vassileva, M. (2006). Microbial solubilization of rock phpsjhate on media containing agro-industrial wastes and effect of the resulting products on plant growth and P uptake. *Plant and Soil, 287*, 77-84.

Vassilev, N., Vassileva, M., Fenice, M. & Federici, F. (1998). Fetilizing effect of microbially treated olive mill wastewaters on *Trifolium* plants. *Bioresource Technology, 66*, 133-137.

Vassilev, N., Vassileva, M., Fenice, M. & Federici, F. (2001). Immobilized cell technology applied in solubilization of insoluble inorganic (rock) phosphate. *Bioresource Technology, 79*, 263-271.

Vimard, B., St-Arnaud, M., Furlan, V. & Fortin, J.A. (1999). Colonization potential of in vitro- produced arbuscular mycorrhizal fungus spores compared with a root-segment inoculum from open pot culture. *Mycorrhiza, 8*, 335-338.

Whitelaw, M.A. (2000). Growth promotion of plants inoculated with phosphate-solubilizing fungi. *Advances in Agronomy, 69*, 99-151.

INDEX

C

D

H

I

N

O

P

Q

R

S

T

Y

Z